INCOMPLETE CATEGORICAL DATA DESIGN

Non-Randomized Response Techniques for Sensitive Questions in Surveys

Chapman & Hall/CRC
Statistics in the Social and Behavioral Sciences Series

Series Editors

Jeff Gill
Washington University, USA

Steven Heeringa
University of Michigan, USA

Wim van der Linden
CTB/McGraw-Hill, USA

J. Scott Long
Indiana University, USA

Tom Snijders
Oxford University, UK
University of Groningen, UK

Aims and scope

Large and complex datasets are becoming prevalent in the social and behavioral sciences and statistical methods are crucial for the analysis and interpretation of such data. This series aims to capture new developments in statistical methodology with particular relevance to applications in the social and behavioral sciences. It seeks to promote appropriate use of statistical, econometric and psychometric methods in these applied sciences by publishing a broad range of reference works, textbooks and handbooks.

The scope of the series is wide, including applications of statistical methodology in sociology, psychology, economics, education, marketing research, political science, criminology, public policy, demography, survey methodology and official statistics. The titles included in the series are designed to appeal to applied statisticians, as well as students, researchers and practitioners from the above disciplines. The inclusion of real examples and case studies is therefore essential.

Published Titles

Analysis of Multivariate Social Science Data, Second Edition
David J. Bartholomew, Fiona Steele, Irini Moustaki, and Jane I. Galbraith

Applied Survey Data Analysis
Steven G. Heeringa, Brady T. West, and Patricia A. Berglund

Bayesian Methods: A Social and Behavioral Sciences Approach, Second Edition
Jeff Gill

Foundations of Factor Analysis, Second Edition
Stanley A. Mulaik

Incomplete Categorical Data Design: Non-Randomized Response Techniques for Sensitive Questions in Surveys
Guo-Liang Tian and Man-Lai Tang

Informative Hypotheses: Theory and Practice for Behavioral and Social Scientists
Herbert Hoijtink

Latent Markov Models for Longitudinal Data
Francesco Bartolucci, Alessio Farcomeni, and Fulvia Pennoni

Linear Causal Modeling with Structural Equations
Stanley A. Mulaik

Multiple Correspondence Analysis and Related Methods
Michael Greenacre and Jorg Blasius

Multivariable Modeling and Multivariate Analysis for the Behavioral Sciences
Brian S. Everitt

Statistical Test Theory for the Behavioral Sciences
Dato N. M. de Gruijter and Leo J. Th. van der Kamp

Chapman & Hall/CRC
Statistics in the Social and Behavioral Sciences Series

INCOMPLETE CATEGORICAL DATA DESIGN

Non-Randomized Response Techniques for Sensitive Questions in Surveys

Guo-Liang Tian

Department of Statistics and Actuarial Science
The University of Hong Kong, Hong Kong, P. R. China

Man-Lai Tang

Department of Mathematics, Hong Kong Baptist University
Kowloon Tong, Hong Kong, P. R. China
Department of Mathematics and Statistics, Hang Seng Management College
Shatin, New Territories, Hong Kong, P. R. China

CRC Press
Taylor & Francis Group
Boca Raton London New York

CRC Press is an imprint of the
Taylor & Francis Group, an **informa** business

A CHAPMAN & HALL BOOK

CRC Press
Taylor & Francis Group
6000 Broken Sound Parkway NW, Suite 300
Boca Raton, FL 33487-2742

First issued in paperback 2019

© 2014 by Taylor & Francis Group, LLC
CRC Press is an imprint of Taylor & Francis Group, an Informa business

No claim to original U.S. Government works

ISBN-13: 978-1-4398-5533-1 (hbk)
ISBN-13: 978-0-367-37962-9 (pbk)

Visit the Taylor & Francis Web site at
http://www.taylorandfrancis.com

and the CRC Press Web site at
http://www.crcpress.com

To Yanli, Margaret, and Adam

To Daisy, Beatrice, and Bryan

Contents

Preface

Acquirement of sensitive information is often needed in a broad range of statistical applications. For instance, some behavioral, epidemiological, public health and social studies may need to solicit information on reproductive history, sexual behavior, abortion, human immunodeficiency virus, acquired immune deficiency syndrome, illegal drug usage, family violence, income, child abuse, employee theft, shoplifting, social security fraud, premature sign-offs on audits, infidelity, driving under the influence, having a baby outside marriage, tax evasion, and cheating in university examinations. When being directly asked these sensitive survey questions, some respondents may refuse to answer, and some may even provide untruthful answers in order to protect their privacy. The problem becomes even more complicated with surveys in diverse populations because of the interaction of sensitivity and respondent diversity. It is, therefore, difficult to draw valid inferences from these inaccurate data that include refusal bias, response bias, and perhaps both. It has long been a challenge to obtain such information while having the privacy of the respondent protected and the resulting data analyzed properly.

Although there are a number of methods (see Barton, 1958) for asking embarrassing questions in non-embarrassing ways, the first ingenious interviewing technique to overcome the above difficulties is the randomized response approach, proposed by Warner (1965), that aims to encourage truthful answers from respondents. The randomized response technique is designed to ask a sensitive question according to the outcome of a randomizing device while the interviewer is blind to the outcome. Since the introduction of the Warner model, voluminous work related to the randomized response technique has been done and can roughly be classified into eight major areas: (a) improvement of the efficiency by refining Warner's randomizing device; (b) extension to sensitive questions with multi-categories; (c) introduction of non-sensitive questions into randomized response models; (d) inclusion of multiple sensitive questions; (e) accounting for non-compliance in the analysis; (f) development of designs for obtaining sensitive quantitative data; (g) development of Bayesian analysis approaches; and (h) comprehensive reviews. Two widely cited books in this field are Fox & Tracy (1986) and Chaudhuri & Mukerjee (1988), which summarized the development of randomized response techniques from 1965 to 1988. A recent follow-up monograph by Chaudhuri (2011) captured some of the developments from early 1990s to 2011. Despite their popularity, randomized response models have long been criticized for their (i) lack of reproducibility; (ii) lack of trust from the interviewees; (iii) higher cost due to the introduction of randomizing devices; and (iv) narrow range of applications.

To overcome some of the aforementioned inadequacies of the randomized response approaches, non-randomized response approaches were first introduced by Swensson (1974) and Takahasi & Sakasegawa (1977), and then were investigated systematically by the authors of this book, their collaborators and other researchers (Jann & Brandenberger, 2012; Jann, Jerke & Krumpal, 2012; Groenitz, 2012). Non-randomized response techniques aim to provide the same privacy protection as randomized response techniques do. However, no randomizing device is needed. Non-randomized response method requires an independent non-sensitive variate such as birth date combined with the sensitive response variable to form an incomplete contingency table, resulting in the reproducibility. The proposed non-randomized response approaches somehow solve issues (i), (iii) and (iv) mentioned above.

This is the first book-length exposition of non-randomized response designs and statistical analyses. The aim of this book is to provide a systematic introduction to non-randomized response techniques for sensitive questions in surveys, which combines the strengths of existing approaches such as randomized response models, incomplete data design, expectation–maximization algorithm, bootstrap method, and data augmentation algorithm. We hope that the systematic introduction will stimulate the application of the non-randomized response approach in surveys and new non-randomized response designs. In particular, we anticipate that this non-randomized response approach would be useful to a wide range of audiences including government agents, academic researchers, and sample survey practitioners.

This book is intended to be written for a one-semester course at the advanced master or PhD level, and as a reference book for researchers, teachers, and applied statisticians. This book is also useful to undergraduates in statistics and to practitioners who need a good understanding of the randomized response approach in surveys with sensitive questions. Knowledge of basic probability, statistics, Bayesian method, and statistical methods for missing data problems is a prerequisite for this book. A brief review on the randomized response models in Chapter 1, along with three appendices on the expectation–maximization and data augmentation algorithms, the inverse Bayes formulae sampling, and some basic statistical distributions make the book quite self-contained. All R codes for examples in this book are available at the following URL:

http://www.saasweb.hku.hk/staff/gltian/

We gratefully acknowledge the support by the Department of Statistics and Actuarial Science of the University of Hong Kong and the Department of Mathematics of Hong Kong Baptist University. We are also thankful to our collaborators on this topic: Prof. Zhi Geng of Peking University, Prof. Ming T. Tan of Lombardi Comprehensive Cancer Center at Georgetown University Medical Center, Dr. Zhen-Qiu Liu of the University of Maryland Greenebaum Cancer Center, Prof. Kam C. Yuen of the University of Hong Kong, Prof. Nian-Sheng Tang of Yunnan University, and Mr. Jun-Wu Yu of Hunan University of Science and Technology. We greatly appreciate the help of our two PhD students Miss Yin Liu of the University of Hong Kong and Miss Qin Wu of Hong Kong Baptist University in preparing the manuscript. The research of the first author was partially supported by a grant (HKU 779210M) from the Research Grant Council of the Hong Kong Special Administrative Region. The research of the second author was partially supported by a grant (HKBU 261508) from the Research Grant Council of the Hong Kong Special Administrative Region.

Hong Kong   Guo-Liang Tian
Hong Kong   Man-Lai Tang
January 31, 2013

CHAPTER 1

Introduction

1.1 Randomized Response Models

Sensitive information is constantly sought in behavioral and social studies with sample surveys containing sensitive questions such as sexual behavior and abortion, human immunodeficiency virus (HIV), acquired immune deficiency syndrome (AIDS), covert or illegal behaviors such as drug use, abuse and violence, and personal income. Some respondents may refuse to answer and even worse provide untruthful answers to maintain their privacy when these questions are asked directly. Furthermore, the sensitivity may be acute in surveys in diverse populations. The final data will inevitably include refusal bias, response bias or both. As a result, it is difficult to make valid inferences based on these inaccurate data.

1.1.1 The Warner model

To overcome the hurdles induced from direct inquiries, Warner (1965) proposed a randomized response technique that allows researchers to not only obtain sensitive information but also protect privacy of respondents. To encourage truthful answers from respondents, the randomized response model offers the respondent a choice of two questions which are complementary to each other. For instance,

(i) I belong to group $\mathcal{A}$ (with probability p of being assigned to respond this statement).

(ii) I do not belong to group $\mathcal{A}$ (with probability $1 - p$ of being assigned to respond this statement).

The respondents are required to give a 'yes' or 'no' answer to either Statement (i) or its complement (ii) depending on the outcome of a randomizing device which is not being revealed to the interviewer. Usually, the probability p of selecting Statement (i) by the randomizing device is designed to be known. Respondents can then answer the sensitive question without revealing their actual answers. It thus reduces the privacy concern and the number of refusals to respond or evasive answers. In practice, we would like to estimate the proportion π of the population belonging to the sensitive group $\mathcal{A}$. Let n' be the number of 'yes' answers obtained from the n respondents selected by the simple random sampling with replacement. The *maximum likelihood estimator* (MLE) of π is shown to be

$$\hat{\pi}_{\mathrm{w}} = \frac{p - 1 + n'/n}{2p - 1}, \qquad p \neq 0.5, \tag{1.1}$$

and its variance is given by

$$\mathrm{Var}(\hat{\pi}_{\mathrm{w}}) = \frac{\pi(1 - \pi)}{n} + \frac{p(1 - p)}{n(2p - 1)^2}. \tag{1.2}$$

The most serious limitation for the Warner's model is its inefficiency. It should be noted that an extra term (i.e., the second term in the right-hand side of (1.2)) is introduced in variance due to the involvement of randomizing device. For any fixed π and fixed sample size n, it is easy to see that $\mathrm{Var}(\hat{\pi}_{\mathrm{w}})$ increases with p when $0 < p < 0.5$, quickly approaches to infinity as $p \to 0.5$, and then quickly decreases with p when $0.5 < p < 1$ (cf. Figure 2.1). However, when p is too small or too large, the privacy of respondents cannot be sufficiently protected (cf. Figure 2.2). Although the privacy of respondents can be sufficiently protected when $p = 0.5$, the variance of $\hat{\pi}_{\mathrm{w}}$ becomes undefined. Therefore, investigators are forced to select a value of p within some sub-interval of $(0, 0.5)$ since $n\mathrm{Var}(\hat{\pi}_{\mathrm{w}})$ is a symmetric function on $p = 0.5$ and thus use an uneven randomizing device (e.g., a biased coin). To reduce the variance of the proportion estimate, investigators have to increase sample size n and it thus results in a significant increase in cost.

1.1.2   Other randomized response models

Since the publication of the Warner model in 1965, voluminous extensions of the randomized response technique have been proposed during the past half century and they can be roughly classified into eight different areas.

(a) Improvement of the efficiency by refining Warner's randomizing device

In the first area, for example, Mangat & Singh (1990) suggested a two-stage randomized response model, which requires the use of two randomizing devices. The conditions when the proposed strategy is more efficient than the Warner's strategy have been obtained for the cases when the respondents are truthful and when they are not completely truthful in their answers. Mangat (1994) considered another randomized response model in which each respondent who is selected in the sample is requested to report yes if he or she belongs to the sensitive group $\mathcal{A}$; otherwise, he or she is instructed to use the Warner device. The proposed model by Mangat (1994) can be shown to be more efficient than both the Warner (1965) and the Mangat & Singh (1990) models. Chang & Liang (1996) developed a new two-stage unrelated randomized response procedure based on the model of Mangat & Singh (1990) and the unrelated question model of Horvitz, Shah & Simmons (1967). Zou (1997) showed that for both two-stage randomized response procedures of Mangat & Singh (1990) and Chang & Liang (1996), there is a simpler single-stage procedure that leads to the same distribution of responses. However, his results cannot be generalized to other randomized response models, as shown by Bhargava & Singh (2002). More recently, Gjestvang & Singh (2006) refined the two-stage randomization by adjusting parameters of the randomizing device that results in a randomized response model which is more efficient than the models by Warner (1965), Mangat & Singh (1990), and Mangat (1994). However, the refined two-stage randomizing device is convoluted and is too complicated to be practical.

(b) Extension to sensitive questions with multi-categories

In the second area, for instance, Abul-Ela, Greenberg & Horvitz (1967) and Bourke (1982) extended the Warner model to the trichotomous case to estimate the proportions of three mutually exclusive groups with at least one sensitive group. Bourke & Dalenius (1973) suggested a Latin square measurement design to extend the Warner design to the multichotomous case. Eriksson (1973) showed how multinomial proportions can be estimated with only one sample using a different randomizing device. Liu, Chow & Mosley (1975) developed a new randomizing device that can be used in the multi-proportions cases. Franklin (1989) considered a dichotomous population with the use of a randomizing device for continuous distributions. The model was further extended to estimate any m proportions ($m > 3$) when all the m group characteristics are mutually exclusive, with at least one and at most $m - 1$ of them being sensitive.

(c) Introduction of non-sensitive questions into randomized response models

The third area is the unrelated question randomized response model which involves non-sensitive question (Horvitz, Shah & Simmons, 1967; Greenberg et al., 1969). They suggested that the respondents might be more cooperative if Statement (ii) in the original Warner model is replaced by, for instance, the following non-sensitive question:

(iii) I was born in the month of April (i.e., I am a member of group $\mathcal{U}$),

which is unrelated to Statement (i). His or her privacy can be protected since the randomizing device is operated by the respondent and the interviewer does not know which question has been answered. Recall that p is the probability of selecting Statement (i). Let π' denote the proportion of individuals in the population who would answer 'yes' to Statement (iii). If π' is known, only one sample is required to estimate the proportion π of the population belonging to the sensitive class $\mathcal{A}$. An unbiased estimator of π is

$$\hat{\pi}_{\mathrm{U}} = \frac{n'/n - (1-p)\pi'}{p} \tag{1.3}$$

with variance

$$\mathrm{Var}(\hat{\pi}_{\mathrm{U}}) = \frac{\pi(1-\pi)}{n} + \frac{(1-p)^2\pi'(1-\pi') + p(1-p)(\pi + \pi' - 2\pi\pi')}{np^2}.$$

When π' is known, the unrelated question randomized response technique is statistically equivalent to the forced response method (Boruch, 1971), which seems to be one of the most widely used randomized response variants. In the forced response model, the respondent is forced by the randomizing device to answer the sensitive question (with probability p) truthfully, or to say 'yes' (with probability θ) or 'no' (with probability $1-p-\theta$), independent of the true answer (Lensvelt-Mulders, Hox & van der Heijden, 2005).

If π' are unknown, two independent samples of size n_1 and n_2 with different probabilities p_1 and p_2 for two randomizing devices are required. Moors (1971) showed that with an optimal allocation of n_1 and n_2, and $p_2 = 0$, the unrelated question model would be more efficient than the Warner model for $p_1 > 0.5$, regardless of the choice of π'. Dowling & Shachtman (1975) proved that $\mathrm{Var}(\hat{\pi}_{\mathrm{U}})$ is less than $\mathrm{Var}(\hat{\pi}_{\mathrm{W}})$ for all π and π', provided that p (or the $\max(p_1, p_2)$ in the two-sample case) is greater than 0.339. Folsom $et\ al.$ (1973) evaluated an alternative two-sample design. Eriksson (1973) and Bourke (1974) generalized the unrelated question model to the situation with m mutually exclusive classes with up to $m-1$ sensitive categories. The basic design uses a deck of cards with each card containing a number of statements. However, all the aforementioned designs require the use of at least one randomizing device.

(d) Inclusion of multiple sensitive questions

The fourth area is the extensions to multiple sensitive questions. Fox & Tracy (1984) considered the estimation of correlation between two sensitive questions. Lakshmi & Raghavarao (1992) also discussed two-by-two contingency table analysis based on binary randomized responses. Christofides (2005) presented a randomized response technique which can be used to estimate the proportion of individuals having two sensitive characteristics at the same time. Kim & Warde (2005) considered a multinomial randomized response model which can handle situations in which untruthful responses may occur. They also derived the Pearson product moment correlation estimator which may be used to quantify the linear relationship between two multinomial variables via a randomized response procedure.

(e) Accounting for noncompliance in the analysis

The fifth area is accounting for noncompliance in the analysis of randomized response data. One difficulty in implementing the randomized response techniques is the choice of an appropriate randomizing device in a self-administered setting. Another challenge in using randomized response models is the possible noncompliance behavior because of respondents' mistrust. To handle noncompliance to the instructions of randomized response techniques, many new models were proposed by various researchers, e.g., Lakshmi & Raghavarao (1992), Clark & Desharnais (1998), Böckenholt & van der Heijden (2007), van den Hout & Klugkist (2009), Ostapczuk *et al.* (2009), Ostapczuk, Musch & Moshagen (2009, 2011), Moshagen (2010), van den Hout, Böckenholt & van der Heijden (2010), Moshagen, Musch & Erdfelder (2012), and so on.

(f) Development of designs for obtaining sensitive quantitative data

In the sixth area, Greenberg *et al.* (1971) extended the unrelated question design for estimating the mean number of abortions in an urban population of women and mean income of heads of households. Warner (1971) established a general linear randomized response model to estimate certain features of the distribution of a random vector with sensitive components. Eriksson (1973) discussed a design for collecting sensitive information on the moments of the distribution of a quantitative variable. Poole (1974) used a multiplication model to estimate the entire distribution of a sensitive quantitative variable, not just the mean and variance. Liu, Chow & Mosley (1975) presented a new randomizing device to obtain discrete quantitative as well as qualitative data on sensitive problems. Pollock & Bek (1976) gave a practical comparison of three (i.e., Greenberg's model, the addition and multiplication models) randomized response models for quantitative data. Horvitz, Greenberg & Abernathy (1976) presented a review on quantitative data designs. Eichhorn & Hayre (1983) studied a multiplicative randomized response method which involves the respondent multiplying his/her sensitive answer by a random realization from a known distribution, and giving the product to the interviewer, who does not know the value of the random realization and thus receives a scrambled response. Singh, Mangat & Singh (1997) were interested in estimating the size and mean of a sensitive quantitative variable for a sub-group of a population.

Gupta, Gupta & Singh (2002) proposed an optional randomized response model to estimate the mean of the stigmatized variable. Grewal, Bansal & Singh (2003) proposed an estimator of population mean of a sensitive quantitative variable using double sampling. Bar-Lev, Bobovitch & Boukai (2004) generalized Eichhorn and Hayre's results and provided an alternative estimator to the mean response which has a uniformly smaller variance as compared to that of Eichhorn & Hayre (1983). Hussain & Shabbir (2007) introduced an unbiased estimator of population mean of a sensitive quantitative variable based on multiple selections of numbers from a scrambling distribution to confound the actual response on sensitive variable with some unrelated variable. Singh, Kim & Grewal (2008) considered the situation if some of the respondents refuse to give scrambled responses related to their sensitive questions and showed that if no scrambling is applied then the proposed method leads to the Rao & Sitter (1995) method of imputation. Other related works include Chaudhuri & Dihidar (2009), Nazuk & Shabbir (2010), Gupta, Shabbir & Sehra (2010), Diana & Perri (2011), Singh & Kim (2011), Chaudhuri (2011, Chapter 7), and so on.

(g) Development of Bayesian analysis approaches

The seventh area is about the development of Bayesian analysis methods. Winkler & Franklin (1979) studied Warner's randomized response model from a Bayesian viewpoint. With a beta prior distribution, the posterior distribution is a mixture of beta distributions. Pitz (1980) considered Simmons' model using the Bayesian method. Spurrier & Padgett (1980) discussed the application of the Bayesian method to the unrelated question model. For a variety of randomized response models, O'Hagan (1987) provided Bayes linear estimators by using a robust, non-parametric, Bayesian information in which only first- and second-order moments are required. Migon & Tachibana (1997) showed how a symbolic computation package could simplify the Bayesian calculation involved in the exact solution and to obtain the numerical approximation through Laplace's method. Unnikrishnan & Kunte (1999) developed a unified model for randomized response sampling strategies of which Warner's and Simmons' as well as polychotomous extensions of these are special cases and also provided Bayesian analysis of these strategies. Bar-Lev, Bobovitch & Boukai (2003) presented a common conjugate prior approach to four randomized response models. DiPietro (2004) described multiple learning opportunities stemming from a classroom project about psychological therapy analyzed through the Bayesian framework. Other related works include Kim, Tebbs & An (2006), van den Hout, Böckenholt & van der Heijden (2010), Song & Kim (2012),

Avetisyan & Fox (2012), and so on.

(h) Comprehensive reviews

The eighth area is about the comprehensive reviews regarding the randomized response techniques, which can be found in Greenberg, Horvitz & Abernathy (1974), Horvitz, Greenberg & Abernathy (1975, 1976), Greenberg, Abernathy & Horvitz (1986), Daniel (1993), Tracy & Mangat (1996), Franklin (1998), Kim & Warde (2004), Kim & Elam (2005), Saha (2007), and in monographs including Cochran (1977, 392–395), Fox & Tracy (1986), Chaudhuri & Mukerjee (1988), Hedayat & Sinha (1991, Chapter 11), Chaudhuri & Stenger (1992, Chapter 10), Lee (1993), and Chaudhuri (2011).

1.1.3 Limitations of the randomized response models

Despite these advances, all randomized response models have several well-recognized inadequacies that have limited their applications.

(a) Lack of reproducibility

A major undesirable feature of randomized response methods is their lack of reproducibility. That is, the same respondent may yield different answers depending on the outcome of the randomizing device in repeated experiments (e.g., repeatedly flip a coin). For example, in the unrelated question design with a coin being the randomizing device, if the outcome is a head, the first question is answered; if the outcome is a tail, the second question is answered. When the result of the first (second) flip is a tail (head), the answer is a 'yes' ('no'). As a result, the interviewers do not know which answer to be collected. Under two trails per respondent (Gould, Shah & Abernathy, 1969), the interviewers at least have three data-gathering schemes: Scheme 1—only the first answer for each respondent is included; Scheme 2—only the second answer of the respondents is collected; Scheme 3—two answers for all respondents are counted. First of all, it is unclear if Scheme 1 and Scheme 2 are statistically equivalent, especially for small sample designs. Next, doubling the number of observations will in general reduce the variance in random samples by one-half (Horvitz, Greenberg & Abernathy, 1976). It seems that increasing the number of trials (e.g., three times or more) will further improve the efficiency. However, how to choose the number of trails in the multiple trial randomized response technique is still a problem to be resolved in practice.

For Warner's original design, it has been proven theoretically by Liu & Chow (1976) that repeated administration of a set of two related questions more than once (say, twice, three times or more) substantially increases the efficiency by reducing the variance of the estimate. However, if the respondent becomes more concerned about his anonymity, then the mean square error may be larger.

(b) Lack of trust

The second inadequacy is the lack of trust from respondents. The setting of the randomizing device is totally controlled by interviewers (i.e., the probabilities of answering Statements (i), (ii) or (iii) are pre-specified by investigators), which makes it difficult to convince the respondents that their privacy is well protected by randomization (Kuk, 1990). Warner (1986), for example, remarked that 'interviewees often act as if they believe the randomizing device simply determines whether they are required to reveal or not to reveal a secret.'

(c) Expensive cost

The third drawback is the expensive cost for the necessity of the usage of randomizing devices. A randomizing device must be provided to the respondent. The device suggested by Warner (1965) is a spinner with an arrow pointer. Greenberg, Abernathy & Horvitz (1986) pointed out the limitation[1] of this device and other randomizing devices including colored plastic balls (Liu & Chow, 1976; Fidler & Kleinknecht, 1977), colored beads (Liu, Chow & Mosley, 1975), coins (Zdep & Rhodes, 1976), dice (Berman, McCombs & Boruch, 1977), and poker chips (Wiseman, Moriarty & Schafer, 1975). For instance, in a self-administered survey, the cost includes the respondent's time and effort, and the devices themselves (especially for a larger sample size). In a telephone survey, the interviewer needs to mail a randomizing device to each respondent, which involves an additional cost such as postage and added record keeping.

(d) Difficulty in understanding

The fourth is the difficulty in understanding randomized response techniques. In the Warner model, it is noteworthy that Statement (ii) is also sensitive because it is simply a complement of Statement (i). In this regard, the interviewee still needs to answer a sensitive question no matter which outcome is being generated by the randomizing device. As pointed out

---

[1]The randomizing device of Warner (1965) is not too satisfactory because the spinner must be perfectly horizontal to be unbiased.

by Franklin (1998), the Warner model implicitly makes an assumption that the respondent is sufficiently cognizant, informed, and educated to recognize and appreciate his or her anonymity. For an audience of lower education level or of less sophistication or in a diverse population as in health disparity studies, whatever explanations, he/she might elect not to reply or might provide an untruthful answer when he/she is being asked to answer the sensitive question.

(e) Narrow range of applications

The fifth is the narrow range of applications. The use of randomized response models has been limited almost exclusively to face-to-face personal interviews. It seems to be infeasible for telephone surveys. It is a common practice in telephone surveys to provide the respondent with a randomizing device by mail prior to the telephone contact. However, as pointed out by Orwin & Boruch (1982), there are numerous drawbacks with this procedure. The first is the additional cost of postage, added record keeping, and the devices themselves. Second, respondents may lose or discard the randomizing device. Third, telephone sampling strategies that capitalize on the non-necessity of linking phone numbers to names, e.g., random digit dialing, cannot be used. Finally, respondents' suspicions about the integrity of the device become rational again. Furthermore, Dalenius (1980) pointed out that most randomized response techniques require randomizing devices and that this presents a serious limitation with respect to telephone surveys. The need for a physical randomizing device in the respondent's presence does present an obstacle.

(f) Difficulty in selection of an appropriate randomizing device

The sixth difficulty is the choice of an appropriate randomizing device in some settings. Coutts & Jann (2011) pointed out that a complicated or novel randomizing device may lead the interviewee to doubt the method itself or, even worse, to feel that they are being tricked by the interviewer into providing information under false pretenses. The result of the randomizing device must be perceived as truly random and unknown to the interviewer in order to convince respondents that their 'yes' answers cannot be interpreted as an admission of guilt.

(g) Difficulty in generalization

Finally, when we estimate the association of two sensitive questions (or one sensitive question and one non-sensitive question), two or more randomizing devices are required if an randomized response technique is applied. This will increase the cognitive load of the randomized response procedure and introduce new sources of error.

1.2   Item Count Techniques

1.2.1   Basic idea for the item count techniques

Item count technique originally proposed by Miller (1984) provides much anonymity and confidentiality. The item count technique is also called the unmatched count technique (Dalton, Wimbush & Daily, 1994; Dalton, Daily & Wimbush, 1997; Coutts & Jann, 2011), the unmatched block design, or block total response (Raghavarao & Federer, 1979). One advantage of the item count technique over the randomized response technique is that no randomized device is required. However, two groups of respondents are generated via randomization.

(a) Description of the item count technique

In an item count design, there are k statements of behaviors which are refereed to the nonkey (or non-sensitive) items and one statement of behavior related to the sensitive characteristic which is refereed to the key (or sensitive) item. All non-sensitive and sensitive items have binary outcomes (e.g., 'yes' or 'no'). The interviewees are randomly assigned to either the control or treatment group. In the control group, respondents will be presented with the k non-sensitive questions only. In the treatment group, the same k non-sensitive questions in the control group together with the sensitive question of interest will be presented to the respondents. Respondents in both groups are then required to report only the number of 'yes' answers to the researchers or interviewers (Droitcour et al., 1991; Tsuchiya, Hirai & Ono, 2007). For example, respondents in the control group will be presented the following questions:

(1)   Do you have a sister?

(2)   Do you like music?

(3)   Is your birthday in January, February or March?

(4)   Have you suffered from athlete's foot?

(5)   Is the last digit of your cell phone an odd number?

According to your case, how many 'yes' answers are there?

Respondents in the treatment group will be presented the following questions:

(1) Do you have a sister?

(2) Do you like music?

(3) Is your birthday in January, February or March?

(4) Have you suffered from athlete's foot?

(5) Is the last digit of your cell phone an odd number?

(6) Have you ever shoplifted?

According to your case, how many 'yes' answers are there?

(b) Parameter estimation

Suppose the answers in the control group are $x_1, \ldots, x_n$, and those in treatment group are $y_1, \ldots, y_m$. Hence, the probability of having the sensitive characteristic (e.g., shoplifted before) can be estimated by

$$\hat{\pi}_{\text{ICT}} = \bar{y} - \bar{x} = \frac{1}{m}\sum_{j=1}^{m} y_j - \frac{1}{n}\sum_{i=1}^{n} x_i \qquad (1.4)$$

and its variance is given by

$$\text{Var}(\hat{\pi}_{\text{ICT}}) = \frac{1}{m-1}\sum_{j=1}^{m}(y_j - \bar{y})^2 + \frac{1}{n-1}\sum_{i=1}^{n}(x_i - \bar{x})^2, \qquad (1.5)$$

where the subscript 'ICT' refers to the 'item count technique.' A $(1-\alpha)100\%$ Wald confidence interval of π can then be constructed by

$$\left[\hat{\pi}_{\text{ICT}} - z_{\alpha/2}\sqrt{\text{Var}(\hat{\pi}_{\text{ICT}})}, \ \hat{\pi}_{\text{ICT}} + z_{\alpha/2}\sqrt{\text{Var}(\hat{\pi}_{\text{ICT}})}\right], \qquad (1.6)$$

where z_α is the upper α-th quantile of the standard normal distribution.

1.2.2   Some applications and generalizations

(a) Some applications

It is noteworthy that all respondents only need to report the total number
of statements that apply without telling which specific statements apply to
them. This technique prohibits interviewers from getting the information
of the sensitive question at the level of individuals. In practice, the interviewees should gain more confidence as their privacy is protected, and they
are willing to provide their truthful answers. Designs based on item count
technique have been adopted in various studies such as drug use (Droitcour
et al., 1991), employee theft for those personnel with access to cash, sup-
plies, merchandise or products easily converted to cash (Wimbush & Dalton,
1997), racial prejudice (Kuklinski, Cobb & Gilens, 1997; Gilens, Sniderman
& Kuklinski, 1998), sexual risk behavior after drinking among college students (LaBrie & Earleywine, 2000), hate crime victimization among college
students (Rayburn, Earleywine & Davison, 2003), counterproductive behavior (Ahart & Sackett, 2004), shoplifting (Tsuchiya, Hirai & Ono, 2007),
eating disordered behavior and attitude (Anderson *et al.*, 2007; Lavender &
Anderson, 2008), and attitude about immigration (Janus, 2010).

(b) Some generalizations

Generalizations of the item count technique have been developed recently
to cope with different scenarios. For example, Tsuchiya (2005) proposed
the cross-based and double-cross-based methods to obtain the domain estimators for the item count technique. Chaudhuri & Christofides (2007)
presented an amendment to the item count technique rendering it well-
equipped with a provision to protect privacy and also a sound theoretical
foundation. Imai (2011) developed multivariate regression models based
on item count technique. Petróczi *et al.* (2011) proposed the single sample
count technique which utilizes known population distributions and embeds
the sensitive question among four unrelated innocuous questions with binomial distribution.

1.2.3   Limitations of the item count techniques

Despite various attractive advantages, existing item count techniques have
several limitations. First, if a design of item count technique consists of k
nonsensitive questions and one sensitive question, respondents who give an
answer of $k + 1$ (i.e., answers to all questions are 'yes') will automatically
expose their sensitive characteristic. In this case, these respondents may

refuse to answer or provide untruthful answers. This definitely weakens the applicability and credibility of item count technique. Second, as pointed out by Chaudhuri & Christofides (2007), the proportion estimate $\hat{\pi}_{\mathrm{ICT}}$ specified by (1.4) may be less than zero. Third, the Wald confidence interval given by (1.6) may be beyond the unit interval $(0, 1)$. Finally, even though (1.6) is within the unit interval $(0, 1)$, its performance was shown to be inferior to other confidence intervals (e.g., the Wilson confidence interval).

1.3 Non-randomized Response Models

Non-randomized response techniques aim to provide the same privacy protection as randomized response techniques do. However, no randomizing device is required.

1.3.1 Swensson's non-randomized response model

(a) Description of the Swensson model

Without the use of any randomizing device, Swensson (1974) proposed a combined-question technique in which two independent random samples of size n_1 and n_2 are required (see also, Horvitz, Greenberg & Abernathy, 1976). The respondents in the first sample are asked to directly reply 'yes' or 'no' to one combined question, e.g.,

(Q1) Are you in the sensitive group $\mathcal{Y}$ (denoted by $Y = 1$) or are you in the non-sensitive group $\mathcal{U}$ (denoted by $U = 1$) but not in $\mathcal{Y}$?

The respondents in the second sample are asked to reply 'yes' or 'no' to another combined question, e.g.,

(Q2) Are you in the sensitive group $\mathcal{Y}$ (denoted by $Y = 1$) or are you neither in the non-sensitive group $\mathcal{U}$ nor in $\mathcal{Y}$?

(b) Parameter estimation

Let $\pi = \mathrm{Pr}(Y = 1)$ and $\theta = \mathrm{Pr}(U = 1)$ denote the respective proportions of the population belonging to the sensitive group $\mathcal{Y}$ and the non-sensitive group $\mathcal{U}$, and λ_1 and λ_2 denote the respective proportions of 'yes' answers in the two samples. Hence, we have

$$\lambda_1 = \pi + \mathrm{Pr}(U = 1, Y = 0) \quad \text{and}$$
$$\lambda_2 = \pi + \mathrm{Pr}(U = 0, Y = 0). \tag{1.7}$$

Since

$$
\begin{aligned}
\lambda_1 + \lambda_2 &= 2\pi + \Pr(U = 1, Y = 0) + \Pr(U = 0, Y = 0) \\
&= 2\pi + \Pr(Y = 0) \\
&= 2\pi + 1 - \pi,
\end{aligned}
$$

we have

$$
\pi = \lambda_1 + \lambda_2 - 1.
$$

The MLE of π and its variance are given by

$$
\hat{\pi}_{\mathrm{S}} = \hat{\lambda}_1 + \hat{\lambda}_2 - 1 = \frac{n_1'}{n_1} + \frac{n_2'}{n_2} - 1 \tag{1.8}
$$

and

$$
\mathrm{Var}(\hat{\pi}_{\mathrm{S}}) = \frac{\lambda_1(1 - \lambda_1)}{n_1} + \frac{\lambda_2(1 - \lambda_2)}{n_2}, \tag{1.9}
$$

respectively, where the subscript 'S' refers to the 'Swensson' model, n_1' and n_2' denote the respective numbers of 'yes' answers in the two samples.

(c) Comparison with the Warner model

Let $n_1 = n_2 = n/2$, $\pi = \theta = 0.1$ and U is independent of Y. From (1.7), we have

$$
\lambda_1 = \pi + \theta(1 - \pi) = 0.19
$$

and

$$
\lambda_2 = \pi + (1 - \theta)(1 - \pi) = 0.91.
$$

From (1.9), we obtain

$$
n\,\mathrm{Var}(\hat{\pi}_{\mathrm{S}}) = 2\lambda_1(1 - \lambda_1) + 2\lambda_2(1 - \lambda_2) = 0.4716.
$$

When $p = 0.7$, from (2.4), we have $n\,\mathrm{Var}(\hat{\pi}_{\mathrm{w}}) = 1.4025$. Thus, the combined-question design is about 3 times as efficient as the Warner design for $p = 0.7$.

1.3.2 Takahasi and Sakasegawa's non-randomized response model

Takahasi & Sakasegawa (1977) proposed another non-randomized response technique by using some auxiliary questions instead of randomizing devices.

(a) The model with independency assumption

An auxiliary question such as "Which do you prefer, 'spring' or 'autumn'?" is asked. After having made a silent answer to this auxiliary question (i.e., not telling the answer to the interviewer) the respondent is asked to say 0 or 1 according to the following list:

(L1) If you prefer 'spring' and you have A, please say 0.

(L2) If you prefer 'spring' and you do not have A, please say 1.

(L3) If you prefer 'autumn' and you have A, please say 1.

(L4) If you prefer 'autumn' and you do not have A, please say 0.

In the list A denotes the sensitive attribute. The aim is to estimate the sensitive proportion π in the population. Let p denote the proportion of the preference of 'autumn' to 'spring'. If p is known and the preference of one season to the other is independent of whether the respondent has A or not (i.e., the independency assumption is valid), then this technique is mathematically identical to the Warner's design. If p is unknown and the independency still holds, then another independent sample is required to estimate the p. However, if p is unknown and the independency cannot be assumed, then the estimation of π becomes impossible.

(b) The model without independency assumption

In the opinion of Suzuki, Takahasi & Sakasegawa (1976), it is very diffi-cult to find any auxiliary question which is independent of some attribute. Thus, Takahasi & Sakasegawa (1977) did not assume the independency and three independent samples of size n_1, n_2 and n_3 are thus required. Each respondent is required to choose one color from red, blue and green (without telling the chosen color to the interviewer) and answer 0 or 1 according to the following lists:

THE LIST FOR THE FIRST SAMPLE:

(1L1) If you choose 'red' and you have A, please say 0.

(1L2) If you choose 'red' and you do not have A, please say 1.

(1L3) If you choose 'blue' and you have A, please say 1.

(1L4) If you choose 'blue' and you do not have A, please say 0.

(1L5) If you choose 'green' and you have A, please say 1.

(1L6) If you choose 'green' and you do not have A, please say 0.

THE LIST FOR THE <u>SECOND</u> SAMPLE:

(2L1) If you choose 'red' and you have A, please say 1.

(2L2) If you choose 'red' and you do not have A, please say 0.

(2L3) If you choose 'blue' and you have A, please say 0.

(2L4) If you choose 'blue' and you do not have A, please say 1.

(2L5) If you choose 'green' and you have A, please say 1.

(2L6) If you choose 'green' and you do not have A, please say 0.

THE LIST FOR THE <u>THIRD</u> SAMPLE:

(3L1) If you choose 'red' and you have A, please say 1.

(3L2) If you choose 'red' and you do not have A, please say 0.

(3L3) If you choose 'blue' and you have A, please say 1.

(3L4) If you choose 'blue' and you do not have A, please say 0.

(3L5) If you choose 'green' and you have A, please say 0.

(3L6) If you choose 'green' and you do not have A, please say 1.

(c) Parameter estimation

Let B_1, B_2 and B_3 denote the color of 'red', 'blue' and 'green', respectively. To estimate the sensitive proportion π in the population, we let $p_i(A)$ denote the proportion of the persons with the attribute A in the population and will choose B_i, $p_i(\bar{A})$ denote the proportion of the persons without the attribute A in the population and will choose B_i, and λ_i be the probability that a respondent in the i-th sample answers 1 for $i = 1, 2, 3$. We have

$$\pi = p_1(A) + p_2(A) + p_3(A),$$
$$1 - \pi = p_1(\bar{A}) + p_2(\bar{A}) + p_3(\bar{A}),$$
$$\lambda_1 = p_1(\bar{A}) + p_2(A) + p_3(A) = \pi + p_1(\bar{A}) - p_1(A),$$
$$\lambda_2 = p_1(A) + p_2(\bar{A}) + p_3(A) = \pi + p_2(\bar{A}) - p_2(A), \quad \text{and}$$
$$\lambda_3 = p_1(A) + p_2(A) + p_3(\bar{A}) = \pi + p_3(\bar{A}) - p_3(A).$$

Since $\lambda_1 + \lambda_2 + \lambda_3 = 3\pi + (1 - \pi) - \pi$, we obtain $\pi = \sum_{i=1}^{3} \lambda_i - 1$. The MLE of π and its variance are respectively given by

$$\hat{\pi}_{\mathrm{TS}} = \sum_{i=1}^{3} \hat{\lambda}_i - 1 = \sum_{i=1}^{3} \frac{n_i'}{n_i} - 1 \tag{1.10}$$

and

$$\mathrm{Var}(\hat{\pi}_{\mathrm{TS}}) = \sum_{i=1}^{3} \frac{\lambda_i(1 - \lambda_i)}{n_i}, \tag{1.11}$$

where the subscript 'TS' refers to the 'Takahasi and Sakasegawa' model, and n_i' is the number of the respondents saying 1 in the i-th sample for $i = 1, 2, 3$.

We note that in Swensson's model or Takahasi and Sakasegawa's model, two or three groups of respondents are needed to be generated by randomization.

1.3.3 Non-randomized response models from a viewpoint of incomplete categorical data design

One important distinction between randomized response models and non-randomized response models to be introduced in this book is that the former usually requires a randomizing device such as a coin or a die which is related to a random variable without reproducibility (see remarks in Section 1.1.3 (a)), while the latter requires an independent non-sensitive variate such as birth date combined with the sensitive response variable to form an incomplete contingency table, resulting in the reproducibility. Some key advantages of the non-randomized methods are summarized as follows:

(1) reproducibility (i.e., the same respondent is expected to give the same answer to the aggregated sensitive and non-sensitive questions);

(2) maximum degree of privacy protection could be achieved by both triangular (in Chapter 3) and parallel (in Chapter 7) models by setting $p = \mathrm{Pr}(W = 1) = 0.5$ and $q = \mathrm{Pr}(U = 1) = 0.5$, where the non-sensitive variate W for both models and the other non-sensitive variate U for the parallel model are uniform.

(3) better cooperation since the non-sensitive question (e.g., the birthday of the respondent) is not controlled by interviewers and respondents are likely to be honest;

(4) significant cost reduction by getting rid of the randomizing device and easy implementation;

(5) natural generalization from a dichotomous situation to a multichoto-
     mous situation within the framework of incomplete contingency table;
     and

(6) broader applications since the non-randomized design is applicable to
     both the face-to-face interview and other forms of surveys.

It is not true that any randomized response model can be easily transformed to a non-randomized response model. Up to now, only the Warner
model and the unrelated question model were successfully transformed to
the corresponding non-randomized crosswise model (Yu, Tian & Tang, 2008)
and the non-randomized parallel model (Tian, 2012), respectively. Next,
although some randomized response models can be transformed to non-
randomized versions, the resulting statistical analysis methods are totally
different. For example, for the randomized unrelated question model with
an unknown $\theta = \Pr(U = 1)$, two independent samples of sizes n_1 and n_2
and two randomizing devices are required, while for its non-randomized version, i.e., the variant of the parallel model in Chapter 10 of this book, only
one sample is needed without using any randomizing device and the corresponding statistical analysis methods are developed based on a trinomial
distribution with two complete observations and one incomplete observation. The second example is as follows. To assess the association of two
sensitive questions with binary outcomes, a randomized response model in
general requires two randomizing devices (Christofides, 2005), while in the
non-randomized hidden sensitivity model (Tian *et al.*, 2007), respondents
only need to answer a non-sensitive question instead of the original two
sensitive questions and the corresponding analysis methods are developed
based on an incomplete 4×4 contingency table. Finally, for other randomized response models (e.g., Kuk, 1990), the corresponding non-randomized
partners are not yet available.

   Other survey techniques with sensitive questions include sealed enve-
lope questionnaires, computer-based questionnaires, nominative techniques
by asking respondents to report on their close friends' socially undesirable
behavior (see, e.g., Bradburn, Sudman & Associates, 1979; Miller, 1983,
1985), negative surveys (Esponda & Guerreroc, 2009; Esponda, Forrest &
Helman, 2009), and so on.

1.4   Scope of the Rest of the Book

Chapter 2 discusses the non-randomized crosswise model. We first introduce
the randomized Warner model to investigate its relative efficiency and degree
of privacy protection. Second, to set the comparison of different models on
the same footing, we present the crosswise model, which is a non-randomized

version of the Warner model. Two asymptotic and two bootstrap confidence intervals for the proportion π of the persons belonging to the sensitive class in the population are presented. An asymptotic property of the MLE (and its modified version) of the π is explored. Third, Bayesian methods for analyzing survey data from the crosswise model are also given. Finally, an induced abortion study and an experimental survey on measuring plagiarism are used to illustrate these methods.

In Chapter 3, we systematically investigate the non-randomized triangular model by introducing the survey design, an alternative formulation of the model, the variance of the estimator, relative efficiency, and degree of privacy protection. We then compare the triangular model with the Warner model theoretically. An alternative derivation of the MLE, two asymptotic confidence intervals, two bootstrap confidence intervals, and some asymptotic properties related to the modified MLE are presented. In addition, Bayesian methods for the triangular model are also discussed. Finally, we analyze three sexual behavior datasets by using the triangular model and corresponding methods.

Chapter 4 introduces sample size formulae for the crosswise and triangular designs based on the power analysis approach. We first consider the triangular model for the one-sample problem. Power functions and their corresponding sample size formulae for the one- and two-sided tests based on the large-sample normal approximation are derived. Evaluation of the performance of the sample size formulae are conducted by comparing exact power with asymptotic power. We then consider the crosswise model for the one-sample problem. We numerically compare the sample sizes required for the crosswise design with those required for the triangular design. Theoretical justification is also provided. Furthermore, we extend the one-sample problem to the two-sample problem. An example based on an induced abortion study in Taiwan is presented to illustrate the presented methods.

In Chapter 5, we introduce a multi-category triangular design for a single sensitive question with multiple outcomes. We show that MLEs of cell probabilities can be obtained via the *expectation–maximization* (EM) algorithm. Asymptotic and bootstrap confidence intervals of the cell probabilities or their functions are then given. Bayesian estimation via the data augmentation algorithm is developed when prior information on the parameters of interest is available. A real dataset from a questionnaire on sexual activities in Korean adolescents is used to illustrate the proposed design and analysis methods.

Chapter 6 introduces a hidden sensitivity model for assessing the association of two sensitive questions with binary outcomes. We present the MLEs of the cell probabilities and the odds ratio for two binary variables associated with the sensitive questions via the EM algorithm. The corresponding standard error estimates are then obtained by bootstrapping methods. A likelihood ratio test and a chi-squared test are developed for testing association between the two binary variables. Simulations are performed to evaluate the empirical type I error rates and powers for the two tests. Bayesian inferences under Dirichlet prior and other priors are investigated. A real data set from an AIDS study is used to illustrate the proposed methodologies.

In Chapter 7, we first present the unrelated question randomized response model to investigate its relative efficiency and degree of privacy protection. Second, we introduce the parallel model, which is a non-randomized version of the unrelated question model. Asymptotic properties of the maximum likelihood estimator (and its modified version) for the proportion of interest are explored. Theoretic comparisons with the crosswise and triangular models show that the parallel model is always more efficient than the two models for most of the possible parameter ranges. And the parallel model is with a wider application range and a better degree of privacy protection. Bayesian methods for analyzing survey data from the parallel model are developed. A real dataset from an 'induced abortion in Mexico' survey, a case study on college students' premarital sexual behavior at Wuhan, and a case study on plagiarism at The University of Hong Kong are used to illustrate these methods.

Sample size formulae for the parallel design are introduced in Chapter 8 by using the power analysis method for both the one- and two-sample problems. Asymptotic power functions and the corresponding sample size formulae for both the one- and two-sided tests based on the large-sample normal approximation are derived. The performance is assessed through comparing the asymptotic power with the exact power and reporting the ratio of the sample sizes with the parallel model and the design of direct questioning. We numerically compare the sample sizes needed for the parallel design with those required for the crosswise and triangular models. Two theoretical justifications are also provided. An example from a survey on 'sexual practices' in San Francisco, Las Vegas, and Portland is used to illustrate the proposed methods.

In Chapter 9, we introduce a multi-category parallel model, where all population categories of interest could be sensitive or one of the categories is totally non-sensitive. Likelihood-based inferences and Bayesian inferences for parameters of interest are developed. In addition, an important special case of the multi-category parallel model is studied to test the association of two sensitive binary variables. Furthermore, theoretic comparisons show that the multi-category parallel model is more efficient than the multi-category triangular model for some cases. An example on the study of association between the number of sex partners and annual income is used to illustrate the methods.

In Chapter 10, we introduce a variant of the parallel model. The survey design and corresponding statistical inferences including likelihood-based methods, bootstrap methods, and Bayesian methods are provided. Theoretical and numerical comparisons showed that the variant of the parallel model over-performs the crosswise and triangular models for most of the possible parameter ranges. An outline for handling the possible noncompliance behavior in the proposed model is provided. An illustrative example from an existing survey on 'sexual practices' in San Francisco, Las Vegas, and Portland is used to demonstrate the proposed statistical analysis methods. Two real surveys on the cheating behavior in examinations at the University of Hong Kong are conducted and are used to illustrate the proposed design and analysis methods.

In Chapter 11, we introduce a combination-questionnaire design to assess the association between one sensitive binary variate (e.g., taking drugs, the number of sex partners) and one non-sensitive binary variate (e.g., health status, cervical cancer). The design consists of a main questionnaire and a supplemental questionnaire without using any randomizing device. The corresponding statistical inference approaches are also presented.

In Appendices A and B, we briefly introduce the EM algorithm, data augmentation algorithm, and the exact *inverse Bayes formula* (IBF) sampling. Appendix C outlines some basic statistical distributions.

CHAPTER 2

The Crosswise Model

In this chapter, we first introduce the randomized Warner model to investigate its point estimation, relative efficiency, and degree of privacy protection. Second, to set the comparison of different models on the same footing, we present a so-called crosswise model, which is a non-randomized version of the original Warner model. Two asymptotic confidence intervals and two bootstrap confidence intervals are provided. Asymptotic property of the modified maximum likelihood estimator for the proportion π of the population belonging to the sensitive class is explored. Third, Bayesian methods for the crosswise model are also given. Finally, an induced abortion study and an experimental survey on measuring plagiarism are used to illustrate these methods.

2.1 The Warner Model

Warner (1965) considered the situation in which respondents from a specific population can be divided into two mutually exclusive groups: one group with a stigmatizing characteristic (e.g., a drug user) and the other group without such a characteristic (e.g., not a drug user). The following two sensitive questions are then presented to respondents:

(a) I am a drug user (i.e., I belong to class $\mathcal{A}$).

(b) I am not a drug user (i.e., I do not belong to class $\mathcal{A}$).

2.1.1 The survey design

Each respondent is directed to answer statement (a) or (b) privately by means of a randomization device such as a dice or a spinner without indicating to the interviewer which question is being answered. Thus, the interviewer receives a response (e.g., 'yes' or 'no') from a respondent without the knowledge of which question is being answered. This alleviates the concerns for privacy and reduces the number of refusals to respond to questions with untruthful answers. Let p denote the probability of assigning statement (a) to a respondent by the randomization device; that is,

$$p = \Pr\{\text{selecting statement (a) by the randomization device}\}.$$

The randomizing device is entirely controlled by the interviewer. In other words, the probability p is chosen by the interviewer as a part of the design and is assumed to be known. In addition, the value of p should be known to all respondents.

2.1.2 Point estimation

Suppose that we want to estimate the proportion π of the population belonging to the sensitive class $\mathcal{A}$, i.e., $\pi = \Pr(\mathcal{A})$. Let n' be the number of 'yes' answers obtained from n respondents. The *maximum likelihood estimator* (MLE) of π can be easily shown to be

$$\hat{\pi}_{\mathrm{W}} = \frac{p - 1 + n'/n}{2p - 1}, \qquad p \neq 0.5, \tag{2.1}$$

provided that $\hat{\pi}_{\mathrm{W}} \in [0, 1]$, where the subscript 'W' refers to the 'Warner' model. Warner (1965) showed that $\hat{\pi}_{\mathrm{W}}$ is unbiased with variance

$$\mathrm{Var}(\hat{\pi}_{\mathrm{W}}) = \mathrm{Var}(\hat{\pi}_{\mathrm{D}}) + \frac{p(1 - p)}{n(2p - 1)^2}, \tag{2.2}$$

where

$$\mathrm{Var}(\hat{\pi}_{\mathrm{D}}) = \frac{\pi(1 - \pi)}{n} \tag{2.3}$$

denotes the variance of $\hat{\pi}_{\mathrm{D}}$ corresponding to the *design of direct questioning* (DDQ). For any fixed π, we note that

$$n\mathrm{Var}(\hat{\pi}_{\mathrm{W}}) = \pi(1 - \pi) + \frac{p(1 - p)}{(2p - 1)^2} \tag{2.4}$$

is an increasing function of p when $0 < p < 0.5$, which quickly approaches to infinity as $p \to 0.5$, and then becomes a decreasing function of p when $0.5 < p < 1$ (see Figure 2.1).

2.1.3 Relative efficiency

The most serious limitation of the Warner model is perhaps its inefficiency when comparing with the DDQ. Note that the second term of (2.2) is introduced due to the randomization device. When $p = 0$ or $p = 1$, the Warner model is reduced to the DDQ. We define the *relative efficiency* (RE) of the Warner design to the DDQ by

$$\mathrm{RE}_{\mathrm{W} \to \mathrm{D}}(\pi, p) \;\; \hat{=} \;\; \frac{\mathrm{Var}(\hat{\pi}_{\mathrm{W}})}{\mathrm{Var}(\hat{\pi}_{\mathrm{D}})}$$

$$= \;\; 1 + \frac{p(1 - p)/(2p - 1)^2}{\pi(1 - \pi)}, \tag{2.5}$$

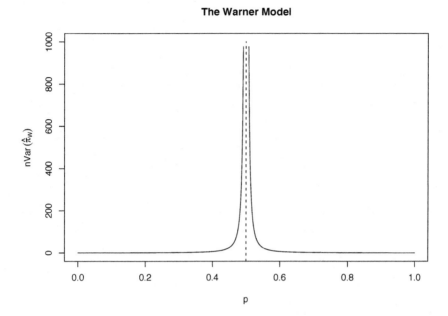

Figure 2.1   Plot of $n\mathrm{Var}(\hat{\pi}_{\mathrm{w}})$ defined in (2.4) against p with $\pi = 0.40$ for the Warner model.

which is free of the sample size (i.e., n).

Let n_{W} and n_{D} be sample sizes required for the Warner design and the DDQ, respectively. To achieve the same estimation precision, we let

$$\frac{\pi(1-\pi)}{n_{\mathrm{W}}} + \frac{p(1-p)}{n_{\mathrm{W}}(2p-1)^2} = \frac{\pi(1-\pi)}{n_{\mathrm{D}}}.$$

From (2.5), it follows that

$$\frac{n_{\mathrm{W}}}{n_{\mathrm{D}}} = \mathrm{RE}_{\mathrm{W}\to\mathrm{D}}(\pi, p).$$

Thus, the relative efficiency is directly related to the ratio of the sample sizes of the two designs.

Table 2.1 shows some values of $\mathrm{RE}_{\mathrm{W}\to\mathrm{D}}(\pi, p)$ for various combinations of π and p. For example, when $\pi = 0.30$ and $p = 0.38$, we have

$$\mathrm{RE}_{\mathrm{W}\to\mathrm{D}}(0.30,\ 0.38) = 20.478,$$

which suggests that the sample size required for the Warner design is about 20 times that required for the DDQ in order to achieve the same estimation precision. To reduce the variability, we have to increase the sample size n, resulting in a possibly significant increase in cost.

Table 2.1 Relative efficiency $\mathrm{RE}_{\mathrm{W}\to\mathrm{D}}(\pi, p)$ for various combinations of π and p

π	p					
	0.20	0.30	0.34	0.38	0.42	0.46
0.05	10.357	28.632	47.135	87.111	201.329	818.105
0.10	5.9383	15.583	25.349	46.447	106.729	432.250
0.15	4.4858	11.294	18.188	33.081	75.6324	305.412
0.20	3.7778	9.2031	14.696	26.564	60.4727	243.578
0.25	3.3704	8.0000	12.688	22.815	51.7500	208.000
0.30	3.1164	7.2500	11.435	20.478	46.3125	185.821
0.35	2.9536	6.7692	10.633	18.979	42.8269	171.604
0.40	2.8519	6.4687	10.131	18.043	40.6484	162.719
0.45	2.7957	6.3030	9.8542	17.526	39.4470	157.818
0.50	2.7778	6.2500	9.7656	17.361	39.0625	156.250

2.1.4 Degree of privacy protection

Intuitively, the optimal *degree of privacy protection* (DPP) is reached at $p = 0.5$, which corresponds to the case of infinite variance (cf. Figure 2.1). When p is either too small or too large, the privacy of respondents cannot be protected sufficiently. Therefore, investigators are forced to select a value of p within some sub-interval of $(0, 0.5)$ since $n\mathrm{Var}(\hat{\pi}_{\mathrm{W}})$ is a symmetric function on $p = 0.5$ and thus adopt an uneven randomizing device (e.g., a biased coin). Since the sensitive information of a respondent regarding his/her membership in the sensitive class $\mathcal{A}$ is characterized through $\mathrm{Pr}(\mathcal{A}|\mathrm{yes})$ and $\mathrm{Pr}(\mathcal{A}|\mathrm{no})$, we define

$$\mathrm{DPP}_{\mathrm{yes}}(\pi, p) \,\hat{=}\, \mathrm{Pr}(\mathcal{A}|\mathrm{yes}) = \frac{\pi p}{\pi p + (1 - \pi)(1 - p)} \qquad (2.6)$$

and

$$\mathrm{DPP}_{\mathrm{no}}(\pi, p) \,\hat{=}\, \mathrm{Pr}(\mathcal{A}|\mathrm{no}) = \frac{\pi(1 - p)}{\pi(1 - p) + (1 - \pi)p} \qquad (2.7)$$

to measure the private information divulged with the Warner model.

Figure 2.2 shows that for a fixed π, $\mathrm{DPP}_{\mathrm{yes}}(\pi, p)$ is a monotonically increasing function of p while $\mathrm{DPP}_{\mathrm{no}}(\pi, p)$ is a monotonically decreasing function of p. In particular, for any $\pi \in (0, 1)$, when $p = 0.5$, we have

$$\mathrm{DPP}_{\mathrm{yes}}(\pi, 0.5) = \mathrm{DPP}_{\mathrm{no}}(\pi, 0.5) = \pi,$$

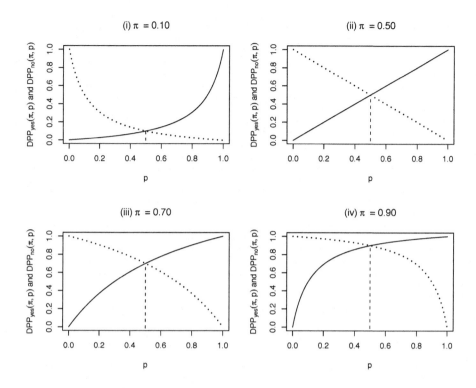

Figure 2.2 Plots of $\mathrm{DPP_{yes}}(\pi, p)$ (solid line) and $\mathrm{DPP_{no}}(\pi, p)$ (dotted line) against p for the Warner model for a fixed π. (i) $\pi = 0.10$; (ii) $\pi = 0.50$; (iii) $\pi = 0.70$; (iv) $\pi = 0.90$.

implying that in this case,

$$\Pr(\mathcal{A}|\mathrm{yes}) = \Pr(\mathcal{A}|\mathrm{no}) = \Pr(\mathcal{A}).$$

In other words, whether the respondent belongs to the sensitive class $\mathcal{A}$ does not depend on the answer 'yes' or 'no', thus arriving at the maximum DPP. However, when $p \in [0,\ 0.50]$, we have $\mathrm{DPP_{yes}}(\pi, p) < \pi$ while $\mathrm{DPP_{no}}(\pi, p) \geqslant \pi$.

2.2 A Non-randomized Warner Model: The Crosswise Model

To compare the randomized Warner model with the non-randomized triangular model (see Chapter 3 of this book) based on the same benchmark, Yu, Tian & Tang (2008) introduced a so-called crosswise model, which can be viewed as a non-randomized version of the original Warner model. Let $\{Y = 1\}$ denote the class of people who possess a sensitive characteristic (e.g., drug-taking) and $\{Y = 0\}$ denote the complementary class. Let W be a non-sensitive dichotomous variate and be independent of Y. The interviewer should select a suitable W so that the proportion

$$p = \Pr(W = 1)$$

should be known in advance or can be estimated easily. Without loss of generality, p is assumed to be known. For example, we may define $W = 1$ if a respondent was born between August and December and $W = 0$ otherwise. Hence, it is reasonable to assume that $p \approx 5/12 = 0.41667$. Our aim is to estimate the proportion

$$\pi = \Pr(Y = 1).$$

2.2.1 The survey design

The interviewer may design a questionnaire in the format as shown on the left-hand side of Table 2.2 and asks each interviewee to put a tick in the upper circle (i.e., $\{Y = 0, \ W = 0\}$) if he/she belongs to one of the two circles or put a tick in the upper square (i.e., $\{Y = 0, \ W = 1\}$) if he/she belongs to one of the two squares. Note that both $\{Y = 0, \ W = 0\}$ and $\{Y = 0, \ W = 1\}$ are non-sensitive. Thus, an interviewee who belongs to the sensitive class (i.e., $\{Y = 1\}$) is not being exposed if a tick is put in the upper circle/square. The corresponding cell probabilities are listed at the right-hand side of Table 2.2. Yu, Tian & Tang (2008) called this the *crosswise* model.

Table 2.2 The crosswise model and the corresponding cell probabilities

Category	$W = 0$	$W = 1$		Category	$W = 0$	$W = 1$	Marginal
$Y = 0$	○	□		$Y = 0$	$(1 - \pi)(1 - p)$	$(1 - \pi)p$	$1 - \pi$
$Y = 1$	□	○		$Y = 1$	$\pi(1 - p)$	πp	π
				Marginal	$1 - p$	p	1

Source: Adapted from Table 2 of Yu, Tian & Tang (2008).
Note: Please put a tick in the upper circle if you belong to one of the two circles or put a tick in the upper square if you belong to one of the two squares.

2.2.2 Connection with the Warner model

Suppose there are a total of n ticks (corresponding to n respondents in the survey without non-response) with n' ticks being put in the upper circle and $n - n'$ ticks being put in the upper square (see Table 2.2). The observed data are denoted by $Y_{\text{obs}} = \{n; \ n', \ n - n'\}$. The probability of putting a tick in the upper circle is then $(1 - \pi)(1 - p) + \pi p$. Letting $\lambda = (1 - \pi)(1 - p) + \pi p$, we have

$$\pi = \frac{p - 1 + \lambda}{2p - 1},$$

where $p = \Pr(W = 1) \neq 0.5$ is known. The likelihood function is proportional to $\lambda^{n'}(1 - \lambda)^{n-n'}$ so that the MLE of λ is given by $\hat{\lambda} = n'/n$. Therefore, the MLE of π is

$$\hat{\pi}_{\mathrm{C}} = \frac{p - 1 + \hat{\lambda}}{2p - 1}, \qquad p \neq 0.5, \tag{2.8}$$

provided that $\hat{\pi}_{\mathrm{C}} \in [0, 1]$, where the subscript 'C' refers to the 'crosswise' model. When $\hat{\pi}_{\mathrm{C}} < 0$ or $\hat{\pi}_{\mathrm{C}} > 1$, we can use the EM algorithm (2.28) and (2.29) with $a = b = 1$ (i.e., if the uniform prior on $[0, 1]$ is adopted, the posterior mode of π is identical to the MLE of π) to find the MLE of π. Since

$$n' \sim \mathrm{Binomial}\,(n, \lambda),$$

we have

$$E(n') = n\lambda \quad \text{and} \quad \mathrm{Var}(n') = n\lambda(1 - \lambda).$$

Hence, $E(\hat{\pi}_{\mathrm{C}}) = \pi$, i.e., $\hat{\pi}_{\mathrm{C}}$ is unbiased, and

$$\begin{aligned} \mathrm{Var}(\hat{\pi}_{\mathrm{C}}) &= \frac{\lambda(1 - \lambda)}{n(2p - 1)^2} \\ &= \frac{\pi(1 - \pi)}{n} + \frac{p(1 - p)}{n(2p - 1)^2}. \end{aligned} \tag{2.9}$$

The MLE in (2.8) and the corresponding variance in (2.9) are identical to (2.1) and (2.2), respectively. Hence, Yu, Tian & Tang (2008) considered the crosswise model as a non-randomized version of the Warner model.

In comparison with the original randomized Warner model, this non-randomized crosswise model has the following advantages:

(1) *Cost reduction* (since no randomization device is required).

(2) *Reproducibility* (i.e., the same respondent is expected to give the same answer by design if the survey is presented to the respondent again).

(3) *Better cooperation* (since the non-sensitive question is not controlled by interviewers and respondents are likely to trust the interviewers).

2.2.3   Two asymptotic confidence intervals

(a) Wald confidence interval

It is easy to show that an unbiased estimate of $\mathrm{Var}(\hat{\pi}_C)$ is given by

$$\overline{\mathrm{Var}}(\hat{\pi}_C) = \frac{\hat{\lambda}(1-\hat{\lambda})}{(n-1)(2p-1)^2}$$

$$= \frac{\hat{\pi}_C(1-\hat{\pi}_C)}{n-1} + \frac{p(1-p)}{(n-1)(2p-1)^2}.$$

By the Central Limit Theorem, $\hat{\pi}_C$ is asymptotically normally distributed, i.e.,

$$(\hat{\pi}_C - \pi)\Big/\sqrt{\overline{\mathrm{Var}}(\hat{\pi}_C)} \sim N(0,1), \quad \text{as } n \to \infty.$$

The $(1-\alpha)100\%$ Wald confidence interval of π is given by

$$[\hat{\pi}_{C,\mathrm{WL}},\ \hat{\pi}_{C,\mathrm{WU}}] = \left[\hat{\pi}_C - z_{\alpha/2}\sqrt{\overline{\mathrm{Var}}(\hat{\pi}_C)},\ \ \hat{\pi}_C + z_{\alpha/2}\sqrt{\overline{\mathrm{Var}}(\hat{\pi}_C)}\right], \quad (2.10)$$

where z_α is the upper α-th quantile of the standard normal distribution.

Remark 2.1   First, it is possible in practice that the lower bound of the Wald confidence interval given by (2.10) is less than zero if the true value of π is near to zero or the upper bound is larger than one if the true value of π is near to one, yielding a useless confidence interval. Second, even $[\hat{\pi}_{C,\mathrm{WL}},\ \hat{\pi}_{C,\mathrm{WU}}]$ is within the unit interval $(0,1)$, its performance was shown (Agresti & Coull, 1998) to be inferior to that of the Wilson (score) confidence interval, which is introduced below.   ¶

(b) Wilson (score) confidence interval

The construction of the $(1-\alpha)100\%$ Wilson (score) confidence interval $[\hat{\pi}_{C,\mathrm{WSL}},\ \hat{\pi}_{C,\mathrm{WSU}}]$ of π is based on

$$1-\alpha = \mathrm{Pr}\left\{\left|\frac{\hat{\pi}_C - \pi}{\sqrt{\mathrm{Var}(\hat{\pi}_C)}}\right| \leqslant z_{\alpha/2}\right\}$$

$$= \mathrm{Pr}\left\{(\hat{\pi}_C - \pi)^2 \leqslant z_{\alpha/2}^2 \mathrm{Var}(\hat{\pi}_C)\right\}$$

$$\stackrel{(2.9)}{=} \mathrm{Pr}\left[\hat{\pi}_C^2 - 2\hat{\pi}_C\pi + \pi^2 \leqslant \frac{z_{\alpha/2}^2}{n}\left\{\pi(1-\pi) + \frac{p(1-p)}{(2p-1)^2}\right\}\right]$$

$$= \mathrm{Pr}\left\{(1+z_*)\pi^2 - (2\hat{\pi}_C + z_*)\pi + \hat{\pi}_C^2 - z_*\varrho \leqslant 0\right\}, \quad (2.11)$$

where

$$z_* \hat{=} \frac{z_{\alpha/2}^2}{n} \quad \text{and} \quad \varrho \hat{=} \frac{p(1-p)}{(2p-1)^2}. \tag{2.12}$$

Solving the quadratic inequality inside the probability in (2.11) yields

$$[\hat{\pi}_{\mathrm{C,WSL}}, \ \hat{\pi}_{\mathrm{C,WSU}}] = \frac{2\hat{\pi}_{\mathrm{C}} + z_* \pm \sqrt{(2\hat{\pi}_{\mathrm{C}} + z_*)^2 - 4(1+z_*)(\hat{\pi}_{\mathrm{C}}^2 - z_*\varrho)}}{2(1+z_*)}, \tag{2.13}$$

which is, in general, within the unit interval $[0,1]$.

Remark 2.2 First, if $\hat{\pi}_{\mathrm{C,WSL}}$ is smaller than zero or $\hat{\pi}_{\mathrm{C,WSU}}$ is greater than one, then the Wilson (score) confidence interval $[\hat{\pi}_{\mathrm{C,WSL}}, \ \hat{\pi}_{\mathrm{C,WSU}}]$ is also useless. Second, it is possible that $[\hat{\pi}_{\mathrm{C,WL}}, \ \hat{\pi}_{\mathrm{C,WU}}]$ is beyond the unit interval $(0,1)$ while $[\hat{\pi}_{\mathrm{C,WSL}}, \ \hat{\pi}_{\mathrm{C,WSU}}]$ is within $(0,1)$. Finally, even $[\hat{\pi}_{\mathrm{C,WL}}, \ \hat{\pi}_{\mathrm{C,WU}}]$ is within $(0,1)$, its performance still depends on the large-sample theory. ¶

2.2.4 Bootstrap confidence intervals

When the sample size n is small to moderate, asymptotic confidence intervals in (2.10) and (2.13) are not reliable. For these cases, the parametric bootstrap approach (Efron & Tibshirani, 1993) can be used to obtain two bootstrap confidence intervals of π.

From the observed data $Y_{\mathrm{obs}} = \{n; \ n', \ n - n'\}$, we first calculate the MLE of λ, i.e., $\hat{\lambda} = n'/n$. Then, we can generate

$$n'^* \sim \mathrm{Binomial}(n, \ \hat{\lambda})$$

to yield $Y_{\mathrm{obs}}^* = \{n; \ n'^*, \ n - n'^*\}$. Based on Y_{obs}^*, we can calculate a bootstrap replication $\hat{\pi}_{\mathrm{C}}^*$ by using the formula (2.8) or the EM algorithm (2.28) and (2.29) with $a = b = 1$. Independently repeating this process G times, we obtain G bootstrap replications $\{\hat{\pi}_{\mathrm{C}}^*(g)\}_{g=1}^G$. Thus, the standard error, $\mathrm{se}(\hat{\pi}_{\mathrm{C}})$, of $\hat{\pi}_{\mathrm{C}}$ can be estimated by the sample standard deviation of the G replications, i.e.,

$$\widehat{\mathrm{se}}(\hat{\pi}_{\mathrm{C}}) = \left[\frac{1}{G-1}\sum_{g=1}^G \left\{\hat{\pi}_{\mathrm{C}}^*(g) - \frac{\hat{\pi}_{\mathrm{C}}^*(1) + \cdots + \hat{\pi}_{\mathrm{C}}^*(G)}{G}\right\}^2\right]^{\frac{1}{2}}.$$

If $\{\hat{\pi}_{\mathrm{C}}^*(g)\}_{g=1}^G$ is approximately normally distributed, a $(1-\alpha)100\%$ bootstrap confidence interval for π is given by

$$\left[\hat{\pi}_{\mathrm{C}} - z_{\alpha/2} \times \widehat{\mathrm{se}}(\hat{\pi}_{\mathrm{C}}), \ \hat{\pi}_{\mathrm{C}} + z_{\alpha/2} \times \widehat{\mathrm{se}}(\hat{\pi}_{\mathrm{C}})\right]. \tag{2.14}$$

If $\{\hat{\pi}_{\mathrm{C}}^*(g)\}_{g=1}^G$ is non-normally distributed, a $100(1-\alpha)\%$ bootstrap confidence interval for π can be obtained by

$$[\hat{\pi}_{\mathrm{C,BL}},\ \hat{\pi}_{\mathrm{C,BU}}], \tag{2.15}$$

where $\hat{\pi}_{\mathrm{C,BL}}$ and $\hat{\pi}_{\mathrm{C,BU}}$ are the $100(\alpha/2)$ and $100(1-\alpha/2)$ percentiles of $\{\hat{\pi}_{\mathrm{C}}^*(g)\}_{g=1}^G$, respectively.

2.2.5 An asymptotic property of the modified MLE

(a) A modified MLE of π

From (2.8), we have $0 \leqslant \hat{\pi}_{\mathrm{C}} \leqslant 1$ if and only if

$$\min(1-p,\ p) \leqslant \hat{\lambda} \leqslant \max(1-p,\ p).$$

Therefore, a modified MLE of π is given by

$$\hat{\pi}_{\mathrm{CM}} = \mathrm{median}\{0, \hat{\pi}_{\mathrm{C}}, 1\} = \begin{cases} 0, & \text{if } \hat{\pi}_{\mathrm{C}} < 0, \\ \hat{\pi}_{\mathrm{C}}, & \text{if } 0 \leqslant \hat{\pi}_{\mathrm{P}} \leqslant 1, \\ 1, & \text{if } \hat{\pi}_{\mathrm{P}} > 1, \end{cases}$$

$$= \begin{cases} 0, & \text{if } 0 \leqslant \dfrac{n'}{n} < \min(1-p,\ p), \\ \hat{\pi}_{\mathrm{C}}, & \text{if } \min(1-p,\ p) \leqslant \dfrac{n'}{n} \leqslant \max(1-p,\ p), \\ 1, & \text{if } \max(1-p,\ p) < \dfrac{n'}{n} \leqslant 1. \end{cases} \tag{2.16}$$

(b) The asymptotic equivalence between $\hat{\pi}_{\mathrm{CM}}$ and $\hat{\pi}_{\mathrm{C}}$

Theorem 2.1 If $0 < \pi < 1$, then $\sqrt{n}(\hat{\pi}_{\mathrm{CM}} - \pi)$ and $\sqrt{n}(\hat{\pi}_{\mathrm{C}} - \pi)$ have the same asymptotic distribution as $n \to \infty$. ¶

Proof. It suffices to show that $\sqrt{n}(\hat{\pi}_{\mathrm{CM}} - \pi) - \sqrt{n}(\hat{\pi}_{\mathrm{C}} - \pi)$ converges to zero in probability as $n \to \infty$, i.e.,

$$\Pr\{|\sqrt{n}(\hat{\pi}_{\mathrm{CM}} - \hat{\pi}_{\mathrm{C}})| > 0\} \to 0, \quad \text{as } n \to \infty. \tag{2.17}$$

Noting that $\hat{\lambda}$ is the MLE of $\lambda = (1-\pi)(1-p) + \pi p$ and

$$\min(1-p,\ p) < \lambda < \max(1-p,\ p)$$

as $0 < \pi < 1$, we naturally obtain $\Pr(|\hat{\lambda} - \lambda| > \varepsilon) \to 0$, as $n \to \infty$, for any $\varepsilon > 0$. We only need to prove

$$\Pr\{|\sqrt{n}(\hat{\pi}_{\mathrm{CM}} - \hat{\pi}_{\mathrm{C}})| > 0\} \leqslant \Pr(|\hat{\lambda} - \lambda| > \varepsilon),$$

or equivalently

$$\{|\sqrt{n}(\hat{\pi}_{\mathrm{CM}} - \hat{\pi}_{\mathrm{C}})| > 0\} \subseteq \{|\hat{\lambda} - \lambda| > \varepsilon\}, \tag{2.18}$$

for any $\varepsilon < \min\{\max(1 - p,\ p) - \lambda, \lambda - \min(1 - p,\ p)\}$. Without loss of generality, we assume $p > 1/2$. We consider three cases.

Case I: $1 - p \leqslant \hat{\lambda} \leqslant p$. From (2.16), we obtain $\hat{\pi}_{\mathrm{CM}} = \hat{\pi}_{\mathrm{C}}$. Therefore, (2.17) follows immediately.

Case II: $\hat{\lambda} < 1 - p$. Now $\hat{\pi}_{\mathrm{CM}} = 0$. If

$$|\sqrt{n}(\hat{\pi}_{\mathrm{CM}} - \hat{\pi}_{\mathrm{C}})| > 0 \quad \Rightarrow \quad |\hat{\lambda} + p - 1| > 0$$

$$\Rightarrow \quad 0 < |\hat{\lambda} + p - 1| = -(\hat{\lambda} + p - 1)$$

$$= -(\hat{\lambda} - \lambda) - \{\lambda - (1 - p)\}$$

$$\Rightarrow \quad |\hat{\lambda} - \lambda| \geqslant -(\hat{\lambda} - \lambda) > \lambda - (1 - p). \tag{2.19}$$

Noting that $\varepsilon < \min\{p - \lambda, \lambda - (1 - p)\}$, we have

$$\{\lambda - (1 - p)\} - \varepsilon > 0. \tag{2.20}$$

By combining (2.19) with (2.20), we obtain

$$|\hat{\lambda} - \lambda| - \varepsilon = |\hat{\lambda} - \lambda| - \{\lambda - (1 - p)\} + \{\lambda - (1 - p)\} - \varepsilon > 0$$

and hence (2.18) follows.

Case III: $\hat{\lambda} > p$. Now $\hat{\pi}_{\mathrm{CM}} = 1$. If

$$|\sqrt{n}(\hat{\pi}_{\mathrm{CM}} - \hat{\pi}_{\mathrm{C}})| > 0 \quad \Rightarrow \quad |p - \hat{\lambda}| > 0$$

$$\Rightarrow \quad 0 < |p - \hat{\lambda}| = -(p - \hat{\lambda})$$

$$= -(\lambda - \hat{\lambda}) - (p - \lambda)$$

$$\Rightarrow \quad |\lambda - \hat{\lambda}| \geqslant -(\lambda - \hat{\lambda}) > p - \lambda. \tag{2.21}$$

Noting that $\varepsilon < \min\{p - \lambda, \lambda - (1 - p)\}$, we have

$$(p - \lambda) - \varepsilon > 0. \tag{2.22}$$

By combining (2.21) and (2.22), we obtain

$$|\hat{\lambda} - \lambda| - \varepsilon = |\hat{\lambda} - \lambda| - (p - \lambda) + (p - \lambda) - \varepsilon > 0.$$

Hence, (2.18) follows. □

2.3 Bayesian Methods for the Crosswise Model

2.3.1 Posterior moments

Let $Y_{\text{obs}} = \{y_{i,\text{C}}: i = 1, \ldots, n\}$ denote the observed data for n respondents, where $y_{i,\text{C}} = 1$ if the i-th respondent puts a tick in the upper circle and $y_{i,\text{C}} = 0$ if the i-th respondent puts a tick in the upper square (see Table 2.2). The likelihood function for π is then given by

$$L_{\text{C}}(\pi|Y_{\text{obs}}) = \prod_{i=1}^{n} \{\pi p + (1-\pi)(1-p)\}^{y_{i,\text{C}}} \{\pi(1-p) + (1-\pi)p\}^{1-y_{i,\text{C}}}$$

$$= \{\pi p + (1-\pi)(1-p)\}^{n'} \{\pi(1-p) + (1-\pi)p\}^{n-n'}, \quad (2.23)$$

where $n' = \sum_{i=1}^{n} y_{i,\text{C}}$. Assume that $\pi \sim \text{Beta}(a, b)$. Thus, the posterior distribution of π is

$$f(\pi|Y_{\text{obs}}) = \frac{\pi^{a-1}(1-\pi)^{b-1} \times L_{\text{C}}(\pi|Y_{\text{obs}})}{c(a, b; \, n', n - n')}, \quad (2.24)$$

where the normalizing constant

$$c(a, b; \, n', n - n') \,\hat{=}\, \frac{p^{n-n'}(1-p)^{n'}}{\Gamma(a + b + n)} \times c^*(a, b; \, n', n - n')$$

and

$$c^*(a, b; \, n', n - n') = \sum_{j_1=0}^{n'} \sum_{j_2=0}^{n-n'} \binom{n'}{j_1} \binom{n-n'}{j_2} \left(\frac{p}{1-p}\right)^{j_1-j_2}$$

$$\times \, \Gamma(a + j_1 + j_2) \Gamma(b + n - j_1 - j_2).$$

Therefore, the r-th posterior moment of π is given by

$$E(\pi^r|Y_{\text{obs}}) = \frac{c^*(a + r, b; \, n', n - n')}{c^*(a, b; \, n', n - n')} \times \frac{\Gamma(a + b + n)}{\Gamma(a + r + b + n)}, \quad r \geqslant 1. \quad (2.25)$$

Remark 2.3 When the sample size n and/or n' are very large, the normalizing constant in (2.24) may be not computable. In such case, we may not obtain the r-th posterior moment of π. ¶

2.3.2 Posterior mode

To derive the posterior mode, we first introduce two latent variables Z_1 and Z_2, where Z_1 denotes the count in cell-$(1, 1)$ while Z_2 denotes the count in cell-$(1,0)$ in Table 2.2. Furthermore, let $\mathbf{z} = (Z_1, Z_2)^\top$ denote the random vector and $\mathbf{z} = (z_1, z_2)^\top$ denote the realization of $\mathbf{z}$. Thus, the complete-data posterior distribution and the conditional predictive distribution are given by

$$f(\pi|Y_{\text{obs}}, \mathbf{z}) = \text{Beta}(\pi|a + z_1 + z_2,\ b + n - z_1 - z_2) \quad \text{and} \tag{2.26}$$

$$
\begin{aligned}
f(\mathbf{z}|Y_{\text{obs}}, \pi) &= f(z_1|Y_{\text{obs}}, \pi) \times f(z_2|Y_{\text{obs}}, \pi) \\
&= \text{Binomial}\left(z_1 \middle| n',\ \frac{\pi p}{\pi p + (1 - \pi)(1 - p)}\right) \\
&\quad \times \text{Binomial}\left(z_2 \middle| n - n',\ \frac{\pi(1 - p)}{\pi(1 - p) + (1 - \pi)p}\right), \tag{2.27}
\end{aligned}
$$

respectively. Using the EM algorithm, the M-step computes the following complete-data posterior mode

$$\tilde{\pi}_C = \frac{a + z_1 + z_2 - 1}{a + b + n - 2} \tag{2.28}$$

while the E-step is to replace $z_1 + z_2$ by the conditional expectation

$$E(Z_1 + Z_2|Y_{\text{obs}}, \pi) = \frac{n'\pi p}{\pi p + (1 - \pi)(1 - p)} + \frac{(n - n')\pi(1 - p)}{\pi(1 - p) + (1 - \pi)p}, \tag{2.29}$$

where $p \neq 1/2$. In fact, when $p = 1/2$, (2.29) becomes

$$E(Z_1 + Z_2|Y_{\text{obs}}, \pi) = n\pi,$$

which does not depend on the observed data Y_{obs} nor n'. In this case, the EM algorithm based on (2.28) and (2.29) converges in one step and we have $\tilde{\pi}_C = (a-1)/(a+b-2)$, which is actually the mode of the prior distribution Beta(a, b).

2.3.3 Generation of i.i.d. posterior samples via the exact IBF sampling

To apply the exact IBF sampling (cf. Appendix B) to the present model, we simply need to identify the conditional support of $\mathbf{z}|(Y_{\text{obs}}, \pi)$. From (2.27), we have

$$\mathcal{S}_{(\mathbf{z}|Y_{\text{obs}})} = \mathcal{S}_{(\mathbf{z}|Y_{\text{obs}}, \pi)} = \{\mathbf{z}_1, \dots, \mathbf{z}_K\}$$

$$= \left\{
\begin{array}{cccc}
(0,0) & (0,1) & \cdots & (0, n - n') \\
(1,0) & (1,1) & \cdots & (1, n - n') \\
\vdots & \vdots & \ddots & \vdots \\
(n',0) & (n',1) & \cdots & (n', n - n')
\end{array}
\right\},$$

where $K = (n' + 1)(n - n' + 1)$. We then calculate $\{\omega_k\}_{k=1}^{K}$ according to (B.2) and (B.3) with $\pi_0 = 0.5$.

2.4   Analyzing the Induced Abortion Data

Liu & Chow (1976) considered an induced abortion study in Taichung City and Taoyuan County, Taiwan (see also Winkler & Franklin, 1979). They adopted the multiple-trial version of the Warner model to increase the efficiency of estimation. Since this book only discusses the single-trial Warner model with the crosswise model as its non-randomized version, we simply use the data from the first trial of each respondent. The target population of interest in this study contains those married women of age 20 to 44 in the South District of Taichung City, Taiwan. The investigators would like to estimate the incidence rate of induced abortions in the target population. With $p = 0.3$, the survey yielded 90 'yes' answers (i.e., $n' = 90$) and 60 'no' answers (i.e., $n = 150$).

Using the likelihood-based method, the proportion of married women of childbearing age who have had induced abortion is estimated to be $\hat{\pi}_{\mathrm{W}} = 0.25$ with estimated variance being (Migon & Tachibana, 1997, p. 406)

$$\widehat{\mathrm{Var}}(\hat{\pi}_{\mathrm{W}}) = 0.01.$$

The resultant 95% Wald confidence interval of π is

$$[0.25 - 1.96\sqrt{0.01}, \ 0.25 + 1.96\sqrt{0.01}] = [0.054, \ 0.446].$$

To illustrate the Bayesian methods presented in Section 2.3, we consider the uniform prior (i.e., $a = b = 1$). Note that $p = 0.3$ and the observed data are

$$Y_{\mathrm{obs}} = \{n; \ n', \ n - n'\} = \{150; \ 90, \ 60\}.$$

Using (2.25), we obtain

$$E(\pi|Y_{\mathrm{obs}}) = 0.2544 \quad \text{and} \quad E(\pi^2|Y_{\mathrm{obs}}) = 0.0742.$$

Thus, $\mathrm{Var}(\pi|Y_{\mathrm{obs}}) = 0.0095$ and the 95% Bayesian credible interval of π based on normality approximation is $[0.0632, 0.4457]$.

Using $\pi^{(0)} = 0.5$ as an initial value, the EM algorithm based on (2.28) and (2.29) converges in 96 iterations. The posterior mode of π is $\tilde{\pi}_{\mathrm{C}} = 0.25$, which is identical to the MLE $\hat{\pi}_{\mathrm{W}}$. Using the exact IBF sampling described in Section 2.3.3, we generate $L = 20{,}000$ i.i.d. posterior samples from $f(\pi|Y_{\mathrm{obs}})$. The histogram based on these samples is plotted in Figure 2.3(ii), which shows that the exact IBF sampling recovers the density completely. The posterior mean, standard deviation and 95% Bayesian credible interval for π are 0.2546, 0.0973 and $[0.0680, \ 0.4490]$, respectively.

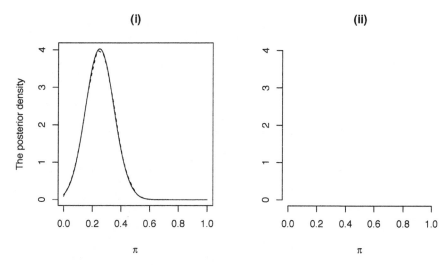

Figure 2.3 Posterior distribution of $\pi = \Pr$(having induced abortion) under the uniform prior (i.e., $a = b = 1$) and $p = \Pr(W = 1) = 0.3$ for the non-randomized crosswise model for $n = 150$ and $n' = 90$. (i) The comparison between the posterior distribution (solid curve) exactly given by (2.24) with the dotted curve estimated by a kernel density smoother based on $L = 20{,}000$ i.i.d. posterior samples generated via the exact IBF sampling. (ii) The histogram based on i.i.d. posterior samples generated via the exact IBF sampling.

2.5 An Experimental Survey Measuring Plagiarism

2.5.1 Survey data

In the Internet and Wikipedia age, universities worry about plagiarism in student homework assignments, projects, and papers. Since plagiarism is a highly sensitive topic, it is possible that some students may misreport their behavior when a direct questioning is applied. Jann & Brandenberger (2012) conducted a survey on plagiarism in student papers in which the crosswise model was implemented and compared to direct questioning. Results from this project were reported by Jann, Jerke & Krumpal (2012). The following two sensitive questions (one is about partial plagiarism and the other is about severe plagiarism) were asked.

(A) "When writing an assignment (e.g., seminar paper, term paper, thesis), have you ever intentionally adopted a passage from someone else's work without citing the original?" (partial plagiarism)

(B) "Did you ever have someone else write a large part of an assignment for you or hand in someone else's work (e.g., from www.hausarbeiten.de) as your own?" (severe plagiarism)

Table 2.3 Survey data for partial plagiarism by using the crosswise model

Type			Frequency
Valid	1	◯ – both 'yes' or both 'no'	198
observation	2	☐ – one 'yes' one 'no'	112
	Sub-total		310
Missing			164
Total			474

Source: Adapted from Table b1 of Jann & Brandenberger (2012).
Note: 198 'yes' to both or 'no' to both the sensitive question (i.e., Question (A))
and the non-sensitive question (i.e., Question (1)). 112 'yes' to Question (A) and
'no' to Question (1) or 'no' to Question (A) and 'yes' to Question (1).

Table 2.4 Survey data for severe plagiarism by using the crosswise model

Type			Frequency
Valid	1	◯ – both 'yes' or both 'no'	230
observation	2	☐ – one 'yes' one 'no'	80
	Sub-total		310
Missing			164
Total			474

Source: Adapted from Table b2 of Jann & Brandenberger (2012).
Note: 230 'yes' to both or 'no' to both the sensitive question (i.e., Question (B))
and the non-sensitive question (i.e., Question (2)). 80 'yes' to Question (B) and
'no' to Question (2) or 'no' to Question (B) and 'yes' to Question (2).

In the direct questioning, students had to answer the above two sensitive
questions. In a non-randomized survey with the crosswise model, the above
sensitive questions were paired with the following two non-sensitive questions and students were asked to provide a joint answer to each pair.

(1) "Is your mother's birthday in January, February, or March?"
     (paired with the partial plagiarism question)

(2) "Is your father's birthday in October, November, or December?"
     (paired with the severe plagiarism question)

Their surveys were carried out from June to July of 2009 among 474
students in three Swiss and German universities, where 111 questionnaires
were distributed at the Swiss Federal Institute of Technology Zurich, 273

Table 2.5 Four 95% confidence intervals of π

Type of CIs	Confidence interval	Width
Wald CI (2.10)	[0.1154586, 0.3297027]	0.2142440
Wilson CI (2.13)	[0.1196295, 0.3323231]	0.2126936
Bootstrap CI (2.14)	[0.1154830, 0.3291879]	0.2137049
Bootstrap CI (2.15)	[0.1193548, 0.3322581]	0.2129032

questionnaires were distributed at the University of Leipzig, and 90 at Ludwig Maximilian University of Munich. We only consider the survey data obtained by using the crosswise model. Tables 2.3 and 2.4 summarize the corresponding survey data. For both surveys on partial plagiarism and severe plagiarism, there are 164 missing values and 310 valid observations.

2.5.2 Analyzing the survey data for partial plagiarism

Let $\pi = \Pr(Y = 1)$ denote the unknown population prevalence for partial plagiarism. The observed data in Table 2.3 can be denoted by

$$Y_{\text{obs}} = \{n; \ n', \ n - n'\} = \{310; \ 198, \ 112\},$$

where $n' = 198$ implies 198 ticks being put in the upper circle and $n - n' = 112$ implies 112 ticks being put in the upper square when the crosswise model is applied (see Table 2.2). In this study, $p = \Pr(W = 1) \approx 0.25$.

From (2.8) and (2.9), the prevalence of partial plagiarism is estimated to be $\hat{\pi}_C = 0.2225806$ and the estimated variance is

$$\widehat{\text{Var}}(\hat{\pi}_C) = 0.002977544.$$

An unbiased estimate of $\text{Var}(\hat{\pi}_C)$ is given by

$$\overline{\text{Var}}(\hat{\pi}_C) = 0.00298718.$$

From (2.10) and (2.13), the 95% Wald and Wilson confidence intervals for π are listed in Table 2.5. We note that the width of the Wilson confidence interval is shorter than that of the Wald confidence interval.

Alternatively, from (2.14) and (2.15), two 95% bootstrap confidence intervals for π with $G = 20{,}000$ bootstrap replications are also reported in Table 2.5, where the width of the second bootstrap confidence interval is shorter than the first bootstrap confidence interval. The corresponding standard error of $\hat{\pi}_C$ is estimated to be

$$\widehat{\text{se}}(\hat{\pi}_C) = 0.05451656.$$

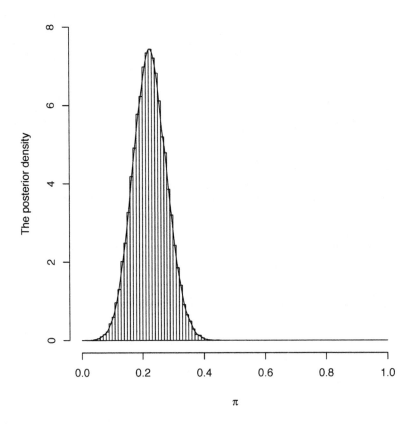

Figure 2.4 A histogram and the corresponding posterior density of $\pi = \Pr$(having partial plagiarism) under the uniform prior (i.e., $a = b = 1$) and $p = \Pr(W = 1) = 0.25$ for the crosswise model for $n = 310$ and $n' = 198$, where the histogram is based on $L = 20{,}000$ i.i.d. posterior samples generated via the exact IBF sampling and the posterior density (solid curve) is estimated by a kernel density smoother based on the same i.i.d. posterior samples.

Now, we consider Bayesian inferences with the uniform prior (i.e., $a = b = 1$) on π. In this case, we find that the normalizing constant in (2.24) is not computable and we cannot obtain the posterior moment of π specified by (2.25).

Using $\pi^{(0)} = 0.5$ as an initial value, the EM algorithm based on (2.28) and (2.29) converges in 84 iterations. The posterior mode of π is $\tilde{\pi}_{\mathrm{C}} = 0.2225806$, which is identical to the MLE $\hat{\pi}_{\mathrm{C}}$. Using the exact IBF sampling described in Section 2.3.3, we generate $L = 20{,}000$ i.i.d. posterior samples from $f(\pi|Y_{\mathrm{obs}})$. The histogram based on these samples and the corresponding posterior density are plotted in Figure 2.4. The resultant posterior mean, standard deviation, and 95% Bayesian credible interval for π are given by 0.2243914, 0.05457976 and [0.1174151, 0.3313678], respectively.

2.5.3   Analyzing the survey data for severe plagiarism

In this subsection, we let $\pi = \Pr(Y = 1)$ denote the unknown population prevalence for severe plagiarism. The observed data in Table 2.4 can be denoted by

$$Y_{\text{obs}} = \{n;\ n',\ n - n'\} = \{310;\ 230,\ 80\},$$

where $n' = 230$ implies 230 ticks being put in the upper circle and $n - n' = 80$ implies 80 ticks being put in the upper square when the crosswise model is applied (see Table 2.2). Again, $p = \Pr(W = 1) \approx 0.25$.

From (2.8) and (2.9), the prevalence of severe plagiarism is estimated to be $\hat{\pi}_{\text{C}} = 0.01612903$ and the estimated variance is 0.002470545. An unbiased estimate of $\text{Var}(\hat{\pi}_{\text{C}})$ is given by $\overline{\text{Var}}(\hat{\pi}_{\text{C}}) = 0.00247854$. From (2.10), the 95% Wald confidence interval of π is

$$[\hat{\pi}_{\text{C,WL}},\ \hat{\pi}_{\text{C,WU}}] = [-0.08144765,\ 0.1137057],$$

which is useless since the lower bound is less than zero. From (2.13), the 95% Wilson confidence interval of π is given by

$$[\hat{\pi}_{\text{C,WSL}},\ \hat{\pi}_{\text{C,WSU}}] = [-0.07495043,\ 0.1190538],$$

which also yields a negative lower bound.

In order to obtain a useful interval estimation of π, we could employ the bootstrap method. From (2.14) and (2.15), two 95% bootstrap confidence intervals for π with $G = 20{,}000$ bootstrap replications are given by

$$[-0.03873587,\ 0.09631006]$$

and

$$[1.082987 \times 10^{-37},\ 0.1129032],$$

respectively. The corresponding standard error of $\hat{\pi}_{\text{C}}$ is estimated by

$$\widehat{\text{se}}(\hat{\pi}_{\text{C}}) = 0.03445049.$$

Alternatively, we may consider Bayesian inferences with the uniform prior (i.e., $a = b = 1$) on π. Here, we also find that the normalizing constant in (2.24) is not computable and we cannot obtain the posterior moment of π specified by (2.25). Using $\pi^{(0)} = 0.05$ as an initial value, the EM algorithm based on (2.28) and (2.29) converges in 727 iterations. The posterior mode of π is $\tilde{\pi}_{\text{C}} = 0.01612903$, which is identical to the MLE $\hat{\pi}_{\text{C}}$. Using the exact IBF sampling described in Section 2.3.3, we generate $L = 20{,}000$ i.i.d. posterior samples from $f(\pi|Y_{\text{obs}})$. The histogram based on these samples and the corresponding posterior density are plotted in Figure 2.5. The resultant posterior mean, standard deviation, and 95%

Bayesian credible interval based on normality approximation for π are given by 0.04748678, 0.03447256 and

$$[0.04748678 - 1.96 \times 0.03447256, \ 0.04748678 + 1.96 \times 0.03447256]$$

$$= \ [-0.02007943, \ 0.11505299],$$

respectively. This Bayesian credible interval with a negative lower bound is not surprising since the posterior density of π is highly skewed toward zero. The 2.5-th and 97.5-th percentiles of the 20,000 i.i.d. posterior samples of π are 0.0019201 and 0.12871372, respectively. Thus, [0.0019201, 0.12871372] is a 95% Bayesian credible interval of π.

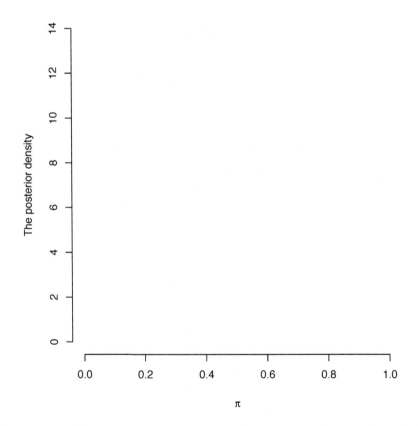

Figure 2.5 A histogram and the corresponding posterior density of $\pi = \Pr($having severe plagiarism$)$ under the uniform prior (i.e., $a = b = 1$) and $p = \Pr(W = 1) = 0.25$ for the crosswise model for $n = 310$ and $n' = 230$, where the histogram is based on $L = 20{,}000$ i.i.d. posterior samples generated via the exact IBF sampling and the posterior density (solid curve) is estimated by a kernel density smoother based on the same i.i.d. posterior samples.

CHAPTER 3

The Triangular Model

In Chapter 2, we introduced the crosswise model which can be regarded as a non-randomized version of the randomized Warner model. Despite the advantages we have described for non-randomized models, the crosswise model still inherits the low efficiency from the randomized Warner model. To improve the relative efficiency, in this chapter we systematically introduce a triangular model, which can be viewed as a variant of the crosswise model (see Yu, Tian & Tang, 2008; Tan, Tian & Tang, 2009; Tang *et al.*, 2012).

3.1 The Triangular Design

Let $\{Y = 1\}$ denote the class of people who possess a sensitive characteristic (e.g., drug-taking, shoplifting, tax evasion, driving under influence, and so on) and $\{Y = 0\}$ denote the complementary class. Let W be a non-sensitive dichotomous variate and be independent of Y. Interviewers should select a suitable W so that the proportion $p = \Pr(W = 1)$ can be estimated easily. Without loss of generality, p is assumed to be known. For example, we may define $W = 1$ if a respondent was born between August and December and $W = 0$ otherwise. Hence, it is reasonable to assume that $p \approx 5/12 = 0.41667$. Our aim is to estimate the proportion $\pi = \Pr(Y = 1)$.

3.1.1 The survey design

Suppose that investigators would like to estimate the proportion of drug users in a specific population. The purpose could be accomplished via the collection of the mixing information of a sensitive question (e.g., are you a drug user?) and an unrelated non-sensitive question (e.g., is your birthday is between August and December?). Interviewers may design a questionnaire in the format as shown on the left-hand side of Table 3.1 and ask each interviewee to put a tick in the circle (i.e., $\{Y = 0, W = 0\}$) if he/she belongs to this circle or put a tick in the upper square (i.e., $\{Y = 0, W = 1\}$) if he/she belongs to one of the three squares. Note that both $\{Y = 0, W = 0\}$ and $\{Y = 0, W = 1\}$ are non-sensitive, and the sensitive class $\{Y = 1\}$ is mixed with another non-sensitive subclass $\{Y = 0, W = 1\}$. Therefore, a respondent who is a drug user can be well covered his/her true identity by

Table 3.1 The triangular model and the corresponding cell probabilities

Category	$W = 0$	$W = 1$	Category	$W = 0$	$W = 1$	Marginal
$Y = 0$	○	□	$Y = 0$	$(1 - \pi)(1 - p)$	$(1 - \pi)p$	$1 - \pi$
$Y = 1$	□	□	$Y = 1$	$\pi(1 - p)$	πp	π
			Marginal	$1 - p$	p	1

Source: Adapted from Table 1 of Yu, Tian & Tang (2008).
Note: Please put a tick in the circle if you belong to this circle or put a tick in the upper square if you belong to one of the three squares.

those non-drug users who are born between August and December, and is possibly willing to put a tick in the upper square. Such a design encourages respondents to not only participate in the survey but also provide their truthful responses. The corresponding cell probabilities are listed at the right-hand side of Table 3.1. Yu, Tian & Tang (2008) called this a triangular model, which belongs to the class of the admissible design defined by Nayak (1994).

3.1.2 Alternative formulation

If $W = 1$ represents that a respondent was born between July and December and $p \approx 1/2$, then we can represent the triangular model in Table 3.1 in the following non-sensitive verbal format:

If you are not a drug user, please truthfully (i.e., according to your actual birthday) put a tick in the following circle (i.e., ○) or square (i.e., □). Otherwise, please put a tick in the following square regardless of your actual birthday.

(1) I was born in the first half of a year, check here   ○

(2) I was born in the second half of a year, check here   □

Obviously, this non-randomized design encourages cooperation from respondents and their sensitive characteristics will not be exposed to others.

In practice, some respondents who are drug users may refuse to provide any answer no matter what survey design is adopted. One immediate advantage with the triangular model is its *robustness* to non-response in the sense that it allows such non-response. For example, for n respondents, we observed n_1 ticks in the circle, n_2 ticks in the square, and n_3 non-responses ($n = n_1 + n_2 + n_3$). This result is equivalent to the observation

$Y_{\text{obs}} = \{n; n_1, n_2 + n_3\}$ (i.e., we observed n_1 ticks in the circle and $n_2 + n_3$ ticks in the upper square) under the assumption that a respondent is always willing to answer the question if he/she is not a drug user.

3.1.3   Variance of the estimator

For the triangular model described in Table 3.1, we define a hidden variable Y_{T} as follows:

$$Y_{\text{T}} = \begin{cases} 1, & \text{if a tick is put in the upper square,} \\ 0, & \text{if a tick is put in the circle,} \end{cases} \tag{3.1}$$

where the subscript 'T' represents the 'triangular' model. Obviously, we have

$$\Pr(Y_{\text{T}} = 1) = \pi + (1 - \pi)p \quad \text{and} \quad \Pr(Y_{\text{T}} = 0) = (1 - \pi)(1 - p),$$

where p is assumed to be known. Let $Y_{\text{obs}} = \{y_{i,\text{T}}: i = 1, \ldots, n\}$ denote the observed data for n respondents with $y_{i,\text{T}} = 1$ if the i-th respondent puts a tick in the upper square and $y_{i,\text{T}} = 0$ if the i-th respondent puts a tick in the circle. The likelihood function for π is

$$L_{\text{T}}(\pi|Y_{\text{obs}}) = \prod_{i=1}^{n} \{\pi + (1 - \pi)p\}^{y_{i,\text{T}}} \{(1 - \pi)(1 - p)\}^{1 - y_{i,\text{T}}}. \tag{3.2}$$

The resultant MLE of π and its corresponding variance are given by

$$\hat{\pi}_{\text{T}} = \frac{\bar{y}_{\text{T}} - p}{1 - p} \quad \text{and} \quad \text{Var}(\hat{\pi}_{\text{T}}) = \text{Var}(\hat{\pi}_{\text{D}}) + \frac{p(1 - \pi)}{n(1 - p)}, \tag{3.3}$$

where $\bar{y}_{\text{T}} = (1/n)\sum_{i=1}^{n} y_{i,\text{T}}$ and $\text{Var}(\hat{\pi}_{\text{D}})$ is defined in (2.3). For any fixed π, we notice that

$$n\,\text{Var}(\hat{\pi}_{\text{T}}) = (1 - \pi)\left(\pi + \frac{p}{1 - p}\right) \tag{3.4}$$

is an increasing function of p when $0 < p < 1$ and approaches to infinity as $p \to 1$ (see Figure 3.1).

3.1.4   Relative efficiency

Similar to (2.5), the RE of the triangular design to the design of direct questioning is

$$\text{RE}_{\text{T}\to\text{D}}(\pi, p) \triangleq \frac{\text{Var}(\hat{\pi}_{\text{T}})}{\text{Var}(\hat{\pi}_{\text{D}})} = 1 + \frac{p/(1 - p)}{\pi}, \tag{3.5}$$

The Triangular Model

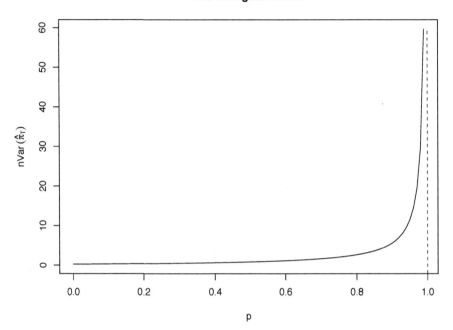

Figure 3.1   Plot of $n\mathrm{Var}(\hat{\pi}_{\mathrm{T}})$ defined in (3.4) against p with $\pi = 0.40$ for the triangular model.

which is also independent of the sample size n. Table 3.2 displays some $\mathrm{RE}_{\mathrm{T}\to\mathrm{D}}(\pi, p)$ for various combinations of π and p. For instance, when $\pi = 0.4$ and $p = 5/12$, we have $\mathrm{RE}_{\mathrm{T}\to\mathrm{D}}(0.4, 5/12) = 2.7857$, indicating that the sample size required for the triangular design is about 2.8 times that required for the direct questioning in order to achieve the same estimation precision.

3.1.5   Degree of privacy protection

Intuitively, the optimal DPP of the triangular model will be achieved at $p \approx 0.5$ when π is very small. Since the privacy information divulged by a respondent regarding his/her membership in the sensitive class $\{Y = 1\}$ is characterized through $\Pr(Y = 1 | Y_{\mathrm{T}} = 0)$ and $\Pr(Y = 1 | Y_{\mathrm{T}} = 1)$, we have

$$\mathrm{DPP}_{\bigcirc}(\pi, p) \triangleq \Pr(Y = 1 | Y_{\mathrm{T}} = 0) = 0 \tag{3.6}$$

and

$$\mathrm{DPP}_{\square}(\pi, p) \triangleq \Pr(Y = 1 | Y_{\mathrm{T}} = 1) = \frac{\pi}{\pi + (1 - \pi)p}. \tag{3.7}$$

In particular, when $p = 0$, we have $\mathrm{DPP}_{\square}(\pi, 0) = 1$, which corresponds to the design of direct questioning. When $p = 1$, we obtain $\mathrm{DPP}_{\square}(\pi, 1) = \pi$,

Table 3.2 Relative efficiency $\mathrm{RE}_{\mathrm{T}\to\mathrm{D}}(\pi, p)$ for various combinations of π and p

π	p						
	0.20	0.30	0.34	0.38	5/12	0.46	0.50
0.05	6.0000	9.5714	11.303	13.258	15.286	18.037	21.000
0.10	3.5000	5.2857	6.1515	7.1290	8.1429	9.5185	11.000
0.15	2.6667	3.8571	4.4343	5.0860	5.7619	6.6790	7.6667
0.20	2.2500	3.1429	3.5758	4.0645	4.5714	5.2593	6.0000
0.25	2.0000	2.7143	3.0606	3.4516	3.8571	4.4074	5.0000
0.30	1.8333	2.4286	2.7172	3.0430	3.3810	3.8395	4.3333
0.35	1.7143	2.2245	2.4719	2.7512	3.0408	3.4339	3.8571
0.40	1.6250	2.0714	2.2879	2.5323	2.7857	3.1296	3.5000
0.45	1.5556	1.9524	2.1448	2.3620	2.5873	2.8930	3.2222
0.50	1.5000	1.8571	2.0303	2.2258	2.4286	2.7037	3.0000

which corresponds to the case that $W = 1$; that is, the respondent has a birthday between January and December. When $p = 0.5$,

$$\mathrm{DPP}_{\square}(\pi, 0.5) = \frac{2\pi}{\pi + 1} > \pi.$$

Figure 3.2 shows that for any fixed π, $\mathrm{DPP}_{\square}(\pi, p)$ is a monotonic decreasing function of p with maximum 1 and minimum π.

3.2 Comparison with the Warner Model

3.2.1 The difference of two variances

Equivalence between the Warner model and the non-randomized crosswise model in efficiency (cf. Section 2.2.2) provides a basis for comparing the Warner model with the non-randomized triangular model. Using the variance criterion, we obtain from (2.2) and (3.3) that

$$\mathrm{Var}(\hat{\pi}_{\mathrm{W}}) - \mathrm{Var}(\hat{\pi}_{\mathrm{T}}) = \frac{p}{n(1 - p)(2p - 1)^2} \times h_{\mathrm{WT}}(p|\pi), \qquad (3.8)$$

where

$$h_{\mathrm{WT}}(p|\pi) = (4\pi - 3)p^2 + (2 - 4\pi)p + \pi$$

is a quadratic function of p for any fixed π ($\pi \neq 3/4$). We have the following result.

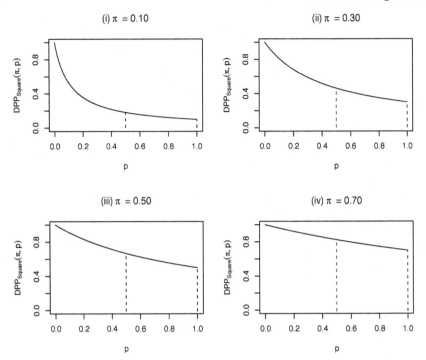

Figure 3.2   Plots of $\mathrm{DPP}_\square(\pi, p)$ defined in (3.7) against p for the triangular model for a fixed π. (i) $\pi = 0.10$; (ii) $\pi = 0.30$; (iii) $\pi = 0.50$; (iv) $\pi = 0.70$.

Theorem 3.1   The triangular model is always more efficient than the Warner model for any $\pi \in (0, 1)$ and $p < 2/3$. Specifically,

(1)   if $\pi = 3/4$, then

$$\mathrm{Var}(\hat{\pi}_\mathrm{W}) \geqslant \mathrm{Var}(\hat{\pi}_\mathrm{T}), \quad \text{for } 0 < p \leqslant \frac{3}{4} \quad (p \neq 0.5). \qquad (3.9)$$

(2)   If $\pi \neq 3/4$, then

$$\mathrm{Var}(\hat{\pi}_\mathrm{W}) \geqslant \mathrm{Var}(\hat{\pi}_\mathrm{T}), \quad \text{for } 0 < p \leqslant p_\pi \quad (p \neq 0.5), \qquad (3.10)$$

where

$$p_\pi = \frac{2\pi - 1 - \sqrt{1 - \pi}}{4\pi - 3}$$

is an increasing function of π and attains its minimum $2/3$ at $\pi = 0$.   ¶

Proof.   (1) If $\pi = 3/4$, then $h_\mathrm{WT}(p|\pi) = 3/4 - p$ is nonnegative when $0 < p \leqslant 3/4$. From (3.8), we obtain (3.9) immediately.
      (2) Case I: $\pi > 3/4$. We have

$$h_\mathrm{WT}(p|\pi) \geqslant 0 \quad \text{for } p \leqslant p_\pi \quad \text{or} \quad p \geqslant p_2,$$

Table 3.3 Relative efficiency $\text{RE}_{\text{W}\to\text{T}}(\pi, p)$ for various combinations of π and p

π	p													
	0.20	0.30	0.34	0.38	0.46	0.49	0.51	0.54	0.56	0.58	7/12	0.60	0.63	0.66
0.05	1.72	2.99	4.17	6.57	45.35	650.6	602.9	33.4	13.6	7.03	6.38	4.10	2.09	1.18
0.10	1.69	2.94	4.12	6.51	45.41	654.4	608.5	33.9	13.9	7.21	6.55	4.23	2.18	1.24
0.20	1.67	2.92	4.11	6.53	46.31	672.9	629.5	35.4	14.6	7.65	6.96	4.52	2.37	1.37
0.30	1.69	2.98	4.20	6.72	48.39	708.1	665.8	37.8	15.7	8.27	7.53	4.93	2.61	1.53
0.40	1.75	3.12	4.42	7.12	51.99	765.4	722.9	41.3	17.2	9.13	8.32	5.47	2.92	1.73
0.50	1.85	3.36	4.80	7.79	57.79	855.7	811.2	46.6	19.5	10.3	9.47	6.25	3.35	2.00
0.60	2.01	3.77	5.45	8.92	67.24	1001	952.2	55.0	23.1	12.3	11.2	7.43	4.00	2.39
0.70	2.29	4.49	6.58	10.9	83.81	1254	1196	69.4	29.3	15.5	14.2	9.41	5.07	3.03
0.80	2.87	5.99	8.93	15.0	117.9	1774	1697	98.7	41.6	22.1	20.2	13.3	7.21	4.29
0.90	4.64	10.5	16.1	27.6	222.0	3358	3219	187.6	79.1	42.1	38.4	25.3	13.5	8.03

where

$$p_2 \triangleq \frac{2\pi - 1 + \sqrt{1 - \pi}}{4\pi - 3}.$$

Note that

$$\frac{\mathrm{d}p_\pi}{\mathrm{d}\pi} = \frac{(\sqrt{1 - \pi} - 2)^2}{2\sqrt{1 - \pi}\,(4\pi - 3)^2} > 0.$$

Hence, p_π is an increasing function of π. It is easy to show that $0.5 < p_\pi < 1 < p_2$. (3.10) follows immediately.

Case II: $\pi < 3/4$. Similarly, we have

$$h_{\text{WT}}(p|\pi) \geqslant 0 \quad \text{for } p_1 \leqslant p \leqslant p_\pi,$$

where

$$p_1 = \frac{-(2\pi - 1 + \sqrt{1 - \pi})}{3 - 4\pi}.$$

Now we have $p_1 < 0 < 1/2 < p_\pi < 1$. Hence, (3.10) follows. $\quad\square$

3.2.2 Relative efficiency of the Warner model to the triangular model

Here, we consider the RE of the Warner model ($p \neq 0.5$) to the triangular model; i.e.,

$$
\begin{aligned}
\text{RE}_{\text{W}\to\text{T}}(\pi, p) &= \frac{\text{Var}(\hat{\pi}_{\text{W}})}{\text{Var}(\hat{\pi}_{\text{T}})} \\
&= \frac{\pi + p(1 - p)/\{(2p - 1)^2(1 - \pi)\}}{\pi + p/(1 - p)},
\end{aligned}
$$

which is independent of the sample size n and depends only on the parameters π and p. We note that interviewers usually select p in $[0.20, 0.46]$ for the Warner model. From Table 3.3, when $0.20 \leqslant p \leqslant 0.46$ (or $0.54 \leqslant p \leqslant 0.66$), the efficiency of the triangular strategy is about 1.7–222 (or 1.2–187) times that of the Warner model. In particular, when $0.49 \leqslant p \leqslant 0.51$ (which are the optimal range for which the privacy of respondents is protected), the efficiency of the triangular strategy is about 600–3358 times that of the Warner model.

3.2.3   Degree of privacy protection

To compare the DPP of the Warner model with that of the triangular model, we consider three cases (i.e., $p = 0.3$, 0.35 and 0.4) which are some practical choices for the Warner model, and two cases (namely $p = 0.5$ and $7/12$) which are two optimal choices for the triangular model. Table 3.4 gives DPPs for various combinations of π and p. From Table 3.4, when comparing the Warner model with $p = 0.35$ with the triangular model with $p = 7/12$, we have

$$\text{DPP}_{\text{yes}} > \text{DPP}_{\bigcirc} \quad \text{and} \quad \text{DPP}_{\text{no}} > \text{DPP}_{\square}$$

for all π in the table. Therefore, the triangular model with $p = 7/12$ has better DPPs than the Warner model with $p = 0.35$. We can reach a similar conclusion when comparing the triangular model with $p = 0.5$ with the Warner model with $p = 0.30$.

Table 3.4   Comparison of DPPs for various combinations of π and p

π	Warner model						Triangular model	
	$p = 0.30$		$p = 0.35$		$p = 0.40$		$p = 0.50$	$p = 7/12$
	DPP_{yes}	DPP_{no}	DPP_{yes}	DPP_{no}	DPP_{yes}	DPP_{no}	$\text{DPP}_{\square}$	$\text{DPP}_{\square}$
0.05	0.022	0.109	0.027	0.089	0.033	0.073	0.095	0.082
0.10	0.045	0.205	0.056	0.171	0.068	0.142	0.181	0.160
0.20	0.096	0.368	0.118	0.317	0.142	0.272	0.333	0.300
0.30	0.155	0.500	0.187	0.443	0.222	0.391	0.461	0.423
0.40	0.222	0.608	0.264	0.553	0.307	0.500	0.571	0.533
0.50	0.300	0.700	0.350	0.650	0.400	0.600	0.666	0.631
0.60	0.391	0.778	0.447	0.735	0.500	0.692	0.750	0.720
0.70	0.500	0.845	0.557	0.813	0.609	0.778	0.823	0.800
0.80	0.632	0.903	0.683	0.881	0.727	0.857	0.889	0.873
0.90	0.794	0.955	0.829	0.944	0.857	0.931	0.947	0.939

Note: DPP_{yes}, DPP_{no}, $\text{DPP}_{\bigcirc}$ and $\text{DPP}_{\square}$ are defined by (2.6), (2.7), (3.6) and (3.7), respectively. In addition, $\text{DPP}_{\bigcirc} = 0$ for all π and p.

3.3 Asymptotic Properties of the MLE

3.3.1 An alternative derivation of the MLE

Suppose that n respondents from a population results in s ticks in the upper square and $n - s$ ticks in the circle (see Table 3.1). By introducing a new parameter $\theta = \pi + (1 - \pi)p$, we have

$$\pi = \frac{\theta - p}{1 - p}. \tag{3.11}$$

The likelihood function is proportional to $\theta^s(1 - \theta)^{n-s}$ and the MLE of θ is given by $\hat{\theta} = s/n$. Therefore, the MLE of π is

$$\hat{\pi}_T = \frac{\hat{\theta} - p}{1 - p}, \tag{3.12}$$

provided that $\hat{\pi}_T \in [0, 1]$. When $\hat{\pi}_T < 0$ or $\hat{\pi}_T > 1$, we can use the EM algorithm (3.32) and (3.33) with $a = b = 1$ (i.e., if the uniform distribution on $[0, 1]$ is adopted as a prior of π, the posterior mode of π is equal to the MLE of π) to find the MLE of π. Note that (3.12) is identical to the $\hat{\pi}_T$ given by (3.3). Since $s \sim \text{Binomial}(n, \theta)$, we have

$$E(s) = n\theta \quad \text{and} \quad \text{Var}(s) = n\theta(1 - \theta).$$

Hence, $E(\hat{\pi}_T) = \pi$; that is, $\hat{\pi}_T$ is unbiased, and

$$\begin{aligned} \text{Var}(\hat{\pi}_T) &= \frac{\theta(1 - \theta)}{n(1 - p)^2} \\ &= \frac{\pi(1 - \pi)}{n} + \frac{p(1 - \pi)}{n(1 - p)}. \end{aligned} \tag{3.13}$$

It is noteworthy that the variance of $\hat{\pi}_T$ can be expressed as the sum of the variance due to sampling and the variance due to the introduction of non-sensitive variable W.

3.3.2 Two asymptotic confidence intervals

It is easy to show that an unbiased estimate of $\text{Var}(\hat{\pi}_T)$ is given by

$$\overline{\text{Var}}(\hat{\pi}_T) = \frac{\hat{\theta}(1 - \hat{\theta})}{(n - 1)(1 - p)^2}. \tag{3.14}$$

When $n \to \infty$, the Central Limit Theorem implies that $\hat{\pi}_T$ is asymptotically normal; that is,

$$(\hat{\pi}_T - \pi) \Big/ \sqrt{\overline{\text{Var}}(\hat{\pi}_T)} \overset{\cdot}{\sim} N(0, 1). \tag{3.15}$$

From (3.15), the common hypothesis testing about π can be easily established. The $(1 - \alpha)100\%$ Wald confidence interval of π is given by

$$[\hat{\pi}_{\mathrm{T,WL}},\ \hat{\pi}_{\mathrm{T,WU}}] = \left[\hat{\pi}_{\mathrm{T}} - z_{\alpha/2}\sqrt{\mathrm{Var}(\hat{\pi}_{\mathrm{T}})},\ \ \hat{\pi}_{\mathrm{T}} + z_{\alpha/2}\sqrt{\mathrm{Var}(\hat{\pi}_{\mathrm{T}})}\right]. \quad (3.16)$$

The $(1 - \alpha)100\%$ Wilson (score) confidence interval $[\hat{\pi}_{\mathrm{T,WSL}},\ \hat{\pi}_{\mathrm{T,WSU}}]$ of π can be constructed based on

$$
\begin{aligned}
1 - \alpha \ \ &= \ \ \mathrm{Pr}\left\{\left|\frac{\hat{\pi}_{\mathrm{T}} - \pi}{\sqrt{\mathrm{Var}(\hat{\pi}_{\mathrm{T}})}}\right| \leqslant z_{\alpha/2}\right\} \\[2mm]
&= \ \ \mathrm{Pr}\left\{(\hat{\pi}_{\mathrm{T}} - \pi)^2 \leqslant z_{\alpha/2}^2 \mathrm{Var}(\hat{\pi}_{\mathrm{T}})\right\} \\[2mm]
&\overset{(3.13)}{=} \ \ \mathrm{Pr}\left[\hat{\pi}_{\mathrm{T}}^2 - 2\hat{\pi}_{\mathrm{T}}\pi + \pi^2 \leqslant \frac{z_{\alpha/2}^2}{n}\left\{\pi(1-\pi) + \frac{p(1-\pi)}{1-p}\right\}\right] \\[2mm]
&= \ \ \mathrm{Pr}\left\{(1 + z_*)\pi^2 - \rho\pi + \hat{\pi}_{\mathrm{T}}^2 - z_* p/(1-p) \leqslant 0\right\}, \quad (3.17)
\end{aligned}
$$

where

$$z_* \ \hat{=}\ \frac{z_{\alpha/2}^2}{n} \quad \text{and} \quad \rho \ \hat{=}\ 2\hat{\pi}_{\mathrm{T}} + \frac{z_*(1 - 2p)}{1 - p}. \quad (3.18)$$

Solving the quadratic inequality inside the probability in (3.17) yields

$$[\hat{\pi}_{\mathrm{T,WSL}},\ \hat{\pi}_{\mathrm{T,WSU}}] = \frac{\rho \pm \sqrt{\rho^2 - 4(1 + z_*)\{\hat{\pi}_{\mathrm{T}}^2 - z_* p/(1-p)\}}}{2(1 + z_*)}, \quad (3.19)$$

which is generally inside $[0, 1]$.

3.3.3   Bootstrap confidence intervals

When the sample size n is small to moderate, asymptotic confidence intervals in (3.16) and (3.19) are not reliable. For such cases, the bootstrap approach can be used to obtain two bootstrap confidence intervals of π.

From the observed data $Y_{\mathrm{obs}} = \{n;\ n - s,\ s\}$, we first calculate the MLE of θ, i.e., $\hat{\theta} = s/n$. Then, we can generate

$$s^* \sim \mathrm{Binomial}(n,\ \hat{\theta})$$

to produce $Y_{\mathrm{obs}}^* = \{n;\ n - s^*,\ s^*\}$. Based on Y_{obs}^*, we can calculate a bootstrap replication $\hat{\pi}_{\mathrm{T}}^*$ using the formula (3.12) or the EM algorithm (3.32) and (3.33) with $a = b = 1$. Independently repeating this process G times, we obtain G bootstrap replications $\{\hat{\pi}_T^*(g)\}_{g=1}^G$. Thus, the standard error, $\mathrm{se}\,(\hat{\pi}_T)$, of $\hat{\pi}_T$ can be estimated by

$$\widehat{\mathrm{se}}(\hat{\pi}_T) = \left[\frac{1}{G-1}\sum_{g=1}^G \left\{\hat{\pi}_T^*(g) - \frac{\hat{\pi}_T^*(1) + \cdots + \hat{\pi}_T^*(G)}{G}\right\}^2\right]^{\frac{1}{2}}.$$

If $\{\hat{\pi}_T^*(g)\}_{g=1}^G$ is approximately normally distributed, a $(1-\alpha)100\%$ bootstrap confidence interval for π is given by

$$\left[\hat{\pi}_T - z_{\alpha/2}\times\widehat{\mathrm{se}}(\hat{\pi}_T),\ \hat{\pi}_T + z_{\alpha/2}\times\widehat{\mathrm{se}}(\hat{\pi}_T)\right]. \tag{3.20}$$

If $\{\hat{\pi}_T^*(g)\}_{g=1}^G$ is non-normally distributed, a $100(1-\alpha)\%$ bootstrap confidence interval for π can be obtained by

$$[\hat{\pi}_{T,BL},\ \hat{\pi}_{T,BU}], \tag{3.21}$$

where $\hat{\pi}_{T,BL}$ and $\hat{\pi}_{T,BU}$ are the $100(\alpha/2)$ and $100(1-\alpha/2)$ percentiles of $\{\hat{\pi}_T^*(g)\}_{g=1}^G$, respectively.

3.3.4 A modified MLE of π

From (3.12), we know that $0 \leqslant \hat{\pi}_T \leqslant 1$ if and only if $p \leqslant \hat{\theta} \leqslant 1$. Therefore, the MLE $\hat{\pi}_T$ can be modified as

$$\hat{\pi}_{TM} = \max(0, \hat{\pi}_T) = \begin{cases} \hat{\pi}_T, & \text{if } p \leqslant \dfrac{s}{n} \leqslant 1, \\[2mm] 0, & \text{if } 0 \leqslant \dfrac{s}{n} < p. \end{cases} \tag{3.22}$$

The following result states that $\hat{\pi}_{TM}$ and $\hat{\pi}_T$ are asymptotically equivalent.

Theorem 3.2 If $0 < \pi < 1$, then $\sqrt{n}(\hat{\pi}_{TM} - \pi)$ and $\sqrt{n}(\hat{\pi}_T - \pi)$ have the same asymptotic distribution for sufficiently large n. ¶

Proof. It suffices to prove that $\sqrt{n}(\hat{\pi}_{TM} - \pi) - \sqrt{n}(\hat{\pi}_T - \pi)$ converges to zero in probability as $n \to \infty$, i.e.,

$$\Pr\{|\sqrt{n}(\hat{\pi}_{TM} - \hat{\pi}_T)| > 0\} \to 0, \quad \text{as } n \to \infty. \tag{3.23}$$

When $p \leqslant \hat{\theta} < 1$, from (3.22), we obtain $\hat{\pi}_{TM} = \hat{\pi}_T$. Therefore (3.23) follows immediately.

Now we consider the case of $0 \leqslant \hat{\theta} < p$. Since $0 < \pi < 1$, from (3.11), we have

$$p < \theta < 1.$$

Noting that $\hat{\theta}$ is the MLE of $\theta = \pi + (1-\pi)p$, we obtain $\Pr(|\hat{\theta}-\theta| > \varepsilon) \to 0$, as $n \to \infty$, for any $\varepsilon > 0$. We only need to show that

$$\Pr\{|\sqrt{n}(\hat{\pi}_{\text{TM}} - \hat{\pi}_{\text{T}})| > 0\} \leqslant \Pr(|\hat{\theta} - \theta| > \varepsilon),$$

or equivalently

$$\{|\sqrt{n}(\hat{\pi}_{\text{TM}} - \hat{\pi}_{\text{T}})| > 0\} \subseteq \{|\hat{\theta} - \theta| > \varepsilon\}, \tag{3.24}$$

for any $\varepsilon < \theta - p$.

Since $0 \leqslant \hat{\theta} < p$, we have $\hat{\pi}_{\text{TM}} = 0$, and if

$$|\sqrt{n}(\hat{\pi}_{\text{TM}} - \hat{\pi}_{\text{T}})| > 0$$

$$\Rightarrow \quad |\hat{\theta} - p| > 0$$

$$\Rightarrow \quad 0 < |\hat{\theta} - p| = -(\hat{\theta} - p) = -(\hat{\theta} - \theta) - (\theta - p)$$

$$\Rightarrow \quad |\hat{\theta} - \theta| \geqslant -(\hat{\theta} - \theta) > \theta - p. \tag{3.25}$$

Since $\varepsilon < \theta - p$, we have

$$(\theta - p) - \varepsilon > 0. \tag{3.26}$$

By combining (3.25) with (3.26), we obtain

$$|\hat{\theta} - \theta| - \varepsilon = |\hat{\theta} - \theta| - (\theta - p) + (\theta - p) - \varepsilon > 0$$

and hence (3.24) follows.   □

3.4   Bayesian Methods for the Triangular Model

In this section, we first derive the exact posterior distribution of π and its explicit posterior moments. We then derive the posterior mode via the EM algorithm (Dempster, Laird & Rubin, 1977) when the posterior distribution of π is highly skewed. Finally, we utilize the exact *inverse Bayes formulae* (IBF) sampling (Tian, Tan & Ng, 2007) to generate i.i.d. posterior samples.

3.4.1   Posterior moments

From (3.2), the likelihood function becomes

$$L_{\text{T}}(\pi|Y_{\text{obs}}) = \{\pi + (1-\pi)p\}^s \{(1-\pi)(1-p)\}^{n-s}, \quad 0 \leqslant \pi \leqslant 1,$$

where $s \mathrel{\hat{=}} \sum_{i=1}^{n} y_{i,\text{T}}$. If we choose the beta distribution $\text{Beta}(a, b)$ as the prior distribution of π, then the posterior distribution of π has the following explicit expression:

$$f(\pi|Y_{\text{obs}}) = \frac{\pi^{a-1}(1-\pi)^{b+n-s-1}\{\pi + (1-\pi)p\}^s}{c_{\text{T}}(a, b; \ s, n-s)}, \tag{3.27}$$

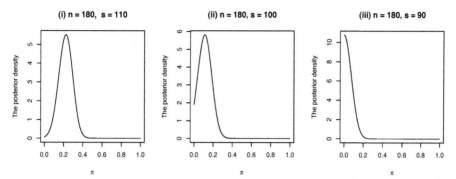

Figure 3.3 Posterior distributions for π under the uniform prior (i.e., $a = b = 1$) and $p = \Pr(W = 1) = 0.5$ for the triangular model. (i) $n = 180$ and $s = 110$; (ii) $n = 180$ and $s = 100$; (iii) $n = 180$ and $s = 90$.

where the normalizing constant is given by

$$
c_{\mathrm{T}}(a, b;\ s, n - s) = \sum_{j=0}^{s} \binom{s}{j} p^{s-j} B(a + j,\ b + n - j). \tag{3.28}
$$

For $a = b = 1$ and $p = \Pr(W = 1) = 0.5$, Figure 3.3 shows the posterior distributions of π for three different combinations of n and s.

When $f(\pi|Y_{\mathrm{obs}})$ is fairly symmetric, the first two posterior moments are good enough to describe the location and discrepancy of the posterior distribution. From (3.27), the r-th posterior moment of π has the following closed-form expression:

$$
E(\pi^r|Y_{\mathrm{obs}}) = \frac{c_{\mathrm{T}}(a + t, b;\ s, n - s)}{c_{\mathrm{T}}(a, b;\ s, n - s)}, \quad r \geqslant 1. \tag{3.29}
$$

3.4.2 Posterior mode

When $f(\pi|Y_{\mathrm{obs}})$ is highly skewed (e.g., see Figure 3.3 (iii)) the posterior mode is usually adopted for describing the location. To derive the mode, we first introduce a latent variable Z, which denotes the number of respondents with the sensitive attribute. Obviously, the number of respondents without the sensitive characteristic is $n - Z$. Let z denote the realization of Z. Thus, the complete-data is $Y_{\mathrm{com}} = \{Y_{\mathrm{obs}}, z\}$. The complete-data posterior distribution and the conditional predictive distribution are given by

$$
f(\pi|Y_{\mathrm{obs}}, z) = \mathrm{Beta}\,(\pi|a + z,\ b + n - z) \tag{3.30}
$$

and

$$
f(z|Y_{\mathrm{obs}}, \pi) = \mathrm{Binomial}\left(z\,\middle|\,s,\ \frac{\pi}{\pi + (1 - \pi)p}\right), \tag{3.31}
$$

respectively. Using the EM algorithm, the M-step computes the complete-
data posterior mode as

$$\tilde{\pi}_{\mathrm{T}} = \frac{a + z - 1}{a + b + n - 2} \tag{3.32}$$

and the E-step is to replace z by the following conditional expectation

$$E(Z|Y_{\mathrm{obs}}, \pi) = \frac{s\,\pi}{\pi + (1 - \pi)p}. \tag{3.33}$$

3.4.3   Generation of i.i.d. posterior samples via the exact IBF sampling

According to the exact IBF sampling presented in Appendix B, to generate
i.i.d. posterior samples we simply need to identify $\mathcal{S}_{(Z|Y_{\mathrm{obs}})}$ and calculate
$\{\omega_k\}_{k=1}^K$. Obviously,

$$\mathcal{S}_{(Z|Y_{\mathrm{obs}})} = \mathcal{S}_{(Z|Y_{\mathrm{obs}}, \pi)} = \{z_1, \ldots, z_K\} = \{0, 1, \ldots, s\},$$

where $K = s + 1$. Setting $\pi_0 = 0.5$, from (B.2) and (B.3), we obtain

$$
\begin{aligned}
q_k(0.5) &= \frac{\binom{s}{z_k} p^{s-z_k} / (1 + p)^s}{0.5^{a+b+n} / B(a + z_k, b + n - z_k)} \\
&\propto \binom{s}{z_k} \frac{\Gamma(a + z_k)\Gamma(b + n - z_k)}{p^{z_k}}
\end{aligned}
$$

and

$$\omega_k = \frac{q_k(0.5)}{\sum_{k'=1}^K q_{k'}(0.5)}$$

for $k = 1, \ldots, K$.

3.5   Analyzing the Sexual Behavior Data

Most studies of sexual behaviors employ conventional self-report surveys.
Researchers have long criticized the validity of these self-reports since sexual
behavior is often highly private. An alternative approach, called *unmatched-
count technique* (UCT), provides participants a chance to answer sensitive
items without directly admitting to the sensitive behavior (Wimbush &
Dalton, 1997). In the UCT method, half of the participants will receive a
set of, for instance, five questions (in which all questions are non-sensitive)
while the other half will receive a set of six questions (in which one of them
is the sensitive question). It should be noted that the five non-sensitive

Table 3.5 Sensitive sexual behavior data

Item	Self-report ($N=102$)			UCT ($N=244$)		
($N=346$)	Rate	Yes	No	Rate	Yes	No
1. Sex without a condom	0.59	60	42	0.70	171	73
2. Drank until intoxication	0.77	79	23	0.70	171	73
3. Sex after drinking	0.48	49	53	0.49	120	124
4. Sex without a condom after drinking	0.36	37	65	0.65	159	85
5. Had sex	0.74	75	27	0.84	205	39

Source: Adapted from Table 1 of LaBrie & Earleywine (2000).

questions are common to all respondents. At the end of the survey, respondents simply indicate the number of statements that are true for them. The base rate estimate for the sensitive item is determined through random assignment of participants and comparisons between the two samples. All samples were obtained via simple random sampling. The main feature of the UCT is that participants do not respond directly to the sensitive item(s).

LaBrie & Earleywine (2000) used an anonymous self-report questionnaire and the UCT to estimate the base rates for some sexual risk behaviors (e.g., having sex without a condom and having sex without a condom after drinking). Three hundred forty-six college students were randomly divided into three groups. Group 1 (102 subjects) received a true/false conventional self-report survey. Groups 2 (122 subjects) and 3 (122 subjects) were UCT protocol groups, with Group 2 receiving Form A and Group 3 receiving Form B (see Appendix B in LaBrie & Earleywine, 2000, for more detail). Their findings are reported in Table 3.5. For example, 36% of the respondents receiving the conventional survey endorsed having sex without a condom after consuming alcohol while the UCT protocol revealed a base rate estimate of 65% for the same behavior. Thus, the anonymous self-report questionnaire revealed only half the percentage of persons engaging in risky sexual behavior after drinking reported by the UCT protocol.

To illustrate the methods in Sections 3.3 and 3.4, for the third sensitive item, we combine the numbers of 'yes' and 'no' with those for the two survey methods, resulting in $n = 346$,

$$n_{yes} = 49 + 120 = 169 \quad \text{and} \quad n_{no} = 53 + 124 = 177.$$

In the triangular model, we further let

$$\pi = \Pr(Y = 1) = \Pr(\text{having sex after drinking})$$

Table 3.6 Four 95% confidence intervals of π

Type of CIs	Confidence interval	Width
Wald CI (3.16)	[0.3994235, 0.5832354]	0.1838119
Wilson CI (3.19)	[0.3945073, 0.5773615]	0.1828542
Bootstrap CI (3.20)	[0.3997872, 0.5832022]	0.1834150
Bootstrap CI (3.21)	[0.3988439, 0.5838150]	0.1849711

and $p = \Pr(W = 1) \approx 0.5$. For the ideal situation (i.e., no sampling errors), the observed counts in the triangle would be

$$s = \frac{n_{\mathrm{no}}}{2} + n_{\mathrm{yes}} \approx 258.$$

Thus, we obtain the observed data $Y_{\mathrm{obs}} = \{n;\ n - s,\ s\} = \{346;\ 88,\ 258\}$. From (3.12) and (3.13), the MLE of π is $\hat{\pi}_{\mathrm{T}} = 0.4913295$ and the estimated variance is

$$\widehat{\mathrm{Var}}(\hat{\pi}_{\mathrm{T}}) = 0.002192472.$$

From (3.14), an unbiased estimate of $\mathrm{Var}(\hat{\pi}_{\mathrm{T}})$ is given by

$$\overline{\mathrm{Var}}(\hat{\pi}_{\mathrm{T}}) = 0.002198827.$$

From (3.16) and (3.19), the 95% Wald confidence interval and Wilson confidence interval of π are listed in Table 3.6. From (3.20) and (3.21), two 95% bootstrap confidence intervals for π with $G = 60{,}000$ bootstrap replications are also reported in Table 3.6. The corresponding standard error of $\hat{\pi}_{\mathrm{T}}$ is estimated by

$$\widehat{\mathrm{se}}(\hat{\pi}_{\mathrm{T}}) = 0.04678954.$$

We note that the width of the Wilson confidence interval is the shortest among the four 95% confidence intervals in Table 3.6.

To illustrate the Bayesian methods presented in Section 3.4, we consider the uniform prior (i.e., $a = b = 1$). Using (3.29), we have

$$E(\pi|Y_{\mathrm{obs}}) = 0.4885057 \quad \text{and} \quad E(\pi^2|Y_{\mathrm{obs}}) = 0.2408194.$$

Thus, the 95% Bayesian credible interval for π based on normality approximation is given by [0.3969598, 0.5800517].

Using $\pi^{(0)} = 0.5$ as an initial value, the EM algorithm (3.32) and (3.33) converges in 30 iterations. The posterior mode of π is $\tilde{\pi}_{\mathrm{T}} = 0.4913295$, which is same as the MLE $\hat{\pi}_{\mathrm{T}}$. Using the exact IBF sampling described in Section 3.4.3, we generate $L = 20{,}000$ i.i.d. posterior samples from $f(\pi|Y_{\mathrm{obs}})$. The histogram based on these samples is plotted in Figure 3.4 (ii), which shows that the exact IBF sampling can recover the density completely. The corresponding posterior mean, standard error, and 95% Bayesian credible interval for π are 0.489817, 0.04711621, and [0.3974693, 0.5821648].

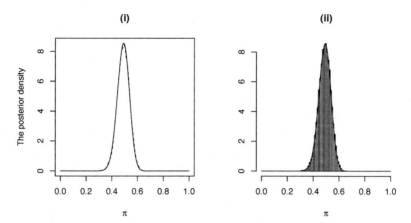

Figure 3.4 Posterior distribution of $\pi = \Pr(\text{having sex after drinking})$ under the uniform prior (i.e., $a = b = 1$) and $p = \Pr(W = 1) \approx 0.5$ for the triangular model for $n = 346$ and $s = 258$. (i) The comparison between the posterior distribution (solid curve) exactly given by (3.27) with the dotted curve estimated by a kernel density smoother based on 20,000 i.i.d. posterior samples generated via the exact IBF sampling. (ii) The histogram based on the 20,000 i.i.d. posterior samples generated via the exact IBF sampling.

3.6 Case Studies on Premarital Sexual Behavior

3.6.1 Questionnaire at Hong Kong Baptist University

Premarital sexual behavior, especially among college students, is a popular issue nowadays, not only among individual citizens, but also for the government and social researchers. However, most people consider it as privacy and feel uncomfortable to disclose information on that. To better protect the respondents' privacy and gather more accurate data, Geng (2011) adopted the non-randomized triangular model to conduct a small study at Hong Kong Baptist University to examine whether the proportions of college students having sex experience are the same for male and female. The question that the last digit of the cell phone number is odd or even is chosen as the non-sensitive question. And $p = \Pr$ (the last digit of a respondent's cell phone number is even) is assumed to be 0.5. The question in the questionnaire is as follows:

- If the last digit of your cell phone number is odd AND you have not had premarital sexual behavior, please check here ○

- Otherwise, please check here □

At the end of the study, Geng (2011) observed that 28 ticks were put in the □ among 45 questionnaires for the male group; while 13 ticks were put in the □ among 30 questionnaires for the female group. Let $\pi = \Pr(Y = 1)$ denote the unknown proportion of having premarital sexual behavior. The observed data in Table 3.1 can be denoted by

$$Y_{\text{obs}} = \{n; \ n - s, \ s\} = \{75; \ 34, \ 41\},$$

where $n = 45 + 30 = 75$, $n - s = 34$ implies 34 ticks being put in the circle and $s = 28 + 13 = 41$ implies 41 ticks being put in the upper square when the triangular model is applied (see Table 3.1).

From (3.12) and (3.13), the MLE of π is $\hat{\pi}_{\text{T}} = 0.09333333$ and the estimated variance is

$$\widehat{\text{Var}}(\hat{\pi}_{\text{T}}) = 0.01321719.$$

From (3.14), an unbiased estimate of $\text{Var}(\hat{\pi}_{\text{T}})$ is given by

$$\overline{\text{Var}}(\hat{\pi}_{\text{T}}) = 0.01339580.$$

From (3.16), the 95% Wald confidence interval of π is

$$[\hat{\pi}_{\text{T,WL}}, \ \hat{\pi}_{\text{T,WU}}] = [-0.1335133, \ 0.32018],$$

which is useless since the lower bound is less than zero. From (3.19), the 95% Wilson confidence interval of π is

$$[\hat{\pi}_{\text{T,WSL}}, \ \hat{\pi}_{\text{T,WSU}}] = [-0.1310325, \ 0.3086041],$$

which also has a negative lower bound.

In order to obtain a useful interval estimation of π, we could employ the bootstrap method. From (3.20) and (3.21), two 95% bootstrap confidence intervals for π with $G = 20{,}000$ bootstrap replications are given by

$$[-0.0777481, \ 0.2896732]$$

and

$$[7.62595 \times 10^{-13}, \ 0.3066667],$$

respectively. The corresponding standard error of $\hat{\pi}_{\text{T}}$ is estimated by

$$\widehat{\text{se}}(\hat{\pi}_{\text{T}}) = 0.09372991.$$

Alternatively, we may consider Bayesian inferences with the uniform prior (i.e., $a = b = 1$) on π. Using (3.29), we have

$$E(\pi|Y_{\text{obs}}) = 0.1325037 \quad \text{and} \quad E(\pi^2|Y_{\text{obs}}) = 0.02471187.$$

Thus, the 95% Bayesian credible interval for π based on normality approximation is given by $[-0.03328312, \ 0.29829051]$.

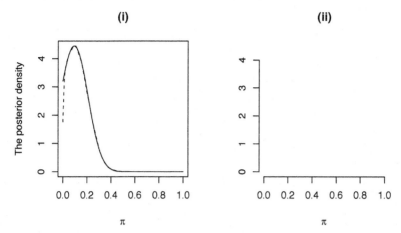

Figure 3.5 Posterior distribution of $\pi = \Pr$(having premarital sexual behavior) under the uniform prior (i.e., $a = b = 1$) and $p = \Pr(W = 1) \approx 0.5$ for the triangular model for $n = 75$ and $s = 41$. (i) The comparison between the posterior distribution (solid curve) exactly given by (3.27) with the dotted curve estimated by a kernel density smoother based on 40,000 i.i.d. posterior samples generated via the exact IBF sampling. (ii) The histogram based on the 40,000 i.i.d. posterior samples generated via the exact IBF sampling.

Using $\pi^{(0)} = 0.5$ as an initial value, the EM algorithm (3.32) and (3.33) converges in 198 iterations. The posterior mode of π is $\tilde{\pi}_{\mathrm{T}} = 0.09333334$, which is same as the MLE $\hat{\pi}_{\mathrm{T}}$. Using the exact IBF sampling described in Section 3.4.3, we generate $L = 40,000$ i.i.d. posterior samples from $f(\pi|Y_{\mathrm{obs}})$. The histogram based on these samples is plotted in Figure 3.5 (ii), which shows that the exact IBF sampling can recover the density completely. The resultant posterior mean, standard error and 95% Bayesian credible interval based on normality approximation for π are given by 0.1320513, 0.08497416 and

$$[0.1320513 - 1.96 \times 0.08497416, \ 0.1320513 + 1.96 \times 0.08497416]$$

$$= \ [-0.03449802, \ 0.29860070],$$

respectively. This Bayesian credible interval with a negative lower bound is not surprising since the posterior density of π is highly skew toward zero. The 2.5-th and 97.5-th percentiles of the 40,000 i.i.d. posterior samples of π are 0.007127091 and 0.319651027, respectively. Thus,

$$[0.007127091, \ 0.319651027]$$

is a 95% Bayesian credible interval of π.

3.6.2   Questionnaire at the Northeast Normal University

The risk of transmission of HIV and other sexually transmitted diseases is higher in sexual relationship with multiple partners and without the use of condoms. Premarital sex often involves multiple partners, and extra-marital sex, by definition, implies multi-partner relationships. Avoidance of multi-partner sexual relationships, use of condoms and sexual abstinence are usually advocated for preventing the spread of HIV and other sexually transmitted diseases. In some Asian countries, premarital sexual activity has long been considered as taboo and becomes a sensitive issue in which we can hardly get reliable answers by direct asking.

To investigate the proportion of college students with premarital sex experience, a survey based on the non-randomized triangular model was conducted in March 2010 among 98 students (48 female and 50 male) at the Northeast Normal University in Changchun, Jilin Province, P. R. China. The respondents were from Faculty of Education, School of Politics and Law, Faculty of Arts, Faculty of Chemistry, School of History and Culture, School of Foreign Language, School of Life Science, School of Physical Education, School of Physics, School of Urban and Environmental Sciences, and School of Mathematics and Statistics. Among the 98 students, 8 were freshmen, 10 were sophomore, 20 were junior, 24 were senior and 36 were postgraduate students. The question in the questionnaire is as follows:

- If the last digit of your cell phone number is odd AND you have not had premarital sexual behavior, please check here   ◯

- Otherwise, please check here   ☐

Thus, it is reasonable to assume the $p = \Pr$ (the last digit of a respondent's cell phone number is even) is around 0.5. At the end of the survey, it was observed that 26 ticks were put in the ☐ for the female group; while 26 ticks were put in the ☐ for the male group. Let $\pi = \Pr\{Y = 1\}$ denote the unknown proportion of having premarital sexual behavior. The observed data can be denoted by $Y_{\mathrm{obs}} = \{n;\ n - s,\ s\} = \{98; 46, 52\}$, where $n = 98$, $n - s = 46$ implies 46 ticks being put in the circle and $s = 26 + 26 = 52$ implies 52 ticks being put in the upper square when the triangular model is applied (see Table 3.1).

From (3.12) and (3.13), the MLE of π is $\hat{\pi}_{\mathrm{T}} = 0.06122449$ and the estimated variance is 0.01016583. From (3.14), an unbiased estimate of $\mathrm{Var}(\hat{\pi}_{\mathrm{T}})$ is given by $\overline{\mathrm{Var}}(\hat{\pi}_{\mathrm{T}}) = 0.01027063$. From (3.16), the 95% Wald confidence interval of π is $[\hat{\pi}_{\mathrm{T,WL}},\ \hat{\pi}_{\mathrm{T,WU}}] = [-0.1374064,\ 0.2598554]$, which is useless since the lower bound is less than zero. From (3.19), the 95% Wilson confidence interval of π is $[\hat{\pi}_{\mathrm{T,WSL}},\ \hat{\pi}_{\mathrm{T,WSU}}] = [-0.1349507,\ 0.2527809]$, which also has a negative lower bound.

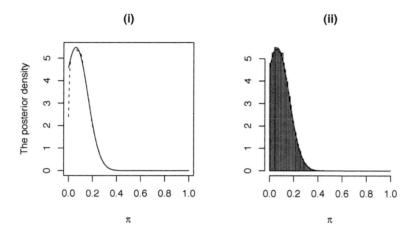

Figure 3.6 Posterior distribution of $\pi = \Pr(\text{having premarital sexual behavior})$ under the uniform prior (i.e., $a = b = 1$) and $p = \Pr(W = 1) \approx 0.5$ for the triangular model for $n = 98$ and $s = 52$. (i) The comparison between the posterior distribution (solid curve) exactly given by (3.27) with the dotted curve estimated by a kernel density smoother based on 40,000 i.i.d. posterior samples generated via the exact IBF sampling. (ii) The histogram based on the 40,000 i.i.d. posterior samples generated via the exact IBF sampling.

To obtain a useful interval estimation of π, we use the bootstrap method. From (3.20) and (3.21), two 95% bootstrap confidence intervals for π with $G = 20,000$ bootstrap replications are given by $[-0.07454688, \ 0.2323579]$ and $[6.170113 \times 10^{-15}, \ 0.2653061]$, respectively. The corresponding standard error of $\hat{\pi}_T$ is estimated by $\widehat{se}(\hat{\pi}_T) = 0.07829205$.

Alternatively, we may consider Bayesian inferences with the uniform prior (i.e., $a = b = 1$) on π. Using (3.29), we have

$$E(\pi|Y_{\text{obs}}) = 0.1056557 \quad \text{and} \quad E(\pi^2|Y_{\text{obs}}) = 0.01617756.$$

Thus, the 95% Bayesian credible interval for π based on normality approximation is given by $[-0.03313729, 0.24444863]$.

Using $\pi^{(0)} = 0.5$ as an initial value, the EM algorithm (3.32) and (3.33) converges in 272 iterations. The posterior mode of π is $\tilde{\pi}_T = 0.06122449$, which is same as the MLE $\hat{\pi}_T$. Using the exact IBF sampling described in Section 3.4.3, we generate $L = 40,000$ i.i.d. posterior samples from $f(\pi|Y_{\text{obs}})$. The histogram based on these samples is plotted in Figure 3.6 (ii), which shows that the exact IBF sampling can recover the density completely. The resultant posterior mean, standard error, and 95% Bayesian credible interval based on normality approximation for π are given by 0.1051566, 0.07037638, and $[-0.03278107, 0.24309435]$, respectively. This Bayesian credible interval with a negative lower bound is not surprising since the posterior density of π is highly skew toward zero. The 2.5-th and 97.5-th percentiles of the 40,000 i.i.d. posterior samples of π are 0.005265239 and 0.263518725, respectively. Thus, [0.005265239, 0.263518725] is a 95% Bayesian credible interval of π.

CHAPTER 4

Sample Sizes for the
Crosswise and Triangular Models

Sample size calculation is an essential component in survey designs on sensitive topics. In this chapter, we will discuss sample size calculation for the crosswise and triangular models based on the power analysis method for both one-sample and two-sample problems. In Section 4.1, we briefly introduce two methods (i.e., the precision analysis method and the power analysis method) for sample size calculations. In Section 4.2, we consider one-sample problem and derive sample size formulae for the triangular model for both one- and two-sided tests based on the large-sample normal approximation. Evaluation of the performance is conducted by comparing exact power with asymptotic power and by calculating n_{T} and the ratio $n_{\mathrm{T}}/n_{\mathrm{D}}$. In Section 4.3, we obtain sample sizes for the crosswise model for a one-sample problem. In Section 4.4, we numerically compare sample sizes required for the crosswise design with those needed for the triangular design. A theoretical justification is also provided. In Section 4.5, we obtain a sample size formula for the triangular model for a two-sample problem. An example about an induced abortion study in Taiwan is presented to illustrate these methods in Section 4.6 (see Tian *et al.*, 2011).

4.1 Precision and Power Analysis Methods

One of the most important steps in sample surveys is the determination of the number of participants. In practice, sample sizes can be determined based on the *precision analysis* method or the *power analysis* method (Chow, Shao & Wang, 2003), which is closely related to the notion of Type I error rate, Type II error rate and power functions in testing hypothesis.

4.1.1 Type I error rate, Type II error rate and power

Assume that $X_1, \ldots, X_n$ is a random sample from a population with probability density function (or probability mass function) $f(x; \theta)$, where $\theta \in \Theta$ and Θ denotes the parameter space. Let $\mathbf{x} = (X_1, \ldots, X_n)^{\top}$ and $\boldsymbol{x} = (x_1, \ldots, x_n)^{\top}$ denote its realization. Consider the following hypotheses:

$$H_0: \theta \in \Theta_0 \quad \text{against} \quad H_1: \theta \in \Theta_1.$$

Let $\mathbb{C}$ and $\mathbb{C}'$ respectively denote the critical region and the acceptance region of a test for testing the null hypothesis H_0 against the alternative hypothesis H_1.

(1) Rejection of the null hypothesis H_0 when it is true is called *Type I error*. The probability of committing a Type I error

$$
\begin{aligned}
\alpha &= \Pr(\text{Type I error}) \\
&= \Pr(\text{rejecting } H_0 | H_0 \text{ is true}) \\
&= \Pr(\mathbf{x} \in \mathbb{C} | \theta \in \Theta_0)
\end{aligned}
$$

is called *Type I error rate*.

(2) Acceptance of the null hypothesis H_0 when it is false is called *Type II error*. The probability of committing a Type II error

$$
\begin{aligned}
\beta &= \Pr(\text{Type II error}) \\
&= \Pr(\text{accepting } H_0 | H_0 \text{ is false}) \\
&= \Pr(\mathbf{x} \in \mathbb{C}' | \theta \in \Theta_1)
\end{aligned}
$$

is called *Type II error rate*.

(3) Power of a test is defined as the probability of correctly rejecting H_0 when the H_0 is false; that is,

$$
\begin{aligned}
\text{Power} &= 1 - \beta \\
&= \Pr(\text{rejecting } H_0 | H_0 \text{ is false}) \\
&= \Pr(\mathbf{x} \in \mathbb{C} | \theta \in \Theta_1).
\end{aligned}
$$

We can summarize these notions into the following table.

Table 4.1 Type I error rate, Type II error rate and power

	H_0 is true ($\theta \in \Theta_0$)	H_0 is false ($\theta \in \Theta_1$)
Reject H_0 ($\mathbf{x} \in \mathbb{C}$)	α	$1 - \beta$ (power)
Accept H_0 ($\mathbf{x} \in \mathbb{C}'$)	$1 - \alpha$ (confidence level)	β

Note: In practice, we usually use 'fail to reject H_0' to replace 'accept H_0.' The confidence level defined by $1-\alpha$ reflects the probability or confidence of not rejecting the true null hypothesis.

4.1.2 Precision analysis

Precision analysis for sample size determination is usually performed by controlling a Type I error rate α (sometimes it is called significance level or size). The confidence level, $1 - \alpha$, then reflects the probability or confidence of not rejecting the true null hypothesis. Due to the duality between a confidence interval and a two-sided hypothesis testing, we may determine a desired sample size by pre-specifying the most tolerable Type I error rate via the confidence interval approach.

For a $(1 - \alpha)100\%$ confidence interval, the precision of the interval depends on its width. The narrower the confidence interval, the more precise the inference provided that the coverage probability is well controlled around the confidence level. Therefore, the precision analysis for sample size determination considers the maximum half width of the $(1 - \alpha)100\%$ confidence interval of the unknown parameter that investigators are willing to accept. Note that the maximum half width of the confidence interval is usually referred as the maximum error of an estimate of the unknown parameter.

4.1.3 Power analysis

Power analysis for sample size determination is performed by controlling a Type II error rate β. Since a Type I error is usually considered to be a more important and serious error that statisticians would like to avoid, a typical approach in hypothesis testing is to control α at an acceptable level and at the same time minimize β by choosing an appropriate sample size. In other words, the H_0 can be tested at a pre-determined nominal level (say, $\alpha = 0.05$) with a desired power (say, $1 - \beta = 0.8$). In addition, investigators are required to specify a meaningful difference for a two-sample design.

4.2 The Triangular Model for One-sample Problem

For the triangular model described in Table 3.1, the MLE of π and its variance are given by (3.3) and (3.13), which can be rewritten as

$$\hat{\pi}_{\mathrm{T}} = \frac{\bar{y}_{\mathrm{T}} - p}{1 - p} \quad \text{and} \quad \mathrm{Var}(\hat{\pi}_{\mathrm{T}}) = \frac{\theta(1 - \theta)}{n(1 - p)^2}, \tag{4.1}$$

where

$$\bar{y}_{\mathrm{T}} = \frac{1}{n} \sum_{i=1}^{n} y_{i,\mathrm{T}} \quad \text{and} \quad \theta = \pi + (1 - \pi)p.$$

As $n \to \infty$, the Central Limit Theorem implies that $\hat{\pi}_T$ is asymptotically normally distributed; that is,

$$\frac{\hat{\pi}_T - \pi}{\sqrt{\mathrm{Var}(\hat{\pi}_T)}} = \frac{n\hat{\pi}_T - n\pi}{\sqrt{n\theta(1-\theta)}/(1-p)} \stackrel{\cdot}{\sim} N(0,1). \qquad (4.2)$$

4.2.1   A one-sided test

To test whether there is a difference between the population proportion (π) with the sensitive characteristic and a pre-specified reference value (π_0), we consider the following hypotheses

$$H_0: \pi = \pi_0 \quad \text{versus} \quad H_1: \pi < \pi_0. \qquad (4.3)$$

Under the null hypothesis H_0, from (4.2), we have

$$\frac{n_T\hat{\pi}_T - n_T\pi_0}{\sqrt{n_T\theta_0(1-\theta_0)}/(1-p)} \stackrel{\cdot}{\sim} N(0,1), \quad \text{as } n_T \to \infty,$$

where n_T denotes the sample size needed for the triangular model for the above one-sided test, and

$$\theta_0 = \pi_0 + (1-\pi_0)p.$$

Thus, we reject the null hypothesis H_0 at the α level of significance when we observe the following event

$$\mathbb{E}_T = \left\{ n_T\hat{\pi}_T \leqslant n_T\pi_0 - z_\alpha\sqrt{n_T\theta_0(1-\theta_0)}\Big/(1-p) \right\}, \qquad (4.4)$$

where z_α is the upper α-th quantile of the standard normal distribution.

Now, we assume that H_1 is true and $\pi = \pi_1$ with $\pi_1 < \pi_0$. Then, the power of the one-sided test is approximately given by

Power (at π_1)

$$= \mathrm{Pr}\,(\text{rejecting } H_0 | \pi = \pi_1)$$

$$= \mathrm{Pr}\left\{ \frac{n_T\hat{\pi}_T - E_{H_1}(n_T\hat{\pi}_T)}{\sqrt{\mathrm{Var}_{H_1}(n_T\hat{\pi}_T)}} \leqslant \frac{n_T\pi_0 - z_\alpha\sqrt{n_T\theta_0(1-\theta_0)}/(1-p) - n_T\pi_1}{\sqrt{n_T\theta_1(1-\theta_1)}/(1-p)} \right\}$$

$$\approx \Phi\left(\frac{\sqrt{n_T}(\pi_0 - \pi_1)(1-p) - z_\alpha\sqrt{\theta_0(1-\theta_0)}}{\sqrt{\theta_1(1-\theta_1)}} \right), \qquad (4.5)$$

where

$$\theta_1 = \pi_1 + (1-\pi_1)p$$

and $\Phi(\cdot)$ denotes the cumulative distribution function of the standard normal distribution. As a result, the sample size needed for achieving a desired power of $1 - \beta$ can be obtained by solving the following equation

$$\sqrt{n_T}\,(\pi_0 - \pi_1)(1 - p) - z_\alpha \sqrt{\theta_0(1 - \theta_0)} = z_\beta \sqrt{\theta_1(1 - \theta_1)},$$

which leads to

$$n_T = \left\{ \frac{z_\alpha \sqrt{\theta_0(1 - \theta_0)} + z_\beta \sqrt{\theta_1(1 - \theta_1)}}{(\pi_0 - \pi_1)(1 - p)} \right\}^2. \tag{4.6}$$

4.2.2 A two-sided test

Now we consider a two-sided test for the following hypotheses

$$H_0: \pi = \pi_0 \quad \text{versus} \quad H_1: \pi \neq \pi_0,$$

with equal allocation $\alpha/2$ to each tail of the rejection region. The relationships among the power, sample size and effect size is approximately given by

$$\text{Power (at } \pi_1) \approx \Phi\left(\frac{\sqrt{n_{T,2}}\,|\pi_0 - \pi_1|(1 - p) - z_{\alpha/2}\sqrt{\theta_0(1 - \theta_0)}}{\sqrt{\theta_1(1 - \theta_1)}} \right).$$

The corresponding sample size formula is

$$n_{T,2} = \left\{ \frac{z_{\alpha/2} \sqrt{\theta_0(1 - \theta_0)} + z_\beta \sqrt{\theta_1(1 - \theta_1)}}{(\pi_0 - \pi_1)(1 - p)} \right\}^2. \tag{4.7}$$

4.2.3 Evaluation of the performance by comparing exact power with asymptotic power

To examine the accuracy of the asymptotic power formula (4.5), we plot in Figure 4.1 the exact as well as asymptotic powers against the sample size n_T under various combinations of (π_0, π_1) at $p = 0.5$ and the significance level $\alpha = 0.05$. The exact power (at π_1) for any particular sample size n_T is

$$\text{Exact power (at } \pi_1) = \sum_{x \in \mathbb{E}_T} \text{Binomial}\,(x | n_T, \, \pi_1 + (1 - \pi_1)p)$$

$$= \sum_{x \in \mathbb{E}_T} \binom{n_T}{x} \theta_1^x (1 - \theta_1)^{n_T - x}, \tag{4.8}$$

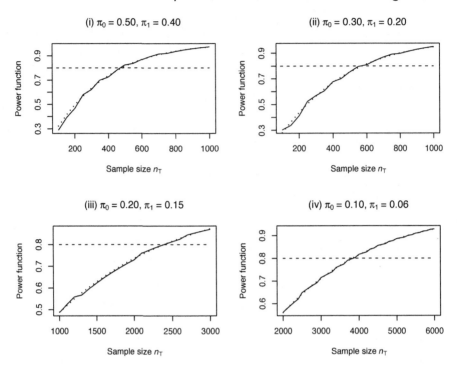

Figure 4.1 Comparisons of the exact power (4.8) (denoted by solid line) with the asymptotic power (4.5) (denoted by dotted line) against the sample size n_{T} under various combinations of (π_0, π_1) at the $p = 0.5$ and $\alpha = 5\%$. (i) $(\pi_0, \pi_1) = (0.50, 0.40)$; (ii) $(\pi_0, \pi_1) = (0.30, 0.20)$; (iii) $(\pi_0, \pi_1) = (0.20, 0.15)$; (iv) $(\pi_0, \pi_1) = (0.10, 0.06)$.

where the rejection region $\mathbb{E}_{\mathrm{T}}$ specified by (4.4) can be rewritten as

$$\mathbb{E}_{\mathrm{T}} = \left\{ X_{\mathrm{T}} \leqslant n_{\mathrm{T}}\theta_0 - z_\alpha \sqrt{n_{\mathrm{T}}\theta_0(1 - \theta_0)} \right\}$$

where

$$X_{\mathrm{T}} = \sum_{i=1}^{n_{\mathrm{T}}} y_{i,\mathrm{T}} \sim \text{Binomial}\,(n_{\mathrm{T}},\, \theta).$$

In general, we find that the asymptotic power function (4.5) provides a satisfactory approximation to the exact power, especially when n_{T} increases.

4.2.4 Evaluation of the performance by calculating n_{T} and $n_{\mathrm{T}}/n_{\mathrm{D}}$

Table 4.2 reports some values of sample size n_{T} given by (4.6) with $\alpha = 5\%$ and 80% power at various combinations of (π_0, π_1, p) and the corresponding

Table 4.2 Sample size n_T for testing H_0: $\pi = \pi_0$ versus H_1: $\pi = \pi_1 < \pi_0$ with $\alpha = 5\%$ and 80% power and the ratio n_T/n_D

π_0	π_1	$p = 0.00$	$p = 0.42$		$p = 0.50$		$p = 0.58$	
		n_D	n_T	n_T/n_D	n_T	n_T/n_D	n_T	n_T/n_D
0.50	0.40	153	389	2.54	483	3.15	614	4.01
	0.35	67	175	2.61	218	3.25	278	4.14
	0.30	37	100	2.70	125	3.37	159	4.29
0.40	0.35	583	1673	2.86	2109	3.61	2719	4.66
	0.30	142	422	2.97	534	3.76	691	4.86
	0.25	61	189	3.09	240	3.93	312	5.11
0.30	0.25	501	1767	3.52	2274	4.53	2983	5.95
	0.20	119	444	3.73	573	4.81	755	6.34
	0.18	81	309	3.81	399	4.92	527	6.50
0.20	0.16	584	2831	4.84	3729	6.38	4986	8.53
	0.13	181	925	5.11	1222	6.75	1637	9.04
	0.10	83	453	5.45	600	7.22	806	9.71
0.10	0.08	1303	11317	8.68	15321	11.7	20927	16.1
	0.06	301	2826	9.38	3834	12.7	5246	17.4
	0.04	121	1254	10.3	1706	14.1	2338	19.3

Source: Table 2 in Tian *et al.* (2011).
Note: n_D is the sample size of the design of direct questioning, given by (4.9).

exact power calculated according to (4.8). Generally, the sample size n_T is sufficient to guarantee the desired power; that is,

$$|\text{actual power} - \text{desired power}| \leqslant 5\%.$$

For a given pair of (π_0, π_1), we can see that n_T is an increasing function of p. In particular, when $p = 0$, the triangular design reduces to the design of direct questioning. Here, we denote the sample size of the design of direct questioning by n_D. In (4.6), setting $p = 0$, we obtain

$$n_D = \left\{ \frac{z_\alpha \sqrt{\pi_0(1 - \pi_0)} + z_\beta \sqrt{\pi_1(1 - \pi_1)}}{\pi_0 - \pi_1} \right\}^2. \tag{4.9}$$

In Table 4.2, we also report the ratios of n_T/n_D for $p = 0.42, 0.5, 0.58$, respectively. For example, when $(\pi_0, \pi_1) = (0.4, 0.3)$ and $p = 0.42$, we have $n_T/n_D = 2.97$, indicating that the sample size required for the triangular design is about 3 times that required for the design of direct questioning in order to achieve the same power for the one-sided test. Here, we choose the non-sensitive dichotomous variate W to be the respondent's birthday duration. For $p = 0.42$ (i.e., $5/12$), 0.50 (i.e., $6/12$) and 0.58 (i.e., $7/12$), they represent $W = 1$ if a respondent was born between January and May, January and June, and January and July, respectively; and $W = 0$ otherwise.

4.3   The Crosswise Model for One-sample Problem

The crosswise model is described in Table 2.2. Let $Y_{\text{obs}} = \{y_{i,\mathrm{C}}: i = 1, \ldots, n\}$ denote the observed data for n respondents, where $y_{i,\mathrm{C}} = 1$ if the i-th respondent puts a tick in the upper circle and $y_{i,\mathrm{C}} = 0$ if the i-th respondent puts a tick in the upper square. The likelihood function for π is then given by

$$L_{\mathrm{C}}(\pi|Y_{\text{obs}}) = \prod_{i=1}^{n} \{(1-\pi)(1-p) + \pi p\}^{y_{i,\mathrm{C}}} \{\pi(1-p) + (1-\pi)p\}^{1-y_{i,\mathrm{C}}}.$$

The MLE of π and its variance (cf. (2.8) and (2.9)) are given by

$$\hat{\pi}_{\mathrm{C}} = \frac{p-1+\bar{y}_{\mathrm{C}}}{2p-1} \quad \text{and} \quad \text{Var}(\hat{\pi}_{\mathrm{C}}) = \frac{\lambda(1-\lambda)}{n(2p-1)^2}, \quad p \neq 0.5, \qquad (4.10)$$

where

$$\bar{y}_{\mathrm{C}} = \frac{1}{n}\sum_{i=1}^{n} y_{i,\mathrm{C}} \quad \text{and} \quad \lambda = (1-\pi)(1-p) + \pi p.$$

4.3.1   A one-sided test

To derive the sample size formula for the crosswise model, we consider the same one-sided hypotheses specified in (4.3). Similar to (4.5) and (4.6), we have

$$\text{Power (at } \pi_1) \approx \Phi\left(\frac{\sqrt{n_{\mathrm{C}}}\,(\pi_0 - \pi_1)|2p-1| - z_\alpha\sqrt{\lambda_0(1-\lambda_0)}}{\sqrt{\lambda_1(1-\lambda_1)}}\right)$$

and

$$n_{\mathrm{C}} = \left\{\frac{z_\alpha\sqrt{\lambda_0(1-\lambda_0)} + z_\beta\sqrt{\lambda_1(1-\lambda_1)}}{(\pi_0 - \pi_1)(2p-1)}\right\}^2. \qquad (4.11)$$

where

$$\lambda_i = (1-\pi_i)(1-p) + \pi_i p, \quad i = 0, 1,$$

and $\pi_1 < \pi_0$. In particular, when $p = 1$ the crosswise design is identical to the design of direct questioning and n_{C} is equal to n_{D} given in (4.9).

4.3.2 Evaluation of the performance by calculating $n_{\rm C}$ and $n_{\rm C}/n_{\rm D}$

Table 4.3 reports some values of sample size $n_{\rm C}$ given in (4.11) with $\alpha = 5\%$ and 80% power under various combinations of (π_0, π_1, p) and the corresponding exact power, which is calculated according to the following formula:

$$\text{Exact power (at } \pi_1) = \sum_{x \in \mathbb{E}_{\rm C}} \binom{n_{\rm C}}{x} \lambda_1^x (1 - \lambda_1)^{n_{\rm C} - x},$$

where the rejection region

$$\mathbb{E}_{\rm C} = \left\{ X_{\rm C} \leqslant n_{\rm C}\lambda_0 - z_\alpha \sqrt{n_{\rm C}\lambda_0(1 - \lambda_0)} \right\}$$

and

$$X_{\rm C} = \sum_{i=1}^{n_{\rm C}} y_{i,\rm C} \sim \text{Binomial}(n_{\rm C}, \lambda).$$

Generally, the sample size $n_{\rm C}$ is sufficient to guarantee the desired power; that is,

$$|\text{actual power} - \text{desired power}| \leqslant 5\%.$$

For a given pair of (π_0, π_1), It is clear that $n_{\rm C}$ is a monotone decreasing function of p. We compare the crosswise design with the design of direct questioning by reporting the ratios of $n_{\rm C}/n_{\rm D}$ for $p = 0.75, 0.70, \ldots, 0.55$, respectively. For example, when $(\pi_0, \pi_1) = (0.4, 0.3)$ and $p = 0.65$, we have $n_{\rm C}/n_{\rm D} = 12$, indicating that the sample size required for the crosswise design is about 12 times that required for the design of direct questioning in order to achieve the same power.

4.4 Comparison for the Crosswise and Triangular Models

4.4.1 Comparison via the calculation of the ratio $n_{\rm C}/n_{\rm T}$

In principle, the optimal degree of privacy protection for the crosswise model is attained at $p = 0.5$, which unfortunately leads to an undefined MLE of parameter of interest and an undefined variance. When p is too small or too large, the privacy of respondents cannot be protected sufficiently. Therefore, investigators have to select p within the interval $[0.42, 0.65]$ as $n\text{Var}(\hat{\pi}_{\rm C})$ is a symmetric function on $p = 0.5$.

In Table 4.4, we compare $n_{\rm C}$ with $n_{\rm T}$. When $p = 0.42$ or 0.58, from Table 4.4, we can see that the efficiency of the triangular strategy is about

Table 4.3 Sample size n_C for testing H_0: $\pi = \pi_0$ versus H_1: $\pi = \pi_1 < \pi_0$ with $\alpha = 5\%$ and 80% power and the ratio n_C/n_D

π_0	π_1	$p = 1$	$p = 0.75$		$p = 0.70$	
		n_D	n_C	n_C/n_D	n_C	n_C/n_D
0.50	0.40	153	617	4.03	964	6.30
	0.35	67	273	4.07	428	6.38
	0.30	37	153	4.13	240	6.48
0.40	0.35	583	2438	4.18	3829	6.56
	0.30	142	606	4.26	954	6.71
	0.25	61	268	4.39	422	6.91
0.30	0.25	501	2356	4.70	3747	7.47
	0.20	119	583	4.89	931	7.82
	0.18	81	404	4.98	645	7.96
0.20	0.16	584	3483	5.96	5657	9.68
	0.13	181	1128	6.23	1838	10.1
	0.10	83	548	6.60	896	10.8
0.10	0.08	1303	12898	9.89	21592	16.6
	0.06	301	3202	10.6	5376	17.9
	0.04	121	1413	11.7	2379	19.7

π_0	π_1	$p = 0.65$		$p = 0.60$		$p = 0.55$	
		n_C	n_C/n_D	n_C	n_C/n_D	n_C	n_C/n_D
0.50	0.40	1716	11.21	3863	25.24	15455	101.01
	0.35	762	11.37	1716	25.61	6868	102.50
	0.30	428	11.56	964	26.05	3863	104.40
0.40	0.35	6835	11.72	15422	26.45	61791	105.98
	0.30	1705	12.00	3852	27.12	15444	108.76
	0.25	756	12.39	1710	28.03	6862	112.49
0.30	0.25	6752	13.47	15339	30.61	61708	123.17
	0.20	1683	14.14	3829	32.17	15422	129.59
	0.18	1167	14.40	2658	32.81	10708	132.19
0.20	0.16	10353	17.72	23770	40.70	96222	164.76
	0.13	3372	18.62	7753	42.83	31411	173.54
	0.10	1648	19.85	3794	45.71	15387	185.38
0.10	0.08	40376	30.98	94044	72.17	383851	294.59
	0.06	10072	33.46	23489	78.03	95941	318.74
	0.04	4466	36.90	10429	86.19	42630	352.31

Source: Table 4 in Tian *et al.* (2011).

Note: n_D is the sample size of the design of direct questioning, given by (4.9).

Table 4.4 The ratio $n_{\mathrm{C}}/n_{\mathrm{T}}$ for testing H_0: $\pi = \pi_0$ versus H_1: $\pi = \pi_1 < \pi_0$ with $\alpha = 5\%$ and 80% power

π_0	π_1	p							
		0.42	0.45	0.49	0.51	0.55	0.58	0.62	0.65
0.50	0.40	14.30	36.53	822.14	779.04	27.79	9.06	3.88	2.24
	0.35	14.12	35.95	810.07	766.67	27.25	8.88	3.80	2.19
	0.30	13.89	35.44	798.35	754.69	26.82	8.73	3.71	2.15
0.40	0.35	13.28	33.72	754.32	711.93	25.24	8.17	3.48	2.00
	0.30	13.15	33.35	744.50	702.54	24.86	8.03	3.41	1.95
	0.25	13.04	32.99	737.04	692.46	24.50	7.90	3.35	1.92
0.30	0.25	12.52	31.61	701.23	658.79	23.12	7.42	3.12	1.78
	0.20	12.45	31.40	694.91	652.66	22.88	7.32	3.08	1.74
	0.18	12.42	31.30	693.31	651.24	22.78	7.28	3.06	1.73
0.20	0.16	12.14	30.47	669.62	626.05	21.73	6.89	2.86	1.61
	0.13	12.13	30.40	667.61	623.78	21.61	6.85	2.84	1.59
	0.10	12.12	30.34	666.10	622.12	21.52	6.81	2.82	1.58
0.10	0.08	12.06	30.05	653.78	607.66	20.82	6.52	2.66	1.47
	0.06	12.07	30.04	653.24	606.93	20.77	6.50	2.65	1.46
	0.04	12.08	30.06	652.71	606.24	20.72	6.48	2.64	1.45

Source: Table 5 in Tian *et al.* (2011).

6–9 or 12–14 times that of the crosswise model. In particular, when $0.49 \leqslant p \leqslant 0.51$ (which is the optimal range for which the privacy of respondents is protected in the triangular model), the efficiency of the triangular strategy is about 600–822 times that of the crosswise model.

4.4.2 A theoretical justification

The above observations are not surprising as we have the following theoretical result.

Theorem 4.1 The triangular model is always more efficient than the crosswise model in the sense that $n_{\mathrm{T}} \leqslant n_{\mathrm{C}}$, for any $\pi_0, \pi_1 \in [0, 0.5]$ and any $p \in [0, 2/3]$. ¶

Proof. From (4.6) and (4.11), we have

$$\frac{n_{\mathrm{T}}}{n_{\mathrm{C}}} = \left(\frac{2p-1}{1-p}\right)^2 \times \left\{\frac{z_\alpha \sqrt{\theta_0(1-\theta_0)} + z_\beta \sqrt{\theta_1(1-\theta_1)}}{z_\alpha \sqrt{\lambda_0(1-\lambda_0)} + z_\beta \sqrt{\lambda_1(1-\lambda_1)}}\right\}^2, \tag{4.12}$$

where

$$\theta_i = \pi_i + (1 - \pi_i)p \quad \text{and} \quad \lambda_i = (1 - \pi_i)(1 - p) + \pi_i p, \quad i = 0, 1.$$

Note that when $0 \leqslant p \leqslant 2/3$, we always have

$$(1 - p)^2 \geqslant (2p - 1)^2,$$

i.e., the first expression on the right-hand side of (4.12) is less than or equal to 1. To obtain $n_{\mathrm{T}} \leqslant n_{\mathrm{C}}$, it suffices to show that $\theta(1 - \theta) \leqslant \lambda(1 - \lambda)$ or equivalently

$$\{\pi + (1-\pi)p\}(1-\pi)(1-p) \leqslant \{(1-\pi)(1-p) + \pi p\}\{\pi(1-p) + (1-\pi)p\}. \quad (4.13)$$

After some simplifications, we can show that (4.13) is equivalent to

$$\pi p \{(2\pi - 1)p - \pi\} \leqslant 0,$$

if $0 \leqslant \pi \leqslant 0.5$. □

4.5 The Triangular Model for Two-sample Problem

When we conduct two independent surveys on a common sensitive question in two different populations/regions (indexed by $k = 1, 2$) using the triangular design presented in Section 3.1.1, it is of design interest to decide the numbers of respondents in two groups in order to compare the proportions (π_k, $k = 1, 2$) of subjects with a sensitive attribute. For a fixed k, we define a hidden variable $Y_{k,\mathrm{T}}$ as follows:

$$Y_{k,\mathrm{T}} = \begin{cases} 1, & \text{if a tick is put in the upper square in the k-th region,} \\ 0, & \text{if a tick is put in the circle in the k-th region,} \end{cases}$$

where $p_k = \mathrm{Pr}(W_k = 1)$, $k = 1, 2$, are assumed to be known but are not necessarily the same. We have $\mathrm{Pr}(Y_{k,\mathrm{T}} = 1) = \pi_k + (1 - \pi_k)p_k$ and $\mathrm{Pr}(Y_{k,\mathrm{T}} = 0) = (1 - \pi_k)(1 - p_k)$. Let $Y_{\mathrm{obs}} = \{y_{ik,\mathrm{T}} : i = 1, \ldots, n_k, \ k = 1, 2\}$ denote the observed data for the $n_1 + n_2$ respondents in the two regions, where $y_{ik,\mathrm{T}} = 1$ (or 0) if the i-th respondent in the k-th region puts a tick in the upper square (or in the circle). Suppose that $\rho = n_1/n_2$. The likelihood function for π_1 and π_2 is then given by

$$L_{\mathrm{T}}(\pi_1, \pi_2 | Y_{\mathrm{obs}}) = \prod_{k=1}^{2} \prod_{i=1}^{n_k} \{\pi_k + (1 - \pi_k)p_k\}^{y_{ik,\mathrm{T}}} \{(1 - \pi_k)(1 - p_k)\}^{1 - y_{ik,\mathrm{T}}}.$$

The resulting MLE of π_k and the corresponding variance are given by

$$\hat{\pi}_k = \frac{\bar{y}_{k,T} - p_k}{1 - p_k} \quad \text{and} \quad \text{Var}(\hat{\pi}_k) = \frac{\vartheta_k(1 - \vartheta_k)}{n_k(1 - p_k)^2}, \quad k = 1, 2,$$

where $\bar{y}_{k,T} = (1/n_k)\sum_{i=1}^{n_k} y_{ik,T}$ and $\vartheta_k = \pi_k + (1 - \pi_k)p_k$. Thus,

$$\widehat{\text{Var}}(\hat{\pi}_k) = \frac{\bar{y}_{k,T}(1 - \bar{y}_{k,T})}{n_k(1 - p_k)^2}$$

is the MLE of $\text{Var}(\hat{\pi}_k)$.

Now, we consider the following two-sided hypotheses

$$H_0: \pi_1 = \pi_2 \quad \text{versus} \quad H_a: \pi_1 \neq \pi_2.$$

The null hypothesis H_0 is rejected at the α level of significance if

$$\left| \frac{\hat{\pi}_1 - \hat{\pi}_2}{\widehat{\text{SE}}_+} \right| > z_{\alpha/2},$$

where

$$\widehat{\text{SE}}_+ = \sqrt{\sum_{k=1}^{2} \widehat{\text{Var}}(\hat{\pi}_k)}$$

denotes the estimate of

$$\text{SE}_+ = \sqrt{\sum_{k=1}^{2} \text{Var}(\hat{\pi}_k)}.$$

Under the alternative hypothesis that $\pi_1 - \pi_2 \neq 0$, the power of the above two-sided test is approximately given by

$$\Phi\left(\frac{|\pi_1 - \pi_2| - z_{\alpha/2} \times \widehat{\text{SE}}_+}{\text{SE}_+} \right).$$

Following Chow, Shao & Wang (2003), we further approximate the power by

$$\Phi\left(\frac{|\pi_1 - \pi_2|}{\text{SE}_+} - z_{\alpha/2} \right).$$

Therefore, the sample size required for achieving a desired power of $1 - \beta$ can be obtained by solving the equation:

$$\frac{|\pi_1 - \pi_2|}{\text{SE}_+} - z_{\alpha/2} = z_\beta.$$

This leads to

$$n_1 = \rho n_2 \tag{4.14}$$

and

$$n_2 = \frac{(z_{\alpha/2} + z_\beta)^2}{(\pi_1 - \pi_2)^2} \left\{ \frac{\vartheta_1(1 - \vartheta_1)}{\rho(1 - p_1)^2} + \frac{\vartheta_2(1 - \vartheta_2)}{(1 - p_2)^2} \right\}. \tag{4.15}$$

4.6 An Example

Liu & Chow (1976) considered an induced abortion study in Taichung City and Taoyuan County, Taiwan (Winkler & Franklin, 1979). Liu and Chow adopted the multiple-trial version of the Warner model to increase the efficiency of estimation. Since the present chapter only involves the single-trial Warner model and the triangular model, we simply use the data from the first trial of each respondent. The target population of interest in this study was married women of age 20 to 44 in the South District of Taichung City, Taiwan. The investigators would like to estimate the incidence rate of induced abortions in the target population. With $p = 0.3$, the survey yielded 90 'yes' answers (i.e., $n\bar{y}_{\mathrm{C}} = \sum_{i=1}^{n} y_{i,\mathrm{C}} = 90$ in (4.10)) and 60 'no' answers (i.e., $n = 150$). From (4.10), the proportion of married women of childbearing age who have had induced abortion is estimated to be $\hat{\pi}_{\mathrm{C}} = 0.25$ with estimated variance being

$$\widehat{\mathrm{Var}}(\hat{\pi}_{\mathrm{C}}) = 0.01.$$

The resulting 95% Wald confidence interval of π is

$$\left[0.25 - 1.96\sqrt{0.01},\ 0.25 + 1.96\sqrt{0.01}\right] = [0.054,\ 0.446].$$

As expected, the survey design using the crosswise/Warner model resulted in a rather wide confidence interval.

Now we consider a survey design to solicit the sensitive information about induced abortion using the triangular model. We first determine the sample size (number of subjects) needed in order to guarantee 80% power using a one-sided test at a significance level of $\alpha = 0.05$ for testing $\pi_0 = 0.37$ against $\pi_1 = 0.27$ with $p = 5/12$. Using the sample size formula (4.6), $n_{\mathrm{T}} = 430$ is required for the triangular design, while the desired sample size is $n_{\mathrm{D}} = 137$ for the design of direct questioning.

Furthermore, we estimate how many subjects are required for comparing induced abortion rates between two regions with 80% power using a two-sided test at a significance level of $\alpha = 0.05$ for testing $\pi_1 = \pi_2$ against $\pi_1 \neq \pi_2$. Assume that true proportions with the sensitive character in the two regions are $\pi_1 = 0.35$ and $\pi_2 = 0.25$, respectively. Using the triangular design with $p_1 = 0.5$ and $p_2 = 0.4$, the sample sizes with $\rho = 1$ (equal allocation) are given by $n_1 = n_2 = 1229$ via (4.14) and (4.15) while the desired sample sizes are $n_1 = n_2 = 326$ for the design of direct questioning.

CHAPTER 5

The Multi-category Triangular Model

In Chapter 3, we introduced a triangular model, which can be applied to surveys for a single sensitive question with two answers. In this chapter, we will generalize this binary triangular model to a non-randomized multi-category triangular model, which can be applied to surveys for a single sensitive question with multiple answers (Tang *et al.*, 2009).

5.1 A Brief Literature Review

Abul-Ela, Greenberg & Horvitz (1967) extended Warner's dichotomous model to a trichotomous model to estimate proportions of three related but mutually exclusive groups of which one or two possess a sensitive characteristic. The model was further extended to estimate proportions of m (> 3) mutually exclusive groups with at least one and at most $m - 1$ of them involving the sensitive attribute. Bourke & Dalenius (1973) considered a Latin square measurement design which was another generalization of the Warner design to the multichotomous case. Their design uses m different possible responses and requires only one sample. An alternative unrelated question randomized response design for estimating the proportions of a population with m ($m > 2$) mutually exclusive sensitive classes (up to $m - 1$ sensitive subclasses) based on only one sample was suggested by Eriksson (1973). Bourke (1974) developed an unrelated question design to estimate the proportions of m mutually exclusive classes, of which up to $m - 1$ include the sensitive attribute. Only one sample is required if the distribution of the unrelated characteristic is known. However, all the aforementioned designs require the use of one or two randomization devices.

From a viewpoint of incomplete categorical data design, Yu, Tian & Tang (2008) proposed a triangular model for a single sensitive question with two answers. Advantages of this non-randomized design include (a) no involvement of randomization device, and (b) reproducibility of response. In many surveys, we are however interested in sensitive questions such as number of sexual partners (0–1, 2–3, $\geqslant 4$) and days of illegal drug usage in last month (0, 1, 2, or $\geqslant 3$). These questions require asking sensitive questions with multiple responses. For these multi-response sensitive questions, Tang *et al.* (2009) proposed a non-randomized multi-category triangular model.

5.2   The Survey Design

5.2.1   Design of questionnaire

Assume that a discrete random variable Y takes values on $\{1,\ldots,m\}$. Furthermore, let $\{Y = 1\}$ denote a non-sensitive class and $\{Y = j\}_{j=2}^{m}$ denote $(m - 1)$ sensitive classes. Let $\pi_j = \Pr(Y = j)$ denote the population proportion for category j. We have $\boldsymbol{\pi} = (\pi_1,\ldots,\pi_m)^{\top} \in \mathbb{T}_m$, where

$$\mathbb{T}_m = \left\{(\pi_1,\ldots,\pi_m)^{\top}\colon\ \pi_j \geqslant 0,\ j = 1,\ldots,m,\ \sum_{j=1}^{m}\pi_j = 1\right\} \quad (5.1)$$

denotes the m-dimensional hyperplane. The objective is to draw statistical inferences on $\{\pi_j\}$.

Investigators may design a questionnaire in the format as shown in Table 5.1 and ask respondents to answer a non-sensitive question Q_{W}, where the non-sensitive question Q_{W} is chosen to have m mutually exclusive possible answers with known probabilities. Assume that W is a non-sensitive variate associated with the Q_{W} and W is independent of Y. In addition, let W also take values on $\{1,\ldots,m\}$ and

$$p_j = \Pr(W = j), \quad j = 1,\ldots,m$$

are known. We will defer the discussion of the choice of Q_{W} in Section 5.2.2. Since Category 1 (i.e., $\{Y = 1\}$) is non-sensitive to any respondent, we have reason to assume that respondents are willing to provide their truthful answers (i.e., putting a tick in Block j for $j = 1,\ldots,m$) according to their true statuses (i.e., $W = 1,\ldots,m$). On the other hand, Categories 2 to m (i.e., $\{Y = 2\},\ldots,\{Y = m\}$) are sensitive to a respondent to some extent, and a truthful answer is unlikely to be obtained directly. In this case, if the respondent belongs to Category j, he/she will be asked to put a tick in Block j ($j = 2,\ldots,m$).

Table 5.2 shows the cell probabilities $\{\pi_j\}$, the observed frequencies $\{n_j\}$ and the unobservable frequencies $\{Z_j\}$. The observed frequency n_2 is the sum of the number of respondents simultaneously belonging to both Category 1 and $\{W = 2\}$ and the number of respondents belonging to Category 2. Similarly, $n_3,\ldots,n_m$ can be interpreted accordingly.

It is noteworthy that the multi-category triangular model (i) prevents respondents from answering any sensitive questions directly; (ii) does not require any randomization device and hence substantially reduces the cost; (iii) can be applied to the face-to-face personal interview, telephone interview, Internet survey and mail questionnaire; and (iv) can obtain exact information on the cell frequency n_1 from Block 1 and hence reduce the variance of the estimator.

Table 5.1 Questionnaire for the multi-category triangular model

Category	$W = 1$	$W = 2$	$\cdots$	$W = m$
1: $\{Y = 1\}$	Block 1: _____	Block 2: _____	$\cdots$	Block m: _____
2: $\{Y = 2\}$	Category 2: Please put a tick in Block 2			
$\vdots$	$\cdots$			
m: $\{Y = m\}$	Category m: Please put a tick in Block m			

Source: Table 1 in Tang *et al.* (2009).
Note: $\{Y = 1\}$ represents the non-sensitive class.

Table 5.2 Cell probabilities, observed and unobservable frequencies for the multi-category triangular model

Category	$W = 1$	$W = 2$	$\cdots$	$W = m$	Total
1: $\{Y = 1\}$	$p_1\pi_1$	$p_2\pi_1$	$\cdots$	$p_m\pi_1$	π_1 (Z_1)
2: $\{Y = 2\}$					π_2 (Z_2)
$\vdots$					$\vdots$
m: $\{Y = m\}$					π_m (Z_m)
Total	$p_1\pi_1$ (n_1)	$p_2\pi_1 + \pi_2 (n_2)$	$\cdots$	$p_m\pi_1 + \pi_m (n_m)$	1 (n)

Source: Table 2 in Tang *et al.* (2009).
Note: $n = \sum_{j=1}^{m} n_j$, $Z_1 = n - \sum_{j=2}^{m} Z_j$, where $(Z_2, \ldots, Z_m)$ are unobservable.

5.2.2 Determination of the non-sensitive question

A good survey method for sensitive questions should be able to protect the privacy of respondents, and to motivate respondents to participate in the study, and to provide truthful answers. In addition, it should be easy to be administrated for both interviewers and respondents. For this purpose, we introduce a non-sensitive question Q_w with m possible answers in the questionnaire. Let W be the corresponding outcome variate of Q_w which is independent of Y.

Let $p_j = \Pr(W = j)$ for $j = 1, \ldots, m$. In practice, we should choose a non-sensitive variate W so that reliable estimates of $\{p_j\}$ can be readily available from census data. Thus, we assume that all $\{p_j\}$ are known (equal or unequal). For example, when $m = 3$, we may let

$W = 1$: if the respondent was born in January–April;

$W = 2$: if the respondent was born in May–August;   and   (5.2)

$W = 3$: if the respondent was born in September–December.

In this case, it is reasonable to assume that each p_j is approximately equal to $1/3$. Similarly, we can define W based on months of birth when $m = 4$ or 6. When $m = 5$, we may define W based on the dates of birth. That is,

$W = 1$: if the respondent was born during dates 1–6;

$W = 2$: if the respondent was born during dates 7–12;

$W = 3$: if the respondent was born during dates 13–18;   (5.3)

$W = 4$: if the respondent was born during dates 19–24;   and

$W = 5$: if the respondent was born during dates 25–30.

When $m = 7$, we might define

$W = 1$: if the respondent was born on Monday;

$W = 2$: if the respondent was born on Tuesday;

$W = 3$: if the respondent was born on Wednesday;

$W = 4$: if the respondent was born on Thursday;

$W = 5$: if the respondent was born on Friday;

$W = 6$: if the respondent was born on Saturday;   and

$W = 7$: if the respondent was born on Sunday.

Unlike randomized response models, the multi-category triangular model requires no randomization device in its operation, and it hence possesses the feature of reproducibility in the sense that the same respondent is expected to give the same answer if the survey is conducted again under the multi-category triangular model. Since interviewers do not know nor control the months or dates of birth of respondents, respondents have confidence to trust the privacy policy and are willing to cooperate. Therefore, it is important to choose W that is independent of the sensitive variate Y.

5.3   Likelihood-based Inferences

5.3.1   MLEs via the EM algorithm

Assume that we carry out a survey with n respondents and we observe n_1 respondents putting a tick in Block 1, n_2 respondents putting a tick in

Block 2, ..., and n_m respondents putting a tick in Block m (see, Table 5.1). Let $Y_{\text{obs}} = \{n; n_1, \ldots, n_m\}$ denote the observed frequencies, where $n = n_1 + \cdots + n_m$. Hence, the observed-data likelihood function of $\boldsymbol{\pi} = (\pi_1, \ldots, \pi_m)^\top$ is

$$L_{\text{MT}}(\boldsymbol{\pi}|Y_{\text{obs}}) = \binom{n}{n_1, \ldots, n_m} (p_1 \pi_1)^{n_1} \prod_{j=2}^{m} (p_j \pi_1 + \pi_j)^{n_j}, \qquad (5.4)$$

where the subscript 'MT' refers to the 'multi-category triangular' model.

We use the EM algorithm (Dempster, Laird & Rubin, 1977) to find the MLEs of $\{\pi_j\}_{j=1}^m$ by introducing a latent vector $\mathbf{z} = (Z_2, \ldots, Z_m)^\top$, where Z_j denotes the number of respondents belonging to the sensitive subclass $\{Y = j\}$, i.e., Category j. Let $\mathbf{z} = (z_2, \ldots, z_m)^\top$ denote the realization of $\mathbf{z}$. We denote the complete data by $\{Y_{\text{obs}}, \mathbf{z}\}$. The complete-data likelihood function for $\boldsymbol{\pi}$ is

$$L_{\text{MT}}(\boldsymbol{\pi}|Y_{\text{obs}}, \mathbf{z}) \propto \prod_{j=1}^{m} \pi_j^{z_j}, \qquad \boldsymbol{\pi} \in \mathbb{T}_m, \qquad (5.5)$$

where $z_1 = n - \sum_{j=2}^m z_j$. Thus, the complete-data MLEs of $\{\pi_j\}_{j=1}^m$ are given by

$$\pi_1 = \frac{n - \sum_{j=2}^m z_j}{n} \quad \text{and} \quad \pi_j = \frac{z_j}{n}, \qquad j = 2, \ldots, m. \qquad (5.6)$$

Since the conditional predictive density is

$$f(\mathbf{z}|Y_{\text{obs}}, \boldsymbol{\pi}) = \prod_{j=2}^{m} f(z_j|Y_{\text{obs}}, \boldsymbol{\pi})$$

$$= \prod_{j=2}^{m} \text{Binomial}\left(z_j \,\middle|\, n_j, \frac{\pi_j}{p_j \pi_1 + \pi_j}\right), \qquad (5.7)$$

the E-step of the EM algorithm is to compute the following conditional expectations

$$E(Z_j|Y_{\text{obs}}, \boldsymbol{\pi}) = \frac{n_j \pi_j}{p_j \pi_1 + \pi_j}, \qquad j = 2, \ldots, m, \qquad (5.8)$$

and the M-step is to replace $\{z_j\}_{j=2}^m$ in (5.6) by the above conditional expectations.

5.3.2 Asymptotic confidence intervals

Let $\boldsymbol{\pi}_{-m} = (\pi_1, \ldots, \pi_{m-1})^{\top}$. The asymptotic variance-covariance matrix of the MLE $\hat{\boldsymbol{\pi}}_{-m}$ is then given by $\mathbf{I}_{\mathrm{obs}}^{-1}(\hat{\boldsymbol{\pi}}_{-m})$, where

$$\mathbf{I}_{\mathrm{obs}}(\boldsymbol{\pi}_{-m}) = -\frac{\partial^2 \ell_{\mathrm{MT}}(\boldsymbol{\pi}|Y_{\mathrm{obs}})}{\partial \boldsymbol{\pi}_{-m} \partial \boldsymbol{\pi}_{-m}^{\top}}$$

denotes the observed information matrix and $\ell_{\mathrm{MT}}(\boldsymbol{\pi}|Y_{\mathrm{obs}}) = \log L_{\mathrm{MT}}(\boldsymbol{\pi}|Y_{\mathrm{obs}})$ is the observed-data log-likelihood function. From (5.4), we have

$$\ell_{\mathrm{MT}}(\boldsymbol{\pi}|Y_{\mathrm{obs}}) = n_1 \log \pi_1 + \sum_{j=2}^{m-1} n_j \log(p_j \pi_1 + \pi_j)$$

$$+ n_m \log(p_m \pi_1 + 1 - \pi_1 - \cdots - \pi_{m-1}).$$

It is easy to show that

$$\frac{\partial \ell_{\mathrm{MT}}(\boldsymbol{\pi}|Y_{\mathrm{obs}})}{\partial \pi_1} = \frac{n_1}{\pi_1} + \sum_{j=2}^{m-1} \frac{n_j p_j}{p_j \pi_1 + \pi_j} + \frac{n_m(p_m - 1)}{p_m \pi_1 + \pi_m},$$

$$\frac{\partial \ell_{\mathrm{MT}}(\boldsymbol{\pi}|Y_{\mathrm{obs}})}{\partial \pi_j} = \frac{n_j}{p_j \pi_1 + \pi_j} - \frac{n_m}{p_m \pi_1 + \pi_m}, \quad 2 \leqslant j \leqslant m-1,$$

and

$$-\frac{\partial^2 \ell_{\mathrm{MT}}(\boldsymbol{\pi}|Y_{\mathrm{obs}})}{\partial \pi_1^2} = \frac{n_1}{\pi_1^2} + \sum_{j=2}^{m-1} \frac{n_j p_j^2}{(p_j \pi_1 + \pi_j)^2} + \frac{n_m(p_m - 1)^2}{(p_m \pi_1 + \pi_m)^2}$$

$$= \frac{n_1}{\pi_1^2} + \sum_{j=2}^{m-1} \phi_j p_j^2 + \phi_m(p_m - 1)^2,$$

$$-\frac{\partial^2 \ell_{\mathrm{MT}}(\boldsymbol{\pi}|Y_{\mathrm{obs}})}{\partial \pi_j^2} = \frac{n_j}{(p_j \pi_1 + \pi_j)^2} + \frac{n_m}{(p_m \pi_1 + \pi_m)^2}$$

$$= \phi_j + \phi_m, \qquad\qquad 2 \leqslant j \leqslant m-1,$$

$$-\frac{\partial^2 \ell_{\mathrm{MT}}(\boldsymbol{\pi}|Y_{\mathrm{obs}})}{\partial \pi_1 \partial \pi_j} = \frac{n_j p_j}{(p_j \pi_1 + \pi_j)^2} - \frac{n_m(p_m - 1)}{(p_m \pi_1 + \pi_m)^2}$$

$$= \phi_j p_j - \phi_m(p_m - 1), \qquad 2 \leqslant j \leqslant m-1,$$

$$-\frac{\partial^2 \ell_{\mathrm{MT}}(\boldsymbol{\pi}|Y_{\mathrm{obs}})}{\partial \pi_j \partial \pi_{j'}} = \frac{n_m}{(p_m \pi_1 + \pi_m)^2} = \phi_m, \quad j \neq j'; \ \ 2 \leqslant j, j' \leqslant m-1,$$

where

$$\phi_j = \frac{n_j}{(p_j \pi_1 + \pi_j)^2}, \quad 2 \leqslant j \leqslant m.$$

Hence, the observed information matrix can be expressed as

$$\mathbf{I}_{\mathrm{obs}}(\boldsymbol{\pi}_{-m}) = \begin{pmatrix} \dfrac{n_1}{\pi_1^2} + \displaystyle\sum_{j=2}^{m-1} \phi_j p_j^2 + \phi_m(p_m - 1)^2, & \boldsymbol{a}^\top \\[2mm] \boldsymbol{a}, & \mathbf{A} \end{pmatrix}, \tag{5.9}$$

where

$$\boldsymbol{a} = \mathrm{diag}(\phi_2, \ldots, \phi_{m-1}) \begin{pmatrix} p_2 \\ \vdots \\ p_{m-1} \end{pmatrix} - \phi_m(p_m - 1)\mathbf{1}_{m-2}$$

is an $(m - 2) \times 1$ vector and

$$\mathbf{A} = \mathrm{diag}(\phi_2, \ldots, \phi_{m-1}) + \phi_m \mathbf{1}_{m-2} \mathbf{1}_{m-2}^\top$$

is an $(m - 2) \times (m - 2)$ matrix.

Let $\mathrm{se}(\hat{\pi}_j)$ denote the standard error of $\hat{\pi}_j$ for $j = 1, \ldots, m - 1$. Note that $\mathrm{se}(\hat{\pi}_j)$ can be estimated by the square root of the j-th diagonal element of $\mathbf{I}_{\mathrm{obs}}^{-1}(\hat{\boldsymbol{\pi}}_{-m})$. We denote the estimated value of $\mathrm{se}(\hat{\pi}_j)$ by $\widehat{\mathrm{se}}(\hat{\pi}_j)$. Thus, a $(1 - \alpha)100\%$ normal-based asymptotic confidence interval for π_j is given by

$$\left[\hat{\pi}_j - z_{\alpha/2} \times \widehat{\mathrm{se}}(\hat{\pi}_j), \ \hat{\pi}_j + z_{\alpha/2} \times \widehat{\mathrm{se}}(\hat{\pi}_j)\right], \quad 1 \leqslant j \leqslant m - 1, \tag{5.10}$$

where z_α is the upper α-th quantile of the standard normal distribution.

Using the delta method (e.g., Casella & Berger, 2002, p. 240), a $(1 - \alpha)100\%$ normal-based asymptotic confidence interval for $\pi_m = 1 - \sum_{j=1}^{m-1} \pi_j$ can be computed by

$$\left[\hat{\pi}_m - z_{\alpha/2} \times \widehat{\mathrm{se}}(\hat{\pi}_m), \ \hat{\pi}_m + z_{\alpha/2} \times \widehat{\mathrm{se}}(\hat{\pi}_m)\right], \tag{5.11}$$

where

$$\widehat{\mathrm{se}}(\hat{\pi}_m) = \left\{ \left(\frac{\partial \pi_m}{\partial \boldsymbol{\pi}_{-m}}\right)^\top \mathbf{I}_{\mathrm{obs}}^{-1}(\boldsymbol{\pi}_{-m}) \left(\frac{\partial \pi_m}{\partial \boldsymbol{\pi}_{-m}}\right) \Big|_{\boldsymbol{\pi}_{-m} = \hat{\boldsymbol{\pi}}_{-m}} \right\}^{\frac{1}{2}}$$

$$= \left\{ \mathbf{1}_{m-1}^\top \mathbf{I}_{\mathrm{obs}}^{-1}(\hat{\boldsymbol{\pi}}_{-m}) \mathbf{1}_{m-1} \right\}^{\frac{1}{2}}. \tag{5.12}$$

5.3.3   Bootstrap confidence intervals

When the sample size n is small to moderate, asymptotic confidence intervals in (5.10) and (5.11) are not reliable. For such cases, the bootstrap approach (Efron & Tibshirani, 1993) can be used to estimate the standard error $\mathrm{se}(\hat{\pi}_j)$ of $\hat{\pi}_j$ for $j = 1, \ldots, m$.

Let $\psi = h(\pi)$ be an arbitrary function of π. Hence, the MLE of ψ is given by $\hat{\psi} = h(\hat{\pi})$, where $\hat{\pi}$ is the MLE of π. Based on the obtained $\hat{\pi}$, we generate

$$(n_1^*, \ldots, n_m^*)^\top \sim \text{Multinomial}(n;\ p_1\hat{\pi}_1, p_2\hat{\pi}_1 + \hat{\pi}_2, \ldots, p_m\hat{\pi}_1 + \hat{\pi}_m). \quad (5.13)$$

For each $Y_{\text{obs}}^* = \{n; n_1^*, \ldots, n_m^*\}$, we can calculate a bootstrap replication $\hat{\psi}^*$ based on $\hat{\pi}^*$. Independently repeating this process G times, we obtain G bootstrap replications $\{\hat{\psi}_g^*\}_{g=1}^G$. Thus, the standard error, $\mathrm{se}(\hat{\psi})$, of $\hat{\psi}$ can be estimated by the sample standard deviation of the G replications, i.e.,

$$\widehat{\mathrm{se}}(\hat{\psi}) = \left\{ \frac{1}{G-1} \sum_{g=1}^G \left(\hat{\psi}_g^* - \frac{\hat{\psi}_1^* + \cdots + \hat{\psi}_G^*}{G} \right)^2 \right\}^{\frac{1}{2}}.$$

If $\{\hat{\psi}_g^*\}_{g=1}^G$ is approximately normally distributed, a $(1-\alpha)100\%$ bootstrap confidence interval for ψ is given by

$$\left[\hat{\psi} - z_{\alpha/2} \times \widehat{\mathrm{se}}(\hat{\psi}),\ \ \hat{\psi} + z_{\alpha/2} \times \widehat{\mathrm{se}}(\hat{\psi}) \right]. \quad (5.14)$$

If $\{\hat{\psi}_g^*\}_{g=1}^G$ is non-normally distributed, a $100(1-\alpha)\%$ bootstrap confidence interval for ψ can be obtained by

$$[\hat{\psi}_{\mathrm{L}},\ \ \hat{\psi}_{\mathrm{U}}], \quad (5.15)$$

where $\hat{\psi}_{\mathrm{L}}$ and $\hat{\psi}_{\mathrm{U}}$ are the $100(\alpha/2)$ and $100(1-\alpha/2)$ percentiles of $\{\hat{\psi}_g^*\}_{g=1}^G$, respectively.

5.4   Bayesian Inferences

The complete-data likelihood function is given by (5.5). If prior information about π is available (say) in the form of a Dirichlet distribution, denoted by $\text{Dirichlet}(a_1, \ldots, a_m)$, then the complete-data posterior distribution is

$$f(\pi|Y_{\text{obs}}, z) = \text{Dirichlet}(\pi|a_1 + z_1, \ldots, a_m + z_m). \quad (5.16)$$

Let the posterior mode of $\boldsymbol{\pi}$ be $\tilde{\boldsymbol{\pi}}$. By the EM algorithm, the complete-data posterior mode is given by

$$\tilde{\pi}_1 = 1 - \sum_{j=2}^{m} \tilde{\pi}_j \quad \text{and} \quad \tilde{\pi}_j = \frac{a_j + z_j - 1}{\sum_{\ell=1}^{m} a_\ell + n - m}, \quad j = 2, \ldots, m. \quad (5.17)$$

The E-step is still given by (5.8) and the M-step is to replace $\{z_j\}_{j=2}^{m}$ in (5.17) by the conditional expectations specified by (5.8).

In particular, when the prior is uniform (i.e., $a_1 = \cdots = a_m = 1$), the posterior mode $\tilde{\boldsymbol{\pi}}$ reduces to the MLE $\hat{\boldsymbol{\pi}}$. In addition, based on (5.7) and (5.16), we can use the data augmentation algorithm to generate posterior samples of $\boldsymbol{\pi}$ by choosing $\boldsymbol{\pi}^{(0)} = (1/m, \ldots, 1/m)^{\top}$ as an initial value (see, Tanner & Wong, 1987).

5.5 Questionnaire on Sexual Activities in Korean Adolescents

Premarital sex and meeting the opposite sex were strictly restricted in traditional Korean society, although many male adolescents might secretly meet and have sex with prostitutes. Due to such social conservativeness, some of the Korean adolescents tend to answer most questions related to sex with reluctance. Anonymity is thus regarded as an important element in ensuring the confidentiality in surveys, and is also frequently perceived as an asset in obtaining information about socially sensitive behaviors like sexual activities.

Youn (2001) conducted an anonymously administered questionnaire on sexual activities in Korean adolescents. The participants consisted of n_{male} = 395 male and n_{female} = 392 female adolescents between the ages of 15 and 22, who had never been married. They were recruited in early 1995 from seven high schools and five universities in the southwestern area of Korea, resulting in a range of students from 9th grade through to college seniors. Half the sample was from the metropolitan Kwangju area with a population of about 1.3 million and the other half was from suburban and rural areas. Three primary objectives of the study are to ask:

(1) How many Korean adolescents in the 1990s were involved in misbehavior relating to sex;

(2) How they perceived their peer sexual activities like coitus, visiting prostitutes or pregnancy; and

(3) How they perceived the coital wishes of date-initiators.

Table 5.3 Numbers of sex partners for Korean adolescents

Category	Number of sex partners	Male	Female
1: $\{Y = 1\}$	0	299	351
2: $\{Y = 2\}$	1	37	28
3: $\{Y = 3\}$	2	22	6
4: $\{Y = 4\}$	3–4	16	5
5: $\{Y = 5\}$	$\geqslant 5$	21	2
Total		$n_{\text{male}} = 395$	$n_{\text{female}} = 392$

Source: Table 3 in Tang *et al.* (2009).

Table 5.4 MLEs and Bayesian estimates of parameters for the sex partner data for Korean adolescents

		MLE	Bayes mean	Bayes std	95% Bayes credible interval	95% Bootstrap CI[†]	95% Bootstrap CI[‡]
M	π_1	0.747	0.735	0.076	[0.578, 0.876]	[0.583, 0.911]	[0.570, 0.899]
	π_2	0.096	0.098	0.029	[0.041, 0.156]	[0.037, 0.156]	[0.038, 0.157]
	π_3	0.058	0.061	0.027	[0.011, 0.117]	[0.002, 0.115]	[0.000, 0.116]
	π_4	0.043	0.047	0.025	[0.004, 0.101]	[−0.012, 0.098]	[0.000, 0.101]
	π_5	0.056	0.058	0.026	[0.009, 0.113]	[0.000, 0.111]	[0.000, 0.114]
F	π_1	0.906	0.818	0.057	[0.692, 0.916]	[0.777, 1.034]	[0.714, 0.966]
	π_2	0.069	0.086	0.027	[0.032, 0.142]	[0.010, 0.127]	[0.018, 0.134]
	π_3	0.013	0.034	0.021	[0.002, 0.083]	[−0.033, 0.059]	[0.000, 0.079]
	π_4	0.010	0.033	0.021	[0.002, 0.081]	[−0.034, 0.054]	[0.000, 0.074]
	π_5	0.003	0.028	0.019	[0.001, 0.072]	[−0.036, 0.041]	[0.000, 0.064]

Source: Table 3 in Tang *et al.* (2009).
Note: M = Male, F = Female.
CI[†] = Bootstrap confidence intervals (5.14) based on normality assumption.
CI[‡] = Bootstrap confidence intervals (5.15) based on non-normality assumption.

To illustrate the design presented in Section 5.2 and the estimation methods in Sections 5.3 and 5.4, we focus on the sensitive question, 'How many coital partners have you had so far?' The numbers of sex partners in Table 5.3 are extracted from Table 2 of Youn (2001). Suppose that we are interested in the true proportions of the $m = 5$ categories shown in Table 5.3. Let $\pi_j = \Pr(Y = j)$ for $j = 1, \ldots, 5$. In this case, we use the non-sensitive variate W defined in (5.3) with $p_1 = \cdots = p_5 \approx 0.2$.

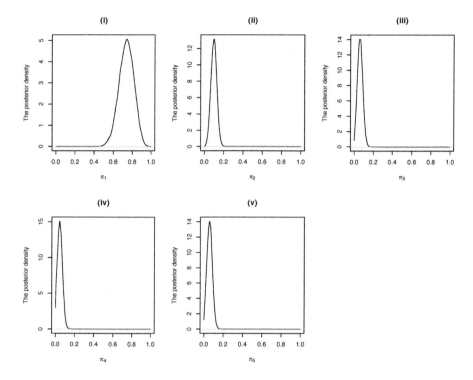

Figure 5.1 The posterior densities of $\{\pi_j\}_{j=1}^5$ are estimated by a kernel density smoother based on the last 20,000 posterior samples generated by the data augmentation algorithm for the male data. (i) $\pi_1 = \Pr(Y = 1)$; (ii) $\pi_2 = \Pr(Y = 2)$; (iii) $\pi_3 = \Pr(Y = 3)$; (iv) $\pi_4 = \Pr(Y = 4)$; (v) $\pi_5 = \Pr(Y = 5)$.

We first consider the male data. With $299/5 \approx 60$, for the ideal situation, the observed counts would be

$$
\begin{aligned}
n_1 &= 59, \\
n_2 &= 60 + 37 = 97, \\
n_3 &= 60 + 22 = 82, \\
n_4 &= 60 + 16 = 76, \quad \text{and} \\
n_5 &= 60 + 21 = 81.
\end{aligned}
$$

Therefore, the observed frequencies are given by

$$
Y_{\text{obs}} = \{n;\, n_1, \ldots, n_5\} = \{395;\, 59, 97, 82, 76, 81\}.
$$

Using $\boldsymbol{\pi}^{(0)} = 0.2 \times \mathbf{1}_5$ as initial values, the EM algorithm based on (5.6) and (5.8) converges in 150 iterations. The resultant MLEs are listed in the third column of Table 5.4.

With the uniform prior (i.e., $a_1 = \cdots = a_5 = 1$), the posterior mode $\tilde{\pi}$ is identical to the MLE $\hat{\pi}$. Based on (5.7) and (5.16), we employ the data augmentation algorithm to generate 40,000 posterior samples and adopt only the last 20,000 samples. The corresponding Bayes estimates (i.e., mean, standard deviation and 95% credible interval) of π are given in columns 4–6 of Table 5.4. Figure 5.1 shows posterior densities of $\{\pi_j\}_{j=1}^{5}$ estimated by a kernel density smoother based on the last 20,000 posterior samples generated by the data augmentation algorithm.

We also report 95% bootstrap confidence intervals for $\{\pi_j\}_{j=1}^{5}$ based on normality assumption and non-normality assumption in columns 7 and 8 of Table 5.4. Here, both the Bayes credible interval and the bootstrap confidence interval based on non-normality assumption indicate that the fourth category has a non-zero proportion while the bootstrap based on normality assumption suggests zero proportion for the fourth category.

For the female data, similarly, the observed frequencies would be

$$Y_{\text{obs}} = \{n;\ n_1, \ldots, n_5\} = \{392;\ 71,\ 98,\ 76,\ 75,\ 72\}.$$

The corresponding MLEs, Bayesian estimates of $\{\pi_j\}_{j=1}^{5}$ are also reported in Table 5.4. Both the Bayes credible interval and the bootstrap confidence interval based on non-normality assumption indicate that the last three categories possess non-zero proportions while the bootstrap based on normality assumption suggest zero proportions for the last three categories. Also, we find that males have more coital partners than females do. This finding is consistent with the general view that men are more involved in sexual activities with multiple partners than women (see, Wiederman, 1997).

CHAPTER 6

The Hidden Sensitivity Model

In previous chapters, we have introduced several non-randomized response models to deal with a single sensitive question with two or multiple answers. In this chapter, we will introduce a hidden sensitivity model for assessing the association of two sensitive questions with binary outcomes.

6.1 Background

Fox & Tracy (1984) considered the estimation of correlation between two sensitive questions. Lakshmi & Raghavarao (1992) discussed a 2 × 2 contingency table based on binary randomized responses. Christofides (2005) presented a randomized response technique for two sensitive characteristics at the same time. However, his procedure requires two randomization devices. Kim & Warde (2005) considered a multinomial randomized response model which can handle untruthful responses. They also derived the Pearson product moment correlation estimator which may be used to quantify the linear relationship between two variables when multinomial response data are observed according to a randomized response procedure. However, all these randomized response procedures make use of randomizing devices which (i) entail extra costs in both efficiency and complexity, (ii) increase the cognitive load of randomized response techniques, and (iii) allow for new sources of error, such as misunderstanding the randomized response procedures or cheating on the procedures (Lensvelt-Mulders et al., 2005).

From a perspective of non-randomized response, Tian et al. (2007) proposed a new survey technique for assessing the association of two binary sensitive variates. To protect respondents' privacy and to avoid the use of any randomization device, they utilized a non-sensitive question in the questionnaire to indirectly obtain a respondent's answer to the two sensitive questions. In Section 6.2, we introduce this survey design. In Section 6.3, we present likelihood-based inferences on parameters of interest. In Section 6.4, we discuss the information loss due to the introduction of a non-sensitive question and the design of cooperative parameters. Some simulation studies are presented in Section 6.5. Bayesian inferences under Dirichlet prior and other priors are given in Sections 6.6 and 6.7. In Section 6.8, a real data set from an AIDS study is used to illustrate these methods.

6.2   The Survey Design

6.2.1   The issue

Let X (e.g., whether or not a drug user) and Y (e.g., whether or not a respondent has AIDS) be two dichotomous variates associated with two sensitive questions, say Q_X and Q_Y. Let $X = 1$ and $Y = 1$ denote the sensitive characteristics of a respondent (e.g., $X = 1$ if a respondent takes drug), and $X = 0$ and $Y = 0$ denote the non-sensitive characteristics of a respondent (e.g., $Y = 0$ if a respondent does not have AIDS). Let $\boldsymbol{\theta} = (\theta_1, \theta_2, \theta_3, \theta_4)^{\top}$, where

$$
\begin{aligned}
\theta_1 &= \Pr(X = 0,\ Y = 0), \\
\theta_2 &= \Pr(X = 0,\ Y = 1), \\
\theta_3 &= \Pr(X = 1,\ Y = 0), \quad \text{and} \\
\theta_4 &= \Pr(X = 1,\ Y = 1).
\end{aligned}
$$

Clearly, we have $\boldsymbol{\theta} \in \mathbb{T}_4$, where $\mathbb{T}_m$ is defined by (5.1). Furthermore, we define

$$
\theta_x = \Pr(X = 1) = \theta_3 + \theta_4
$$

and

$$
\theta_y = \Pr(Y = 1) = \theta_2 + \theta_4.
$$

A commonly used quantity for measuring the association of two binary variates is the odds ratio

$$
\psi = \frac{\theta_1 \theta_4}{\theta_2 \theta_3}.
$$

The objective of this chapter is to make statistical inferences on $\boldsymbol{\theta}$, θ_x, θ_y and ψ.

6.2.2   The design of questionnaire

A good survey method for sensitive questions should be able to

(1)   protect the privacy of respondents,

(2)   motivate the respondents to participate in the study, and

(3)   provide truthful answers.

In addition, it should also be easy to be administrated for both interviewers and respondents. For this purpose, we introduce a non-sensitive question, say Q_W, with four possible answers in the questionnaire.

Table 6.1 Questionnaire for the hidden sensitivity model

Category	$W = 1$	$W = 2$	$W = 3$	$W = 4$
I: $\{X = 0, Y = 0\}$	Block 1: ___	Block 2: ___	Block 3: ___	Block 4: ___
II: $\{X = 0, Y = 1\}$	Category II: Please put a tick in Block 2			
III: $\{X = 1, Y = 0\}$	Category III: Please put a tick in Block 3			
IV: $\{X = 1, Y = 1\}$	Category IV: Please put a tick in Block 4			

Source: Table I in Tian *et al.* (2007).

(a) Choice of a non-sensitive variate

Assume that W is the outcome variate of Q_{w}, and W is independent of both X and Y. Let $p_i = \Pr(W = i)$ for $i = 1, \ldots, 4$. For example, let $W = i$ if the respondent was born in the i-th quarter, where

$$
\begin{aligned}
\text{the first quarter} &= \{\text{January, February, March}\}, \\
\text{the second quarter} &= \{\text{April, May, June}\}, \\
\text{the third quarter} &= \{\text{July, August, September}\}, \quad \text{and} \\
\text{the fourth quarter} &= \{\text{October, November, December}\}.
\end{aligned}
$$

Thus it is reasonable to assume that each p_i is approximately equal to $1/4$. Since interviewers do not know nor control the date of birth, respondents have reasons to trust the privacy policy and are willing to cooperate. In this case, it is also reasonable to assume that W is independent of the two sensitive binary variates X and Y.

Alternatively, we can consider $W = i$ to represent that a respondent lives in the i-th region. For instance, $W = 1$, 2, 3 and 4 if he/she lives in the north, south, east and west of certain region, respectively. If there are more than four sub-regions, one can always merge adjacent sub-regions to form a total of four sub-regions. We can then estimate the corresponding $\{p_i\}$ from geographic surveys or census data.

(b) Estimation of $\{p_i\}$

Design of $\{p_i\}$ will be discussed in Section 6.4. In practice, we should choose a non-sensitive variate W so that reliable estimates of $\{p_i\}$ can be readily available from some geographic surveys or census data. In this chapter, we assume that all $\{p_i\}$ are known.

Table 6.2 Cell probabilities, observed counts and unobservable counts for the hidden sensitivity model

Category	$W = 1$	$W = 2$	$W = 3$	$W = 4$	Total
I	$p_1\theta_1$	$p_2\theta_1$	$p_3\theta_1$	$p_4\theta_1$	$\theta_1\ (Z_1)$
II					$\theta_2\ (Z_2)$
III					$\theta_3\ (Z_3)$
IV					$\theta_4\ (Z_4)$
Total	$p_1\theta_1(n_1)$	$p_2\theta_1 + \theta_2(n_2)$	$p_3\theta_1 + \theta_3(n_3)$	$p_4\theta_1 + \theta_4(n_4)$	$1\ (n)$

Source: Table II in Tian *et al.* (2007).
Note: $n = \sum_{i=1}^{4} n_i$, $Z_1 = n - (Z_2 + Z_3 + Z_4)$, where (Z_2, Z_3, Z_4) are unobservable.

(c) The hidden sensitivity model

Table 6.1 shows the questionnaire, under which each respondent is asked to answer the non-sensitive question Q_w. On the one hand, since Category I, i.e., $\{X = 0,\ Y = 0\}$, is non-sensitive to any respondent, we have reason to assume that each respondent is willing to provide his/her truthful answer (i.e., putting a tick in Block i with $i = 1, \ldots, 4$) according to his/her true status (i.e., $W = 1, \ldots, 4$).

On the other hand, Categories II, III and IV(i.e., $\{X = 0,\ Y = 1\}$, $\{X = 1,\ Y = 0\}$ and $\{X = 1,\ Y = 1\}$) are sensitive to respondents and truthful answers are unlikely to be obtained. In this case, if a respondent belongs to Category II (III or IV), he/she will be asked to put a tick in Block 2 (3 or 4, respectively).

Tian *et al.* (2007) called this technique the hidden sensitivity model since the truthful sensitive characteristics of all respondents will be hidden. Table 6.2 shows the cell probabilities $\{\theta_i\}$, the observed frequencies $\{n_i\}$ and the unobservable frequencies $\{Z_i\}$. The observed frequency n_2 is the sum of the frequency of respondents belonging to Block 2 and the frequency of those belonging to Category II. Similarly, n_3 and n_4 can be interpreted accordingly.

6.3   Likelihood-based Inferences

In this section, the MLEs of $\boldsymbol{\theta}$ and ψ are derived by using the EM algorithm and the estimated standard error are obtained by the bootstrap approach. A likelihood ratio test and a chi-squared test are presented for testing association between the two sensitive variables.

6.3.1 MLEs via the EM algorithm

Let $Y_{\text{obs}} = \{n; \; n_1, \ldots, n_4\}$ denote the observed counts, where $n = \sum_{i=1}^{4} n_i$. Hence, the observed-data likelihood function for $\boldsymbol{\theta} \in \mathbb{T}_4$ is given by

$$L_{\text{H}}(\boldsymbol{\theta}|Y_{\text{obs}}) = \binom{n}{n_1, \ldots, n_4} (p_1\theta_1)^{n_1} \prod_{i=2}^{4} (p_i\theta_1 + \theta_i)^{n_i}, \qquad (6.1)$$

where the subscript 'H' denotes the 'hidden' sensitivity model. We treat the observed counts n_2, n_3 and n_4 as incomplete data and use the EM algorithm (Dempster, Laird & Rubin, 1977) to find the MLE $\hat{\boldsymbol{\theta}}$ of $\boldsymbol{\theta}$ (Bourke & Moran, 1988). The counts $\{Z_i\}$ in Table 6.2 can be viewed as missing data. Briefly, Z_1, Z_2, Z_3 and Z_4 represent the counts of the respondents belonging to Categories I, II, III and IV, respectively. Here, we denote the missing data by $Y_{\text{mis}} = \{z_2, z_3, z_4\}$ and the complete data by $Y_{\text{com}} = \{Y_{\text{obs}}, Y_{\text{mis}}\}$. Notice that all $\{p_i\}$ are known. Consequently, the complete-data likelihood function is

$$L_{\text{H}}(\boldsymbol{\theta}|Y_{\text{com}}) \propto \prod_{i=1}^{4} \theta_i^{z_i}, \qquad \boldsymbol{\theta} \in \mathbb{T}_4,$$

where $z_1 = n - (z_2 + z_3 + z_4)$. Hence, the complete-data MLEs for $\boldsymbol{\theta}$ are given by

$$\theta_i = \frac{z_i}{n}, \qquad i = 1, \ldots, 4. \qquad (6.2)$$

Given $(Y_{\text{obs}}, \boldsymbol{\theta})$, Z_i follows the binomial distribution with parameters n_i and $\theta_i/(p_i\theta_1 + \theta_i)$, $i = 2, 3, 4$. Therefore, the E-step of the EM computes the following conditional expectations

$$E(Z_i|Y_{\text{obs}}, \boldsymbol{\theta}) = \frac{n_i\theta_i}{p_i\theta_1 + \theta_i}, \qquad i = 2, 3, 4, \qquad (6.3)$$

$$E(Z_1|Y_{\text{obs}}, \boldsymbol{\theta}) = n - \sum_{i=2}^{4} \frac{n_i\theta_i}{p_i\theta_1 + \theta_i},$$

and the M-step updates (6.2) by replacing $\{z_i\}$ with the conditional expectations.

6.3.2 Bootstrap confidence intervals

Once we obtain the MLE $\hat{\boldsymbol{\theta}}$ via the EM algorithm (6.2) and (6.3), we can immediately compute the MLE of the odds ratio by

$$\hat{\psi} = \frac{\hat{\theta}_1\hat{\theta}_4}{\hat{\theta}_2\hat{\theta}_3}.$$

The bootstrap approach (Efron & Tibshirani, 1993) can be used to estimate the standard error of $\hat{\psi}$. Based on the obtained $\hat{\boldsymbol{\theta}}$, we generate

$$(n_1^*, \ldots, n_4^*)^\top \sim \text{Multinomial}\,(n;\ p_1\hat{\theta}_1,\ p_2\hat{\theta}_1 + \hat{\theta}_2,\ p_3\hat{\theta}_1 + \hat{\theta}_3,\ p_4\hat{\theta}_1 + \hat{\theta}_4).$$

Once we obtain $Y_{\text{obs}}^* = \{n;\ n_1^*, \ldots, n_4^*\}$, we can calculate a bootstrap replication $\hat{\psi}^*$ via the EM algorithm (6.2) and (6.3). Independently repeating this process G times, we obtain G bootstrap replications $\{\hat{\psi}_g^*\}_{g=1}^G$. Consequently, the standard error, $\text{se}(\hat{\psi})$, of $\hat{\psi}$ can be estimated by the sample standard deviation of the G replications, i.e.,

$$\widehat{\text{se}}(\hat{\psi}) = \left\{ \frac{1}{G-1} \sum_{g=1}^G \left(\hat{\psi}_g^* - \frac{\hat{\psi}_1^* + \cdots + \hat{\psi}_G^*}{G} \right)^2 \right\}^{\frac{1}{2}}. \tag{6.4}$$

If $\{\hat{\psi}_g^*\}_{g=1}^G$ is approximately normally distributed, a $(1-\alpha)100\%$ bootstrap confidence interval for ψ is given by

$$\left[\hat{\psi} - z_{\alpha/2} \times \widehat{\text{se}}(\hat{\psi}),\ \ \hat{\psi} + z_{\alpha/2} \times \widehat{\text{se}}(\hat{\psi}) \right], \tag{6.5}$$

where z_α is the upper α-th quantile of $N(0,1)$. Alternatively, if $\{\hat{\psi}_g^*\}_{g=1}^G$ is non-normally distributed, a $(1-\alpha)100\%$ bootstrap confidence interval of ψ is given by

$$[\hat{\psi}_{\text{L}},\ \hat{\psi}_{\text{U}}], \tag{6.6}$$

where $\hat{\psi}_{\text{L}}$ and $\hat{\psi}_{\text{U}}$ are the $100(\alpha/2)$ and $100(1-\alpha/2)$ percentiles of $\{\hat{\psi}_g^*\}_{g=1}^G$, respectively.

6.3.3   Testing of association

Suppose that we want to test the hypotheses

$$H_0\colon \psi = 1 \quad \text{versus} \quad H_1\colon \psi \neq 1.$$

(a) The likelihood ratio test

The likelihood ratio statistic can be used to test if association (or the dependence) exists between X and Y and is given by

$$\Lambda_1 = -2\{\ell_{\text{H}}(\hat{\boldsymbol{\theta}}_{\text{R}}|Y_{\text{obs}}) - \ell_{\text{H}}(\hat{\boldsymbol{\theta}}|Y_{\text{obs}})\},$$

where $\hat{\boldsymbol{\theta}}_{\mathrm{R}}$ denotes the restricted MLE of $\boldsymbol{\theta}$ under H_0, $\hat{\boldsymbol{\theta}}$ denotes the MLE of $\boldsymbol{\theta}$, which can be obtained by the EM algorithm (6.2) and (6.3), and $\ell_{\mathrm{H}}(\boldsymbol{\theta}|Y_{\mathrm{obs}}) = \log L_{\mathrm{H}}(\boldsymbol{\theta}|Y_{\mathrm{obs}})$. Under H_0: $\theta_1\theta_4 = \theta_2\theta_3$, we have

$$\theta_1 = (1-\theta_x)(1-\theta_y), \qquad \theta_2 = (1-\theta_x)\theta_y, \tag{6.7}$$

$$\theta_3 = \theta_x(1-\theta_y), \qquad \theta_4 = \theta_x\theta_y, \tag{6.8}$$

and the complete-data likelihood function becomes

$$L_{\mathrm{H}}(\theta_x, \theta_y | Y_{\mathrm{com}}, H_0) \propto \{(1-\theta_x)(1-\theta_y)\}^{z_1}\{(1-\theta_x)\theta_y\}^{z_2}$$
$$\times \{\theta_x(1-\theta_y)\}^{z_3}(\theta_x\theta_y)^{z_4}.$$

Thus, the corresponding M-step becomes

$$\hat{\theta}_{x,\mathrm{R}} = \frac{z_3 + z_4}{n}, \qquad \hat{\theta}_{y,\mathrm{R}} = \frac{z_2 + z_4}{n}, \tag{6.9}$$

and the E-step computes the conditional expectation (6.3) by replacing $\{\theta_i\}$ with (6.7) and (6.8). Once we obtain $\hat{\theta}_{x,\mathrm{R}}$ and $\hat{\theta}_{y,\mathrm{R}}$, the restricted MLE $\hat{\boldsymbol{\theta}}_{\mathrm{R}}$ can be computed from equation (6.7) and (6.8). Under H_0, Λ_1 asymptotically follows the chi-squared distribution with one degree of freedom.

(b) The chi-squared test

Alternatively, we can consider the chi-squared statistic to test H_0. Let $\boldsymbol{\lambda} = (\lambda_1, \ldots, \lambda_4)^{\mathsf{T}}$, where $\lambda_1 = p_1\theta_1$ and $\lambda_i = p_i\theta_1 + \theta_i$ for $i = 2, 3, 4$. In matrix notation, we have

$$\boldsymbol{\lambda} = \mathbf{P}\boldsymbol{\theta} = \begin{pmatrix} p_1 & 0 & 0 & 0 \\ p_2 & 1 & 0 & 0 \\ p_3 & 0 & 1 & 0 \\ p_4 & 0 & 0 & 1 \end{pmatrix} \boldsymbol{\theta}. \tag{6.10}$$

Under H_0, the chi-squared statistic

$$\Lambda_2 = \sum_{i=1}^{4} \frac{(n_i - n\hat{\lambda}_{i,\mathrm{R}})^2}{n\hat{\lambda}_{i,\mathrm{R}}}$$

asymptotically follows the chi-squared distribution with one degree of freedom, where

$$\hat{\boldsymbol{\lambda}}_{\mathrm{R}} = (\hat{\lambda}_{1,\mathrm{R}}, \ldots, \hat{\lambda}_{4,\mathrm{R}})^{\mathsf{T}} = \mathbf{P}\hat{\boldsymbol{\theta}}_{\mathrm{R}}$$

is the restricted estimator of $\boldsymbol{\lambda}$ under H_0.

6.4   Information Loss and Design Consideration

In this section, we discuss the information loss due to the introduction of the non-sensitive question. We also discuss the design of the cooperative parameters.

6.4.1   Information loss due to the introduction of the non-sensitive variate

From (6.10), we obtain $\boldsymbol{\theta} = \mathbf{P}^{-1}\boldsymbol{\lambda}$. From (6.1), the MLE of $\boldsymbol{\lambda}$ is given by

$$\hat{\boldsymbol{\lambda}} = (\hat{\lambda}_1, \ldots, \hat{\lambda}_4)^\top = \left(\frac{n_1}{n}, \ldots, \frac{n_4}{n}\right)^\top$$

so that an alternative estimator of $\boldsymbol{\theta}$ is given by

$$\hat{\boldsymbol{\theta}}_v = \mathbf{P}^{-1}\hat{\boldsymbol{\lambda}}. \tag{6.11}$$

It should be noted that it is possible that $\hat{\boldsymbol{\theta}}_v \notin \mathbb{T}_4$. For instance, if

$$p_1 = \cdots = p_4 = 0.25 \quad \text{and} \quad (n_1, \ldots, n_4)^\top = (12, \ 10, \ 15, \ 13)^\top,$$

then

$$\hat{\boldsymbol{\theta}}_v = (0.96, \ -0.04, \ 0.06, \ 0.02)^\top \notin \mathbb{T}_4.$$

In this chapter, we say an estimator $\hat{\boldsymbol{\theta}}_v$ is *valid* [1] if $\hat{\boldsymbol{\theta}}_v \in \mathbb{T}_4$. Obviously, if $\hat{\boldsymbol{\theta}}_v$ is a valid estimate then $\hat{\boldsymbol{\theta}}_v = \hat{\boldsymbol{\theta}}$. In the subsequent discussion, we focus on those valid estimates.

Since

$$E(\hat{\boldsymbol{\theta}}_v) = \mathbf{P}^{-1}E(\hat{\boldsymbol{\lambda}}) = \mathbf{P}^{-1}\boldsymbol{\lambda} = \boldsymbol{\theta},$$

$\hat{\boldsymbol{\theta}}_v$ is unbiased. From (6.11), the covariance matrix of $\hat{\boldsymbol{\theta}}_v$ is

$$\text{Cov}(\hat{\boldsymbol{\theta}}_v) = \frac{\mathbf{C} + \mathbf{D}}{n},$$

where

$$\mathbf{C} = \mathbf{P}^{-1}\text{diag}(\boldsymbol{\lambda})(\mathbf{P}^{-1})^\top - \text{diag}(\boldsymbol{\theta}) \quad \text{and} \quad \mathbf{D} = \text{diag}(\boldsymbol{\theta}) - \boldsymbol{\theta}\boldsymbol{\theta}^\top.$$

---

[1]Bourke (1982) used the term 'admissible.' Since 'admissible' has its special meaning in decision theory, we use the term 'valid' instead of 'admissible' in this book to avoid unnecessary confusion.

Let M_r $(r = 1, \ldots, 4)$ denote the principal minor of order r for the matrix $\mathbf{C}$. It is easy to obtain that

$$
\begin{aligned}
M_1 &= \frac{(1 - p_1)\theta_1}{p_1} > 0, \\
M_2 &= \frac{p_2(1 - p_1 - p_2)\theta_1^2}{p_1} > 0, \\
M_3 &= \frac{p_2 p_3 p_4 \theta_1^3}{p_1} > 0, \quad \text{and} \\
M_4 &= |\mathbf{C}| = 0.
\end{aligned}
$$

Thus, $\mathbf{C}$ is a non-negative definite matrix.

Note that $\mathrm{Cov}(\hat{\boldsymbol{\theta}}_v)$ is written as a sum of two components. The first component $\mathbf{C}/n$ represents the increase in the covariance due to the introduction of the non-sensitive question Q_{w} and the second component $\mathbf{D}/n$ denotes the covariance matrix if the sensitive questions are directly asked without the use of Q_{w}. By introducing the non-sensitive question Q_{w}, the hidden sensitivity model hence reduces false responses and increases the cooperation from respondents for answering sensitive questions at the expense of some information loss (i.e., an increase of $\mathbf{C}/n$ in the covariance matrix of estimates).

6.4.2 Design of the cooperative parameters

If $p_1 = 1$ and $p_2 = p_3 = p_4 = 0$, then the hidden sensitivity model is equivalent to directly asking respondents the sensitive questions and it will somehow lose the cooperation from respondents. In this case, the parameters $\{p_i\}$ can be used to measure respondents' cooperation and they are called the *cooperative parameters*. In this subsection, we investigate the design of $\{p_i\}$. For this purpose, we first demonstrate that the conventional A-optimality criterion, which is widely used in experimental designs, is not applicable to the hidden sensitivity model. We then propose a feasible criterion for determining $\{p_i\}$ in terms of the proportion of information loss that the researcher can tolerate.

It is natural to find $\{p_i\}$ minimizing $\mathrm{Cov}(\hat{\boldsymbol{\theta}}_v)$. There are several criteria for minimizing a matrix, including A-, D- and E-optimality criteria (Atkinson & Donev, 1992). Here, we consider the A-optimality criterion that minimizes $\mathrm{tr}(\mathbf{C})$. It is easy to show that

$$
\mathrm{tr}(\mathbf{C}) = \frac{\theta_1\{(1 - p_1^2) + \sum_{i=2}^{4} p_i^2\}}{p_1} \geqslant 0.
$$

Thus, $\mathrm{tr}(\mathbf{C})$ achieves its minimum value zero at $\boldsymbol{p}_{\min} = (1, 0, 0, 0)^{\top}$. As mentioned above, this is not applicable to the hidden sensitivity model.

Let

$$\rho(\boldsymbol{p}) = \frac{\mathrm{tr}(\mathbf{C})}{n \ \mathrm{tr}\{\mathrm{Cov}(\hat{\boldsymbol{\theta}}_v)\}}$$

denote the proportion of information loss due to the introduction of the non-sensitive question Q_{w} with probability distribution $\boldsymbol{p}$. Suppose that α_0 is the least proportion of information loss that the researcher must sacrifice to encourage respondents' cooperation and that β_0 is the largest proportion of information loss that the researcher may tolerate. Hence, we can consider the criterion:

$$\alpha_0 \leqslant \rho(\boldsymbol{p}) \leqslant \beta_0,$$

where $0 < \alpha_0 < \beta_0 < 1$ are known constants pre-specified by the interviewer. The inequalities are equivalent to

$$\frac{\alpha_0(1 - \sum_{i=1}^{4} \theta_i^2)}{(1 - \alpha_0)\theta_1} \leqslant \frac{1 - p_1^2 + \sum_{i=2}^{4} p_i^2}{p_1} \leqslant \frac{\beta_0(1 - \sum_{i=1}^{4} \theta_i^2)}{(1 - \beta_0)\theta_1},$$

which however do not depend on the sample size n. In other words, any $\boldsymbol{p}$ belonging to this interval can be used as the cooperative parameters.

6.5 Simulation Studies

6.5.1 Comparison of the likelihood ratio test with the chi-squared test

To investigate the performance of the likelihood ratio statistic Λ_1 and the chi-squared statistic Λ_2, we conduct simulation studies to evaluate the empirical Type I error rates (or the actual significance levels) and powers of the two statistics. For a given θ_1, we consider the following three combinations of θ_2, θ_3 and θ_4:

$$\text{Scenario 1:} \quad (\theta_2, \theta_3, \theta_4) = (2, 3, 1)\frac{1 - \theta_1}{6}, \quad \psi = \frac{\theta_1}{1 - \theta_1};$$

$$\text{Scenario 2:} \quad (\theta_2, \theta_3, \theta_4) = (5, 4, 1)\frac{1 - \theta_1}{10}, \quad \psi = \frac{\theta_1}{2(1 - \theta_1)}; \quad (6.12)$$

$$\text{Scenario 3:} \quad (\theta_2, \theta_3, \theta_4) = (8, 11, 1)\frac{1 - \theta_1}{20}, \quad \psi = \frac{5\theta_1}{22(1 - \theta_1)}.$$

The sample sizes used in our simulation studies are given by $n = 50$, 100 $(100)1000$. To compare the Type I error rates (i.e., with $\theta_1\theta_4/(\theta_2\theta_3) = \psi$ being 1), we choose $\theta_1 = 0.50$ for Scenario 1, $\theta_1 = 2/3$ for Scenario 2 and $\theta_1 = 22/27$ for Scenario 3. For the comparisons of powers (i.e., with ψ being non-unity), the selected θ_1 and the corresponding ψ are listed in Table 6.3.

Table 6.3 The values of θ_1 and the corresponding values of ψ for the three scenarios

Scenario	θ_1					
	0.200	0.400	0.600	0.800	0.900	0.950
Scenario 1: ψ	0.250	0.667	1.500	4.000	9.000	19.00
Scenario 2: ψ	0.125	0.333	0.750	2.000	4.500	9.500
Scenario 3: ψ	0.057	0.152	0.340	0.900	2.045	4.320

Source: Table III in Tian *et al.* (2007).

For any given combination (n, θ_1), we independently generate

$$(n_1^{(\ell)}, \ldots, n_4^{(\ell)})^\top \sim \text{Multinomial}\left(n; \frac{\theta_1}{4}, \frac{\theta_1}{4} + \theta_2, \frac{\theta_1}{4} + \theta_3, \frac{\theta_1}{4} + \theta_4\right), \quad (6.13)$$

for $\ell = 1, \ldots, L\,(L = 5{,}000)$, where only $p_i = 0.25$, $i = 1, \ldots, 4$, are considered. All hypothesis testings are conducted at level 0.05. Let r_j be the number of rejections of the null hypothesis (i.e., H_0: $\psi = 1$) by the statistics $\Lambda_j\,(j = 1, 2)$. Hence, the actual significance level can be estimated by r_j/L with $\psi = 1$ and the power of the test statistic $\Lambda_j\,(j = 1, 2)$ can be estimated by r_j/L with $\psi \neq 1$.

Figure 6.1 gives the comparisons of Type I error rates between the likelihood ratio and chi-squared tests for three scenarios. In general, we find that the chi-squared test can well control its Type I error rates around the pre-chosen nominal level while the likelihood ratio test may be relatively conservative (see, Figures 6.1 (i) and 6.1 (ii)). As expected, when sample size increases the Type I error rates for both tests become closer to the pre-chosen nominal level except for the likelihood ratio test at large θ_1 value (see, Figure 6.1 (iii)).

Figure 6.2 gives the comparisons of powers between the likelihood ratio and chi-squared tests for different cases. It is interesting to note that both tests perform quite similarly for small ψ (e.g., ≤ 4). On the other hand, the chi-squared test is slightly powerful than the likelihood ratio test for moderate to large ψ (e.g., > 4) and small to moderate sample sizes (e.g., ≤ 200). As expected, when sample size increases we cannot distinguish the power performance for both tests.

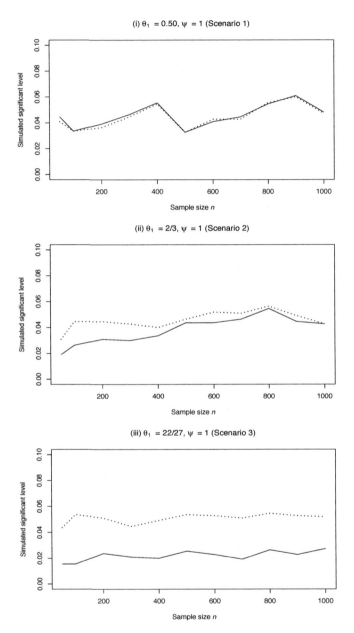

Figure 6.1 Comparisons of Type I error rates between the likelihood ratio test (solid line) and the chi-squared test (dotted line): (i) $\theta_1 = 0.50$, $\psi = 1$ (Scenario 1); (ii) $\theta_1 = 2/3$, $\psi = 1$ (Scenario 2); (iii) $\theta_1 = 22/27$, $\psi = 1$ (Scenario 3).

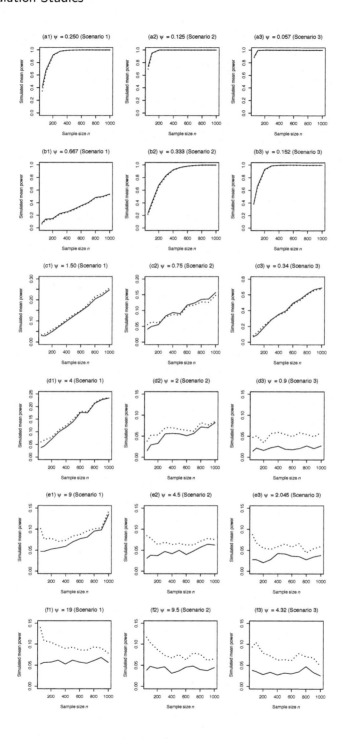

Figure 6.2 Comparisons of powers between the likelihood ratio test (solid line) and the chi-squared test (dotted line).

6.5.2   The probability of obtaining valid estimates

If $\hat{\boldsymbol{\theta}}_v$ defined by (6.11) is a valid estimate for any sample size n and cell probability vector θ, then the EM algorithm developed in Section 6.3.1 is not necessary since (6.11) becomes very simple. However, Bourke (1982) claimed that the larger the sample size n, the smaller the probability of obtaining invalid estimate from (6.11). It seems that

$$\lim_{n \to \infty} \Pr(\hat{\boldsymbol{\theta}}_v \in \mathbb{T}_4) = 1.$$

However, this is not true for some cases, for instance, when θ_1 is very large. Probably, Bourke (1982) overlooked the fact that $\Pr(\hat{\boldsymbol{\theta}}_v \in \mathbb{T}_4)$ depends on both n and θ. This section aims to investigate the relationship between $\Pr(\hat{\boldsymbol{\theta}}_v \in \mathbb{T}_4)$ and n for any given θ via the Monte Carlo simulations. Here, we only consider the case in which all $p_i = 0.25$, $i = 1, \ldots, 4$.

In practice, the proportion of respondents having the sensitive characteristics is usually very small while the proportion of respondents having the non-sensitive characteristics is very large. Therefore, we consider $\theta_1 = 0.70, 0.80, 0.90, 0.92, 0.94, 0.96, 0.98, 0.99$ and $n = 50, 100(200)900$, $1000(500)10000$. For any fixed θ_1, the other three components $(\theta_2, \theta_3, \theta_4)$ are determined by the three scenarios defined in (6.12). For any given combination (n, θ_1), we independently generate $(n_1^{(\ell)}, \ldots, n_4^{(\ell)})^\top$ from (6.13) for $\ell = 1, \ldots, L$ ($L = 10{,}000$). Based on (6.11), we calculate

$$\hat{\boldsymbol{\theta}}_v^{(\ell)} = \mathbf{P}^{-1} \left(n_1^{(\ell)}/n, \ldots, n_4^{(\ell)}/n \right)^\top$$

and estimate $\Pr(\hat{\boldsymbol{\theta}}_v \in \mathbb{T}_4)$ by

$$\widehat{\Pr}(\hat{\boldsymbol{\theta}}_v \in \mathbb{T}_4) = \frac{\sum_{\ell=1}^{L} I(\hat{\boldsymbol{\theta}}_v^{(\ell)} \in \mathbb{T}_4)}{L},$$

where $I(\cdot)$ denotes the indicator function.

Figure 6.3 plots $\widehat{\Pr}(\hat{\boldsymbol{\theta}}_v \in \mathbb{T}_4)$ against the sample size n for all three scenarios with various θ_1. From each plot, we can conclude that $\widehat{\Pr}(\hat{\boldsymbol{\theta}}_v \in \mathbb{T}_4)$ is generally a monotonic increasing function of n, and approximately,

$$\widehat{\Pr}(\hat{\boldsymbol{\theta}}_v \in \mathbb{T}_4) \text{ under Scenario 1} \ \geqslant \ \widehat{\Pr}(\hat{\boldsymbol{\theta}}_v \in \mathbb{T}_4) \text{ under Scenario 2}$$
$$\geqslant \ \widehat{\Pr}(\hat{\boldsymbol{\theta}}_v \in \mathbb{T}_4) \text{ under Scenario 3}.$$

In addition, when θ_1 is not too large (e.g., $\theta_1 = 0.70$ or 0.80), for Scenario 1, we have

$$\lim_{n \to \infty} \Pr(\hat{\boldsymbol{\theta}}_v \in \mathbb{T}_4) \to 1,$$

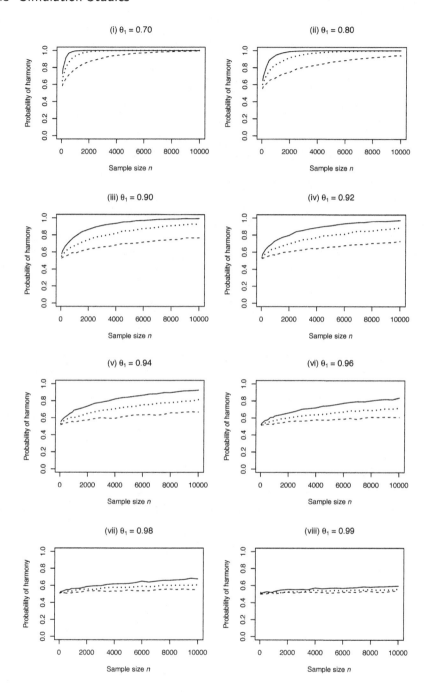

Figure 6.3   The relationship between the estimated probability of obtaining valid estimates $\widehat{\Pr}(\hat{\boldsymbol{\theta}}_v \in \mathbb{T}_4)$ and the sample size n for three scenarios with various θ_1. Scenario 1 (solid line): $(\theta_2, \theta_3, \theta_4) = (2, 3, 1)(1 - \theta_1)/6$. Scenario 2 (dotted line): $(\theta_2, \theta_3, \theta_4) = (5, 4, 1)(1 - \theta_1)/10$. Scenario 3 (dashed line): $(\theta_2, \theta_3, \theta_4) = (8, 11, 1)(1 - \theta_1)/20$.

which is consistent with the claim of Bourke (1982). However, when θ_1 is very large (e.g., $\theta_1 = 0.99$), Figure 6.3(h) shows that

$$\lim_{n \to \infty} \Pr(\hat{\boldsymbol{\theta}}_v \in \mathbb{T}_4) \leqslant 0.6$$

for all three scenarios.

6.6 Bayesian Inferences under Dirichlet Prior

6.6.1 Posterior moments

Assume that the observed-data likelihood function $L_{\mathrm{H}}(\boldsymbol{\theta}|Y_{\mathrm{obs}})$ is given by (6.1). A natural prior for $\boldsymbol{\theta}$ is the Dirichlet distribution Dirichlet$(\boldsymbol{a})$ with $\boldsymbol{a} = (a_1, \ldots, a_4)^{\mathsf{T}}$. Thus, the posterior distribution of $\boldsymbol{\theta} \in \mathbb{T}_4$ has the following closed-form expression:

$$f(\boldsymbol{\theta}|Y_{\mathrm{obs}}) = c_{\mathrm{H}}^{-1}(\boldsymbol{a}, \boldsymbol{n}) \left(\prod_{i=1}^{4} \theta_i^{a_i - 1} \right) \times \theta_1^{n_1} \prod_{i=2}^{4} (p_i \theta_1 + \theta_i)^{n_i}, \qquad (6.14)$$

where $\boldsymbol{n} = (n_1, \ldots, n_4)^{\mathsf{T}}$, $n = \sum_{i=1}^{4} n_i = \boldsymbol{n}^{\mathsf{T}} \mathbf{1}_4$, the normalizing constant

$$c_{\mathrm{H}}(\boldsymbol{a}, \boldsymbol{n}) = \frac{c_{\mathrm{H}}^*(\boldsymbol{a}, \boldsymbol{n})}{\Gamma(\sum_{i=1}^{4} a_i + n)}$$

and

$$c_{\mathrm{H}}^*(\boldsymbol{a}, \boldsymbol{n}) = \sum_{j_2=0}^{n_2} \sum_{j_3=0}^{n_3} \sum_{j_4=0}^{n_4} \left\{ \Gamma(a_1 + n_1 + j_2 + j_3 + j_4) \prod_{\ell=2}^{4} \binom{n_\ell}{j_\ell} \Gamma(a_\ell + n_\ell - j_\ell) p_\ell^{j_\ell} \right\}.$$

Let $\boldsymbol{r} = (r_1, \ldots, r_4)^{\mathsf{T}}$. The posterior moment of $\boldsymbol{\theta}$ is given by

$$E\left(\theta_1^{r_1} \theta_2^{r_2} \theta_3^{r_3} \theta_4^{r_4} \middle| Y_{\mathrm{obs}}\right) = \frac{c_{\mathrm{H}}^*(\boldsymbol{a} + \boldsymbol{r}, \boldsymbol{n})}{c_{\mathrm{H}}^*(\boldsymbol{a}, \boldsymbol{n})} \cdot \frac{\Gamma(\sum_{i=1}^{4} a_i + n)}{\Gamma(\sum_{i=1}^{4}(a_i + r_i) + n)}.$$

Hence, the posterior moments of θ_x, θ_y and ψ can be readily expressed as

$$\left\{ \begin{array}{rcl} E(\theta_x|Y_{\mathrm{obs}}) &=& E(\theta_3|Y_{\mathrm{obs}}) + E(\theta_4|Y_{\mathrm{obs}}), \\[4pt] E(\theta_x^2|Y_{\mathrm{obs}}) &=& E(\theta_3^2|Y_{\mathrm{obs}}) + 2E(\theta_3\theta_4|Y_{\mathrm{obs}}) + E(\theta_4^2|Y_{\mathrm{obs}}), \\[4pt] E(\theta_y|Y_{\mathrm{obs}}) &=& E(\theta_2|Y_{\mathrm{obs}}) + E(\theta_4|Y_{\mathrm{obs}}), \\[4pt] E(\theta_y^2|Y_{\mathrm{obs}}) &=& E(\theta_2^2|Y_{\mathrm{obs}}) + 2E(\theta_2\theta_4|Y_{\mathrm{obs}}) + E(\theta_4^2|Y_{\mathrm{obs}}), \\[4pt] E(\psi|Y_{\mathrm{obs}}) &=& E(\theta_1\theta_2^{-1}\theta_3^{-1}\theta_4|Y_{\mathrm{obs}}), \\[4pt] E(\psi^2|Y_{\mathrm{obs}}) &=& E(\theta_1^2\theta_2^{-2}\theta_3^{-2}\theta_4^2|Y_{\mathrm{obs}}). \end{array} \right.$$

6.6.2 Posterior mode

To derive the posterior mode of $\boldsymbol{\theta}$, we treat the observed counts n_2, n_3 and n_4 as incomplete data and the counts Z_2, Z_3 and Z_4 as missing data (see Table 6.2). Let $\boldsymbol{z} = (z_2, z_3, z_4)^\top$ and $z_1 = n - z_2 - z_3 - z_4$. Thus, the complete-data posterior distribution and the conditional predictive distribution are given by

$$f(\boldsymbol{\theta}|Y_{\text{obs}}, \boldsymbol{z}) = \text{Dirichlet}\,(\boldsymbol{\theta}|a_1 + z_1, \dots, a_4 + z_4) \qquad (6.15)$$

and

$$f(\boldsymbol{z}|Y_{\text{obs}}, \boldsymbol{\theta}) = \prod_{i=2}^{4} \text{Binomial}\left(z_i \middle| n_i, \ \frac{\theta_i}{p_i\theta_1 + \theta_i}\right), \qquad (6.16)$$

respectively. Based on the EM algorithm, the M-step finds the complete-data posterior mode by

$$\tilde{\theta}_i = \frac{a_i + z_i - 1}{\sum_{\ell=1}^{4} a_\ell + n - 4}, \quad i = 2, 3, 4, \quad \tilde{\theta}_1 = 1 - \tilde{\theta}_2 - \tilde{\theta}_3 - \tilde{\theta}_4, \qquad (6.17)$$

and the E-step replaces $\{z_i\}$ by the conditional expectations

$$E(Z_i|Y_{\text{obs}}, \boldsymbol{\theta}) = \frac{n_i\theta_i}{p_i\theta_1 + \theta_i}, \quad i = 2, 3, 4. \qquad (6.18)$$

6.6.3 Generation of posterior samples via the DA algorithm

Based on (6.15) and (6.16), we can use the DA algorithm (Tanner & Wong, 1987) to generate posterior samples of $\boldsymbol{\theta}$. We may choose $\boldsymbol{\theta}^{(0)} = 0.25 \times \mathbf{1}_4$ as the initial value.

6.7 Bayesian Inferences under Other Priors

In the previous section, we consider the Dirichlet distribution as the prior of $\boldsymbol{\theta}$. It is well known that the covariance structure associated with the Dirichlet distribution is completely non-positive. Obviously, those cases that possess, for instance, positive covariance structures cannot be modeled by the Dirichlet prior. To explore the essence of the Dirichlet prior, we first transform the original parameter space $\Theta = \{\theta_1, \theta_2, \theta_3\}$ into an orthogonal parameter space, say $\Theta_x = \{\theta_x, \xi, \eta\}$, and present an equivalent prior distribution for $(\theta_x, \xi, \eta)^\top$ when $(\theta_1, \theta_2, \theta_3)^\top$ follows a Dirichlet distribution. Next, we introduce three new joint priors for $(\theta_x, \xi, \eta)^\top$ for modeling

(1) independence structure with restrictions,

(2) negative correlation structure, and

(3) positive correlation structures.

Finally, for the positive correlation structure, we derive the corresponding posterior moments for the parameters of interest via the importance sampling and generate posterior samples via the DA algorithm.

6.7.1 Orthogonal parameter space

Note that $\theta_x = \Pr(X = 1)$ is the marginal probability of X. Let $\xi = \Pr(Y = 0|X = 0)$ and $\eta = \Pr(Y = 1|X = 1)$ be the corresponding conditional probabilities. Table 6.4 illustrates the fundamental relationship between the two parameter spaces $\Theta = \{\theta_1, \theta_2, \theta_3\}$ and $\Theta_x = \{\theta_x, \xi, \eta\}$.

Table 6.4 Parameter spaces Θ and Θ_x and their cell probabilities

Category	$Y = 0$	$Y = 1$	Marginal	Category	$Y = 0$	$Y = 1$	Marginal
$X = 0$	θ_1	θ_2	$\theta_1 + \theta_2$	$X = 0$	$(1-\theta_x)\xi$	$(1-\theta_x)(1-\xi)$	$1 - \theta_x$
$X = 1$	θ_3	θ_4	$\theta_3 + \theta_4$	$X = 1$	$\theta_x(1-\eta)$	$\theta_x\eta$	θ_x
Space Θ			1	Space Θ_x			1

Source: Table 5 in Tian *et al.* (2009).

It is noteworthy that the following one-to-one transformation

$$\theta_x = \theta_3 + \theta_4, \qquad \xi = \frac{\theta_1}{\theta_1 + \theta_2} \quad \text{and} \quad \eta = \frac{1 - \theta_1 - \theta_2 - \theta_3}{1 - \theta_1 - \theta_2} \qquad (6.19)$$

maps the original parameter space Θ into the orthogonal parameter space (i.e., an unit cube in $\mathbb{R}^3$) Θ_x. The corresponding Jacobian is given by

$$|J| = \left| \frac{\partial(\theta_1, \theta_2, \theta_3)}{\partial(\theta_x, \xi, \eta)} \right| = \theta_x(1 - \theta_x).$$

We have the following result.

Theorem 6.1 If $(\theta_1, \theta_2, \theta_3)^\top$ follows Dirichlet $(a_1, a_2, a_3; a_4)$ prior in space Θ, then the equivalent prior distribution of $(\theta_x, \xi, \eta)^\top$ in Θ_x is given by

$$\theta_x \sim \text{Beta}(a_3 + a_4, a_1 + a_2),$$

$$\xi \sim \text{Beta}(a_1, a_2), \quad \text{and} \qquad (6.20)$$

$$\eta \sim \text{Beta}(a_4, a_3) \quad \text{or} \quad 1 - \eta \sim \text{Beta}(a_3, a_4),$$

where θ_x, ξ and η are mutually independent.   ¶

6.7.2 Joint prior for modeling independence with constraints

Sometimes, prior information on θ_x, ξ and η are available in the form of constraints. For example, let $X = 1$ if a person has annual income being greater than or equal to $100,000; $X = 0$ otherwise, and $Y = 1$ if a person travels at least once a year; $Y = 0$ otherwise. Thus, we have

$$\eta = \Pr(Y = 1 | \text{the annual income} \geqslant \$100{,}000)$$

and

$$\xi = \Pr(Y = 0 | \text{the annual income} < \$100{,}000).$$

In general, the possibility of traveling every year is positively related to annual income. Therefore, it is reasonable to impose the following constraints on η and ξ:

$$\eta \geqslant 1 - \eta \quad \text{and} \quad \xi \geqslant 1 - \xi, \tag{6.21}$$

i.e., $\eta \geqslant 0.5$ and $\xi \geqslant 0.5$. Let $\mathrm{TBeta}(a, b;\ L, U)$ denote the truncated beta distribution defined on the interval $(L,\ U)$. Hence, an alternative joint prior for $(\theta_x, \xi, \eta)^\top$ to (6.20) is

$$\theta_x \sim \mathrm{Beta}\,(a_3 + a_4,\ a_1 + a_2),$$

$$\xi \sim \mathrm{TBeta}(a_1,\ a_2;\ 0.5, 1), \quad \text{and} \tag{6.22}$$

$$\eta \sim \mathrm{TBeta}(a_4,\ a_3;\ 0.5, 1),$$

where θ_x, ξ and η are independent. In other words, the joint prior (6.22) is adequate for modeling the assumption of independence between ξ and η with constraints (6.21).

6.7.3 Joint prior for modeling negative correlation structure

In some applications, the assumption of independence between ξ and η may not be adequate while the negative correlation structure appears to be more practical. One possible way for modeling negative correlation structure is to consider the following inequality constraint:

$$\xi \geqslant 1 - \eta.$$

Define $\xi^* = 1 - \xi$ and $\eta^* = 1 - \eta$. From this inequality constraint, we obtain

$$\xi^* \geqslant 0, \quad \eta^* \geqslant 0 \quad \text{and} \quad \xi^* + \eta^* \leqslant 1. \tag{6.23}$$

Naturally, a Dirichlet prior can be assigned to $(\xi^*, \eta^*)^\top$. It is well known that the components of a Dirichlet random vector are negatively correlated, and so are ξ and η.

6.7.4   Joint prior for modeling positive correlation structure

Now, we consider the case that ξ and η are positively correlated. The first problem is to identify an appropriate prior distribution. The second problem is to compute the corresponding posterior moments for the parameters of interest. Here, we introduce a positively correlated bivariate-beta distribution as the joint prior of $(\xi, \eta)^\top$. We then employ the importance sampling to calculate the posterior moments and the DA algorithm to obtain posterior samples.

(a) Positively correlated bivariate-beta distribution

A random vector $\mathbf{w} = (W_1, W_2)^\top$ is said to follow a *positively correlated bivariate-beta distribution* if the conditional distribution of $W_1|\tau$ and the conditional distribution of $W_2|\tau$ are independent, and

$$W_i|\tau \sim \text{Beta}\,(\gamma_i \tau,\ \gamma_i(1-\tau)), \quad \gamma_i > 0, \quad i = 1, 2,$$

where $\tau \sim \text{Beta}\,(a,\ b)$ and $a,\ b > 0$. We will write $\mathbf{w} \sim \text{PCBBeta}(\gamma_1, \gamma_2;\ a, b)$ (see, e.g., Albert & Gupta, 1983; 1985). Theorem 6.2 below gives the joint density of $\mathbf{w}$ and an algorithm for generating the random vector $\mathbf{w}$.

Theorem 6.2   If $\mathbf{w} = (W_1, W_2)^\top \sim \text{PCBBeta}(\gamma_1, \gamma_2;\ a, b)$, then

(1)   The density of $\mathbf{w}$ is

$$f(\mathbf{w}) = \int_0^1 \frac{\tau^{a-1}(1-\tau)^{b-1}}{B(a,b)} \prod_{i=1}^2 \frac{w_i^{\gamma_i\tau-1}(1-w_i)^{\gamma_i(1-\tau)-1}}{B(\gamma_i\tau, \gamma_i(1-\tau))}\, d\tau. \quad (6.24)$$

(2)   A random sample of $\mathbf{w}$ can be generated as follows: First generate a random variable τ from $\text{Beta}(a,b)$, and then independently generate W_i from $\text{Beta}(\gamma_i\tau,\ \gamma_i(1-\tau))$ for $i = 1, 2$.   ¶

Theorem 6.3   If $\mathbf{w} = (W_1, W_2)^\top \sim \text{PCBBeta}(\gamma_1, \gamma_2;\ a, b)$, then the correlation coefficient of W_1 and W_2 is

$$\rho(W_1, W_2) = \sqrt{\frac{(\gamma_1 + 1)(\gamma_2 + 1)}{(\gamma_1 + a + b + 1)(\gamma_2 + a + b + 1)}}. \quad (6.25)$$

¶

Proof. Let μ and σ^2 denote the mean and variance of τ, respectively. We have

$$\mu = \frac{a}{a+b} \quad \text{and} \quad \sigma^2 = \frac{ab}{(a+b+1)(a+b)^2}.$$

Using the rule of conditional expectation, we obtain for $i = 1, 2$,

$$E(W_i) = E\{E(W_i|\tau)\} = E(\tau) = \mu,$$

$$E(W_i^2) = E\{E(W_i^2|\tau)\} = E\left(\frac{\gamma_i \tau^2 + \tau}{\gamma_i + 1}\right) = \frac{\gamma_i(\sigma^2 + \mu^2) + \mu}{\gamma_i + 1},$$

$$E(W_1 W_2) = E\{E(W_1 W_2|\tau)\} = E(\tau^2) = \sigma^2 + \mu^2, \quad \text{and}$$

$$\text{Var}(W_i) = \frac{\gamma_i \sigma^2 + \mu - \mu^2}{\gamma_i + 1}.$$

Hence,

$$\rho(W_1, W_2) = \frac{E(W_1 W_2) - E(W_1) E(W_2)}{\sqrt{\text{Var}(W_1) \cdot \text{Var}(W_2)}}$$

$$= \sqrt{\frac{(\gamma_1 + 1)(\gamma_2 + 1)}{(\gamma_1 + (\mu - \mu^2)/\sigma^2)(\gamma_2 + (\mu - \mu^2)/\sigma^2)}}.$$

Note that

$$\frac{\mu - \mu^2}{\sigma^2} = a + b + 1,$$

we obtain (6.25) immediately. □

Theorem 6.3 shows that a PCBBeta distribution can be used to quantify positive correlations between two continuous random variables. In particular, if $\gamma_1 = \gamma_2 = \gamma$ and $a = b = 1$, then

$$\rho(W_1, W_2) = \frac{\gamma + 1}{\gamma + 3} = 1 - \frac{2}{\gamma + 3},$$

which is an increasing function of γ. The larger the pre-specified value of γ, the stronger the correlation between W_1 and W_2.

(b) Computing posterior moments via importance sampling

Substituting (6.19) into (6.1), we can rewrite the likelihood function as

$$L_{\text{H}}(\theta_x, \xi, \eta | Y_{\text{obs}}) = (1 - \theta_x)^{n_1 + n_2} \times h(\theta_x, \xi, \eta), \tag{6.26}$$

where

$$h(\theta_x, \xi, \eta) = \xi^{n_1}\{p_2\xi + (1 - \xi)\}^{n_2}\{p_3(1 - \theta_x)\xi + \theta_x(1 - \eta)\}^{n_3}$$
$$\times \{p_4(1 - \theta_x)\xi + \theta_x\eta\}^{n_4}.$$

Motivated by (6.20) and (6.24), we may consider the following distributions as the joint prior for $(\theta_x, \xi, \eta)^\top$ if ξ and η are believed to be positively correlated

$$\theta_x \sim \text{Beta}(\alpha, \beta), \qquad (\xi, \eta)^\top \sim \text{PCBBeta}(\gamma, \gamma; 1, 1), \qquad (6.27)$$

and they are independent. The resultant posterior distribution is

$$f(\theta_x, \xi, \eta | \text{data}) = \frac{\theta_x^{\alpha-1}(1 - \theta_x)^{n_1+n_2+\beta-1} \times f(\xi, \eta) \times h(\theta_x, \xi, \eta)}{c_{\text{H}}(\alpha, \beta, \gamma, \boldsymbol{n})}.$$

By importance sampling, we obtain

$$c_{\text{H}}(\alpha, \beta, \gamma, \boldsymbol{n}) \approx \frac{B(\alpha, n_1 + n_2 + \beta)}{L} \sum_{\ell=1}^{L} h(\theta_x^{(\ell)}, \xi^{(\ell)}, \eta^{(\ell)}),$$

where $\{\theta_x^{(\ell)}\}_{\ell=1}^L$ is a sample of size L from $\text{Beta}(\alpha, \ n_1 + n_2 + \beta)$ and $\{\xi^{(\ell)}, \eta^{(\ell)}\}_{\ell=1}^L$ is a sample of size L from $\text{PCBBeta}(\gamma, \gamma; 1, 1)$ by using Theorem 6.2 (2).

The parameters of interest can be expressed as

$$\theta_x = \theta_x, \quad \theta_y = (1 - \theta_x)(1 - \xi) + \theta_x\eta \quad \text{and} \quad \delta = \frac{\xi\eta}{(1 - \xi)(1 - \eta)}.$$

Therefore, the posterior moments of θ_x can be readily calculated via the importance sampling. However, the calculation of posterior moments such as $E(\theta_y^r | Y_{\text{obs}})$ and $E(\psi^r | Y_{\text{obs}})$ requires the evaluation of the expression

$$\{(1 - \theta_x)(1 - \xi) + \theta_x\eta\}^r \times h(\theta_x, \xi, \eta)$$

L times at

$$(\theta_x^{(\ell)}, \xi^{(\ell)}, \eta^{(\ell)}), \quad \ell = 1, \dots, L.$$

To this end, the DA algorithm can be adopted and we will discuss the algorithm as follows.

(c) Generating posterior samples via the DA algorithm

The likelihood function (6.26) and the prior assumption (6.27) can be reformulated in terms of a hierarchical model with three stages.

Stage 1.   The observed data Y_{obs} is augmented with latent data $\boldsymbol{z} = (z_2, z_3, z_4)^{\top}$ so that the complete-data likelihood is

$$L_{\text{H}}(\theta_x, \xi, \eta | Y_{\text{obs}}, \boldsymbol{z}) = \theta_x^{z_3 + z_4}(1 - \theta_x)^{n - z_3 - z_4}$$

$$\times \xi^{z_1}(1 - \xi)^{z_2} \times \eta^{z_4}(1 - \eta)^{z_3},$$

where $z_1 = n - z_2 - z_3 - z_4$. Similar to (6.16), the conditional predictive distribution is

$$f(\boldsymbol{z} | Y_{\text{obs}}, \theta_x, \xi, \eta)$$

$$= \text{Binomial}\left(z_2 \middle| n_2, \ \frac{1 - \xi}{p_2 \xi + 1 - \xi}\right)$$

$$\times \text{Binomial}\left(z_3 \middle| n_3, \ \frac{\theta_x(1 - \eta)}{p_3(1 - \theta_x)\xi + \theta_x(1 - \eta)}\right)$$

$$\times \text{Binomial}\left(z_4 \middle| n_4, \ \frac{\theta_x \eta}{p_4(1 - \theta_x)\xi + \theta_x \eta}\right). \tag{6.28}$$

Stage 2.   Given a hyperparameter τ, the joint prior is a product of independent beta distributions:

$$f(\theta_x, \xi, \eta | \tau) = f(\theta_x) \times f(\xi | \tau) \times f(\eta | \tau)$$

$$= \text{Beta}(\theta_x | \alpha, \ \beta) \times \text{Beta}(\xi | \gamma \tau, \ \gamma(1 - \tau))$$

$$\times \text{Beta}(\eta | \gamma \tau, \ \gamma(1 - \tau)).$$

Stage 3.   Let the prior distribution of τ be $U(0, 1)$, namely, $f(\tau) = 1$ for $\tau \in (0, 1)$. We write the joint distribution of the complete-data and parameters as

$$L_{\text{H}}(\theta_x, \xi, \eta | Y_{\text{obs}}, \boldsymbol{z}) \times f(\theta_x, \xi, \eta | \tau) \times f(\tau).$$

Thus, we have

$$f(\theta_x, \xi, \eta | Y_{\text{obs}}, \boldsymbol{z}, \tau)$$

$$= \text{Beta}(\theta_x | z_3 + z_4 + \alpha, \ n - z_3 - z_4 + \beta)$$

$$\times \text{Beta}(\xi | z_1 + \gamma \tau, \ z_2 + \gamma[1 - \tau])$$

$$\times \text{Beta}(\eta | z_4 + \gamma \tau, \ z_3 + \gamma[1 - \tau]) \tag{6.29}$$

and

$$f(\tau | Y_{\text{obs}}, \xi, \eta) \propto \frac{(\xi \eta)^{\gamma \tau}\{(1 - \xi)(1 - \eta)\}^{\gamma(1 - \tau)}}{\{B(\gamma \tau, \gamma(1 - \tau))\}^2}, \tag{6.30}$$

where $\tau \in (0, 1)$.

Sampling from (6.28) and (6.29) is quite straightforward. Note that (6.30) is an un-normalized one-dimensional density function defined on $(0, 1)$. The grid-point method (see, e.g., Gelman *et al.*, 1995, p. 302) can be used to generate random samples from this distribution. The implementation of the Gibbs sampling and the calculation of arbitrary expectations of interest were presented thoroughly in Gelfand & Smith (1990) and Arnold (1993), and are hence omitted here.

6.8   Analyzing HIV Data in an AIDS Study

Strauss, Rindskopf & Falkin (2001) reported an HIV data set which examined the relationship between self-reported HIV status and history of sex exchange for drugs and money. All participants were drug dependent women offenders who were mandated to treatment through the criminal justice system of New York City. The data were collected as part of an evaluation study of four drug treatment programs, respectively classified as prison-based, jailed-based, community-based residential, and community-based outpatient. The data reflected baseline responses from 325 clients interviewed at the four treatment programs from May, 1995 through December, 1996. Notice that there are incomplete data for 83 subjects. Table 6.5 gives the cross-classification of history of sex exchange (no or yes, denoted by $X = 0$ or $X = 1$) and HIV status (negative or positive, denoted by $Y = 0$ or $Y = 1$) as reported by the women. The objective is to examine if association exists between sex exchange and HIV status. Obviously, both questions (i.e., sex history and HIV status) are highly sensitive questions to respondents.

Table 6.5 HIV data from Strauss, Rindskopf & Falkin (2001)

History of sex exchange	HIV status		
	$Y = 0$ (HIV$-$)	$Y = 1$ (HIV$+$)	Missing
$X = 0$ (no)	108 (m_1, θ_1)	18 (m_2, θ_2)	44
$X = 1$ (yes)	93 (m_3, θ_3)	23 (m_4, θ_4)	39

Source: Table 7 in Tian *et al.* (2009).
Note: X denotes history of sex exchange and Y denotes HIV status.

6.8.1   Likelihood-based methods

To construct a benchmark for comparison, we first use the routine method to analyze the HIV data by discarding the missing data. The MLEs, standard errors, two 95% bootstrap confidence intervals based on the normality and

Table 6.6 Estimates of parameters for complete cross-classification HIV data

Parameter	MLE	std	95% bootstrap CI†	95% bootstrap CI‡
θ_1	0.4463	0.0317	[0.3840, 0.5085]	[0.3842, 0.5082]
θ_2	0.0743	0.0168	[0.0412, 0.1075]	[0.0413, 0.1074]
θ_3	0.3843	0.0308	[0.3238, 0.4447]	[0.3264, 0.4462]
θ_4	0.0951	0.0188	[0.0581, 0.1319]	[0.0578, 0.1322]
ψ	1.4839	0.6080	[0.2921, 2.6755]	[0.7534, 3.1148]

Note: CI† = Bootstrap confidence intervals (6.5) based on normality assumption. CI‡ = Bootstrap confidence intervals (6.6) based on non-normality assumption.

non-normality assumptions for the cell probabilities and the odds ratio are reported in Table 6.6. Note that the differences between the two bootstrap confidence intervals for $\{\theta_i\}_{i=1}^4$ are very small, while those for ψ are very large. This is not surprising since the bootstrap densities of $\{\hat\theta_i\}_{i=1}^4$ are quite symmetric while the bootstrap density of $\hat\psi$ is highly skewed. Therefore, bootstrap confidence intervals based on the non-normality assumption are recommended.

For illustration, we let $W = i$ if the respondent's mother (or the respondent herself/himself) was born in the i-th quarter, and it is thus reasonable to assume that

$$p_i = \Pr(W = i) = 0.25, \quad i = 1, \ldots, 4,$$

and W is independent of the two sensitive questions. For the ideal situation (i.e., no sampling errors), the observed counts would be

$$n_1 = \frac{m_1}{4} = 27,$$

$$n_2 = 27 + m_2 = 45,$$

$$n_3 = 27 + m_3 = 120, \quad \text{and}$$

$$n_4 = 27 + m_4 = 50.$$

To consider sampling errors, we first generate 50 i.i.d. random samples from Multinomial$(108; 0.25 \times \mathbf{1}_4)$ and then average these counts, yielding

$$(n_1', \ldots, n_4')^\top = (28, \ 26, \ 26, \ 28)^\top.$$

Therefore, we obtain the observed counts

$$Y_{\text{obs}} = \{n; \ n_1, \ldots, n_4\} = \{242; \ 28, \ 44, \ 119, \ 51\}.$$

Table 6.7 Posterior modes and estimates of parameters for the HIV data

Parameter	Posterior mode	Bayes mean	Bayes std	95% Bayes credible intervals
θ_1	0.4628	0.4635	0.0757	[0.3208, 0.6172]
θ_2	0.0661	0.0680	0.0317	[0.0105, 0.1339]
θ_3	0.3760	0.3723	0.0426	[0.2877, 0.4538]
θ_4	0.0950	0.0960	0.0346	[0.0300, 0.1648]
ψ	1.7692	3.6939	31.431	[0.5733, 11.932]

Source: Table 8 in Tian *et al.* (2009).

Using $\boldsymbol{\theta}^{(0)} = 0.25 \times \mathbf{1}_4$ as the initial values, the EM algorithm in (6.2) and (6.3) converges in 50 iterations. The resultant MLEs are

$$\hat{\boldsymbol{\theta}} = (0.4628,\ 0.0661,\ 0.3761,\ 0.0950)^\top \quad \text{and} \quad \hat{\psi} = 1.7692,$$

which are very close to the MLEs in Table 6.6. Based on (6.4) and (6.6), we generate $G = 10{,}000$ bootstrap samples to estimate the standard errors of $\{\hat{\theta}_i\}$ and $\hat{\psi}$ which are given by 0.0794, 0.0337, 0.0438, 0.0354 and 11.7544, respectively. The corresponding 95% non-normal-based bootstrap confidence intervals are [0.2975, 0.6028], [0.0116, 0.1446], [0.2977, 0.4710], [0.0294, 0.1653] and [0.4656, 9.9934].

To test the null hypothesis

$$H_0\text{: } \psi = 1 \quad \text{against} \quad H_1\text{: } \psi \neq 1,$$

we need to obtain the restricted MLE $\hat{\boldsymbol{\theta}}_{\mathrm{R}}$. Using $(\theta_x^{(0)}, \theta_y^{(0)}) = (0.5,\ 0.5)$ as the initial value, the EM algorithm in (6.3) and (6.9) converges to

$$\hat{\theta}_{x,\mathrm{R}} = 0.482477 \quad \text{and} \quad \hat{\theta}_{y,\mathrm{R}} = 0.179607$$

in 45 iterations. From (6.7), the restricted MLEs are given by

$$\hat{\boldsymbol{\theta}}_{\mathrm{R}} = (0.4245,\ 0.0929,\ 0.3958,\ 0.0866)^\top.$$

The log-likelihood ratio statistic Λ_1 is equal to 1.0191 with p-value being 0.3127 while the chi-square statistic Λ_2 is equal to 1.0257 with p-value being 0.3112. All results suggest that there is no association between sex exchange and HIV status based on the current data.

In addition, from (6.11) we have

$$\hat{\boldsymbol{\lambda}} = (0.11570,\ 0.18182,\ 0.49174,\ 0.21074)^\top$$

and

$$\hat{\boldsymbol{\theta}}_v = (0.4628,\ 0.0661,\ 0.3761,\ 0.0950)^\top \in \mathbb{T}_4,$$

which indicates that $\hat{\boldsymbol{\theta}}_v$ is valid.

6.8.2 Bayesian methods

If Dirichlet $(1,1,1,1)$ (i.e., the uniform distribution on $\mathbb{T}_4$) is adopted as the prior distribution of $\boldsymbol{\theta}$, then the posterior modes of $\boldsymbol{\theta}$ are equal to the corresponding MLEs. Using $\boldsymbol{\theta}^{(0)} = 0.25 \times \mathbf{1}_4$ as initial values, the EM algorithm based on (6.17) and (6.18) converges in 100 iterations. The posterior modes of $\boldsymbol{\theta}$ and odds ratio ψ are listed in the second column of Table 6.7. Based on (6.15) and (6.16), we employ the DA algorithm to generate 1,000,000 posterior samples and only use the second half of the samples. The Bayes estimates of $\boldsymbol{\theta}$ and ψ are given in Table 6.7. Since the Bayes credible interval of ψ includes the value of 1, we have reason to believe that there is no association between sex exchange and HIV status.

The posterior densities of the $\{\theta_i\}_{i=1}^4$ and ψ estimated by a kernel density smoother are plotted in Figures 6.4 and 6.5.

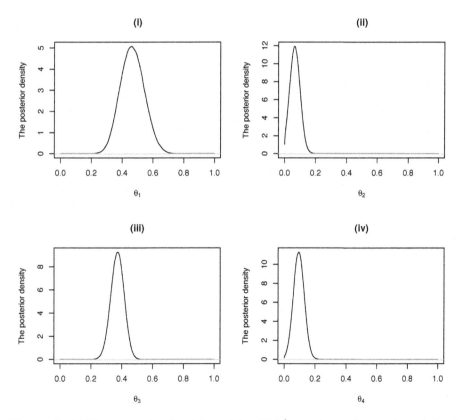

Figure 6.4 The posterior densities of the $\{\theta_i\}_{i=1}^4$ estimated by a kernel density smoother based on the second 500,000 posterior samples of $\boldsymbol{\theta}$ generated by the data augmentation algorithm when the prior distribution is Dirichlet $(1,1,1,1)$. (i) The posterior density of θ_1; (ii) The posterior density of θ_2; (iii) The posterior density of θ_3; (iv) The posterior density of θ_4.

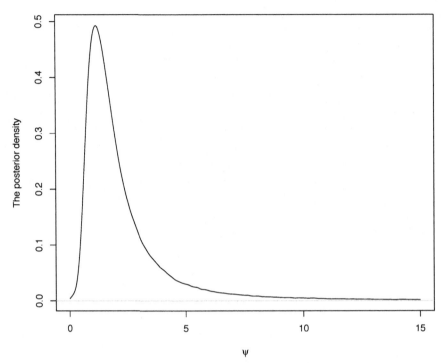

Figure 6.5 The posterior density of the odds ratio ψ estimated by a kernel density smoother based on the second 500,000 posterior samples of $\boldsymbol{\theta}$ generated by the data augmentation algorithm when the prior distribution is Dirichlet $(1, 1, 1, 1)$.

CHAPTER 7

The Parallel Model

Amongst various randomized response alternatives of the original Warner design, we note that the randomized unrelated question design is the most widely used by investigators in practice. For example, the method has been applied to

(i) estimate the frequency of the illegitimacy of offspring (Greenberg *et al.*, 1969);

(ii) estimate the incidence of induced abortions (Abernathy, Greenberg & Horvitz, 1970; Shimizu & Bonham, 1978).

(iii) investigate the connection between driving after drinking and automobile accidents (Folsom *et al.*, 1973);

(iv) estimate the proportion of drug use (Goodstadt & Gruson, 1975).

In this chapter, we first introduce the randomized unrelated question model to investigate its relative efficiency and degree of privacy protection in Section 7.1.

The crosswise and triangular models require that one category (denoted by $\{Y = 0\}$) is non-sensitive. Thus, they cannot be applied to situation where two categories (the other category denoted by $\{Y = 1\}$) are sensitive like income, the number of sex partners, disloyal or loyal to his/her boss and so on. For example, let $Y = 0$ ($Y = 1$) if the number of sex partners $\leqslant 3$ ($\geqslant 4$). Next, the two models still have a lower efficiency for some cases. Third, it was shown that the triangular model cannot sufficiently protect respondents' privacy. Hence, these limitations of the existing nonrandomized response models motivate us to introduce a so-called parallel model in Section 7.2, which is a non-randomized version of the unrelated question model. Asymptotic properties of the maximum likelihood estimator (and its modified version) for the proportion π of the population belonging to the sensitive class are also explored. Theoretical comparisons with the existing non-randomized crosswise and triangular models are presented in Sections 7.3 and 7.4, respectively. Bayesian methods for analyzing survey data from the parallel model are given in Section 7.5. In Section 7.6, a real dataset from a survey on induced abortion in Mexico is used to illustrate these methods. Two case studies are provided in Sections 7.7 and 7.8. A discussion is given in Section 7.9.

7.1   The Unrelated Question Model

The unrelated question randomized response model was first suggested by
Walt R. Simmons (cf. Horvitz, Shah & Simmons 1967) and its theoretical framework was developed by Greenberg *et al.* (1969). In the unrelated
question model, one sensitive question and one non-sensitive question are
presented to an interviewee:

(a)   I am a drug user (i.e., I am a member of group $\mathcal{Y}$).

(b)   My birthday is in the first half of a month (i.e., I am a member of
      group $\mathcal{U}$).

Note that Statement (b) is unrelated to Statement (a) and is completely
innocuous.

7.1.1   The survey design

Each interviewee is asked to reply with only a 'yes' or 'no' answer to Statement (a) or (b). The selection of Statement (a) or (b) is made by a randomization device such as a coin, a deck of cards, or a sealed clear plastic
box containing colored beads. In this way, the respondent's status is not
revealed to the interviewer provided that the latter cannot observe the randomization process in the device. His or her privacy can be protected since
the randomization device is operated by the respondent and the interviewer
does not know which question is being answered by the respondent.

Let p denote the probability of assigning Statement (a) to a respondent
by the randomization device, i.e.,

$$p = \Pr\{\text{selecting Statement (a) by the randomization device}\}.$$

Here, p is designed to be known. In other words, the randomization device
is controlled by the interviewer. Furthermore, let q denote the proportion
of individuals in the population who would answer 'yes' to Statement (b),
i.e.,

$$q = \Pr\{\text{answering 'yes' to Statement (b)}\} = \Pr(\mathcal{U}).$$

Greenberg *et al.* (1969) considered two cases where q is unknown and q is
known. If q is *unknown*, two independent samples of size n_1 and n_2 with
different probabilities p_1 and p_2 for two RDs are required. In what follows,
we focus only on the case of q being known.

7.1.2   Estimation

If q is known, only one sample is required to estimate the proportion π of the population belonging to the sensitive class $\mathcal{Y}$, i.e.,

$$\pi = \Pr\{\text{answering 'yes' to Statement (a)}\} = \Pr(\mathcal{Y}).$$

If λ denotes the probability that a 'yes' answer is reported, then

$$
\begin{aligned}
\lambda \;=\;& \Pr\{\text{answering 'yes' to (a)}\} \times \Pr\{\text{selecting Statement (a)}\} \\
& + \Pr\{\text{answering 'yes' to (b)}\} \times \Pr\{\text{selecting Statement (b)}\} \\
=\;& \pi p + q(1-p)
\end{aligned}
\tag{7.1}
$$

so that $\pi = \{\lambda - q(1-p)\}/p$.

Let n' denote the number of 'yes' reported from n respondents. Hence, $n' \sim \text{Binomial}(n, \lambda)$, $E(n') = n\lambda$ and $\text{Var}(n') = n\lambda(1-\lambda)$. Since the MLE of λ is $\hat{\lambda} = n'/n$, the MLE of π is given by

$$\hat{\pi}_{\mathrm{U}} = \frac{\hat{\lambda} - q(1-p)}{p} = \frac{n'/n - q(1-p)}{p}, \tag{7.2}$$

provided that $\hat{\pi}_{\mathrm{U}} \in [0,1]$, where the subscript 'U' refers to the 'unrelated' question model. It is easy to show that $\hat{\pi}_{\mathrm{U}}$ is unbiased and

$$
\begin{aligned}
\text{Var}(\hat{\pi}_{\mathrm{U}}) \;=\;& \frac{\text{Var}(\hat{\lambda})}{p^2} = \frac{\text{Var}(n')}{(np)^2} = \frac{\lambda(1-\lambda)}{np^2} \\[2mm]
\overset{(7.1)}{=}\;& \frac{\pi(1-\pi)}{n} + \frac{p(1-p)(1-2q)\pi + q(1-p)(1-q+qp)}{np^2} \\[2mm]
=\;& \text{Var}(\hat{\pi}_{\mathrm{D}}) + \frac{(1-p)f_{\mathrm{U}}(q|\pi, p)}{np^2},
\end{aligned}
\tag{7.3}
$$

where $\text{Var}(\hat{\pi}_{\mathrm{D}})$ denotes the variance of the corresponding MLE of π in the *design of direct questioning* (DDQ) and

$$
\begin{aligned}
f_{\mathrm{U}}(q|\pi, p) \;\hat{=}\;& p(1-2q)\pi + q(1-q+qp) \\
=\;& (p-1)q^2 + (1-2\pi p)q + \pi p.
\end{aligned}
\tag{7.4}
$$

It is easy to verify that $f_{\mathrm{U}}(q|\pi, p) = f_{\mathrm{U}}(1-q|1-\pi, p)$.

On the one hand, for any fixed π and q, we note that

$$n\,\text{Var}(\hat{\pi}_{\mathrm{U}}) = \pi(1-\pi) + \frac{(1-p)f_{\mathrm{U}}(q|\pi, p)}{p^2} \tag{7.5}$$

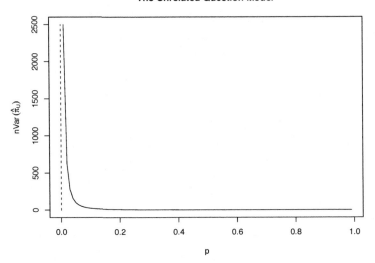

Figure 7.1 Plot of $n\mathrm{Var}(\hat\pi_{\mathrm{U}})$ defined by (7.5) against p with $\pi = 0.3$ and $q = 0.5$ for the unrelated question model.

is a decreasing function of p when $p \in (0,1)$ (see Figure 7.1). We can see that $n\mathrm{Var}(\hat\pi_{\mathrm{U}}) \to \infty$ as $p \to 0$ and $\mathrm{Var}(\hat\pi_{\mathrm{U}}) \to \mathrm{Var}(\hat\pi_{\mathrm{D}})$ as $p \to 1$.

On the other hand, for any fixed π and p, let

$$0 = \frac{\mathrm{d}f_{\mathrm{U}}(q|\pi, p)}{\mathrm{d}q} = 2(p-1)q + 1 - 2\pi p,$$

we have

$$q = \begin{cases} 0, & \text{if } \pi p \geqslant 0.5, \\[2mm] \dfrac{1 - 2\pi p}{2(1-p)}, & \text{if } \pi p < 0.5 \text{ and } (1-\pi)p < 0.5, \\[2mm] 1, & \text{if } (1-\pi)p \geqslant 0.5. \end{cases}$$

Therefore, we obtain the following results:

1°   If $\pi p \geqslant 0.5$, then (7.5) is a decreasing function of q, see Figure 7.2 (i);

2°   If $\pi p < 0.5$ and $(1-\pi)p < 0.5$, then $n\mathrm{Var}(\hat\pi_{\mathrm{U}})$ is a quadratic function of q and reaches its maximum value $1/(4p^2)$ at $q = (1 - 2\pi p)/\{2(1-p)\}$, see Figure 7.2 (ii);

3°   If $(1 - \pi)p \geqslant 0.5$, then $n\mathrm{Var}(\hat\pi_{\mathrm{U}})$ is an increasing function of q, see Figure 7.2 (iii).

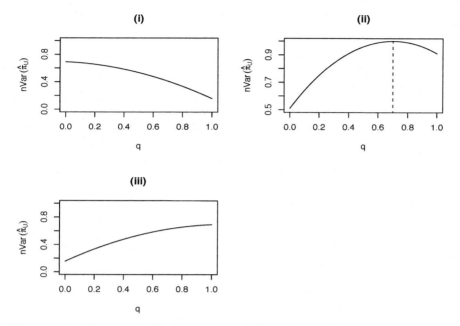

Figure 7.2 Plot of $n\mathrm{Var}(\hat{\pi}_{\mathrm{U}})$ defined by (7.5) against q for the unrelated question model. (i) $\pi p \geqslant 0.5$; (ii) $\pi p < 0.5$ and $(1-\pi)p < 0.5$; (iii) $(1-\pi)p \geqslant 0.5$.

7.1.3   Relative efficiency

When $p = 1$, the unrelated question model is reduced to the DDQ. The RE of the unrelated question design to the DDQ is given by

$$\mathrm{RE}_{\mathrm{U}\to\mathrm{D}}(\pi, q, p) = \frac{\mathrm{Var}(\hat{\pi}_{\mathrm{U}})}{\mathrm{Var}(\hat{\pi}_{\mathrm{D}})}$$

$$= 1 + \frac{(1-p)f_{\mathrm{U}}(q|\pi, p)}{\pi(1-\pi)p^2},$$

which is free of the sample size n. It is easy to verify that

$$\mathrm{RE}_{\mathrm{U}\to\mathrm{D}}(1-\pi, 1-q, p) = \mathrm{RE}_{\mathrm{U}\to\mathrm{D}}(\pi, q, p).$$

Table 7.1 reports some values of $\mathrm{RE}_{\mathrm{U}\to\mathrm{D}}(\pi, q, p)$ for various combinations of π, q and p. For example, when $\pi = 0.10$, $q = 0.50$ and $p = 0.60$, we have

$$\mathrm{RE}_{\mathrm{U}\to\mathrm{D}}(0.10, 0.50, 0.60) = 5.9383,$$

which implies that the sample size required for the unrelated question design is about 6 times that required for the DDQ in order to achieve the same estimation precision.

Table 7.1 Relative efficiency $\mathrm{RE}_{\mathrm{U}\to\mathrm{D}}(\pi, q, p)$ for various combinations of π, q and p

π	q	p				
		1/3	0.40	0.50	0.60	2/3
0.05	1/3	34.450	22.578	13.046	7.9916	5.8538
0.10		19.024	12.666	7.5432	4.8134	3.6543
0.20		11.555	7.8750	4.8889	3.2840	2.5972
0.30		9.3598	6.4762	4.1217	2.8460	2.2963
0.50		8.5556	6.0000	3.8889	2.7284	2.2222
0.70		10.629	7.4286	4.7566	3.2693	2.6138
0.90		24.950	17.111	10.506	6.7888	5.1358
0.95		47.081	32.052	19.362	12.202	9.0117
0.05	0.50	43.105	28.631	16.789	10.356	7.5789
0.10		23.222	15.583	9.3333	5.9383	4.4722
0.20		13.500	9.2031	5.6875	3.7778	2.9531
0.30		10.524	7.2500	4.5714	3.1164	2.4881
0.50		9.0000	6.2500	4.0000	2.7778	2.2500
0.70		10.524	7.2500	4.5714	3.1164	2.4881
0.90		23.222	15.583	9.3333	5.9383	4.4722
0.95		43.105	28.631	16.789	10.356	7.5789
0.05	2/3	47.081	32.052	19.362	12.202	9.0117
0.10		24.950	17.111	10.506	6.7888	5.1358
0.20		14.055	9.7500	6.1389	4.1173	3.2222
0.30		10.629	7.4286	4.7566	3.2693	2.6138
0.50		8.5556	6.0000	3.8889	2.7284	2.2222
0.70		9.3598	6.4762	4.1217	2.8460	2.2963
0.90		19.024	12.666	7.5432	4.8134	3.6543
0.95		34.450	22.578	13.046	7.9916	5.8538

Note: $\mathrm{RE}_{\mathrm{U}\to\mathrm{D}}(1 - \pi, 1 - q, p) = \mathrm{RE}_{\mathrm{U}\to\mathrm{D}}(\pi, q, p)$.

7.1.4 Degree of privacy protection

Since the sensitive information of a respondent regarding his/her membership in sensitive class $\mathcal{Y}$ is characterized through $\Pr(\mathcal{Y}|\text{yes})$ and $\Pr(\mathcal{Y}|\text{no})$, we use

$$\mathrm{DPP}_{\text{yes}}(\pi, q, p) \,\hat{=}\, \Pr(\mathcal{Y}|\text{yes}) = \frac{\pi p}{\pi p + q(1 - p)} \qquad (7.6)$$

and

$$\mathrm{DPP}_{\text{no}}(\pi, q, p) \,\hat{=}\, \Pr(\mathcal{Y}|\text{no}) = 0 \qquad (7.7)$$

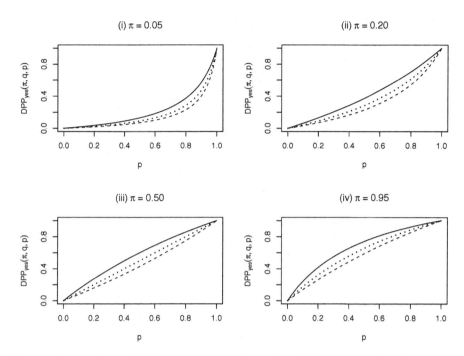

Figure 7.3 Plots of $\mathrm{DPP}_{\mathrm{yes}}(\pi, q, p)$ defined by (7.6) against p for the unrelated question model with a fixed π and three different values of q, where the solid line is corresponding to $q = 1/3$; the dotted line is corresponding to $q = 0.5$; and the dashed line is corresponding to $q = 2/3$. (i) $\pi = 0.05$; (ii) $\pi = 0.20$; (iii) $\pi = 0.50$; (iv) $\pi = 0.95$.

to measure the private information divulged with the unrelated question model. Either $\mathrm{DPP}_{\mathrm{yes}}(\pi, q, p)$ or $\mathrm{DPP}_{\mathrm{no}}(\pi, q, p)$ is called the *degree of privacy protection* (DPP) for the unrelated question model. In particular, when $p = 1$, we have $\mathrm{DPP}_{\mathrm{yes}}(\pi, q, 1) = 1$, which is corresponding to the design of direct questioning.

For any fixed π and q, $\mathrm{DPP}_{\mathrm{yes}}(\pi, q, p)$ is a monotonically increasing function of p. Figure 7.3 shows three curves (corresponding to $q = 1/3$, 0.5 and 2/3) of $\mathrm{DPP}_{\mathrm{yes}}(\pi, q, p)$ against p with a fixed π, where $\pi = 0.05$, 0.20, 0.50 and 0.95, respectively.

In addition, for any fixed π and p, $\mathrm{DPP}_{\mathrm{yes}}(\pi, q, p)$ is a monotonically decreasing function of q, we have

$$\frac{\pi p}{\pi p + 1 - p} = \mathrm{DPP}_{\mathrm{yes}}(\pi, 1, p)$$

$$\leqslant \mathrm{DPP}_{\mathrm{yes}}(\pi, q, p)$$

$$\leqslant \mathrm{DPP}_{\mathrm{yes}}(\pi, 0, p) = 1.$$

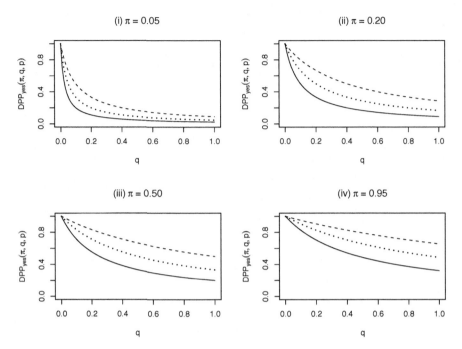

Figure 7.4 Plots of $\mathrm{DPP}_{\mathrm{yes}}(\pi, q, p)$ defined by (7.6) against q for the unrelated question model with a fixed π and three different values of p, where the solid line is corresponding to $p = 1/3$; the dotted line is corresponding to $p = 0.5$; and the dashed line is corresponding to $p = 2/3$. (i) $\pi = 0.05$; (ii) $\pi = 0.20$; (iii) $\pi = 0.50$; (iv) $\pi = 0.95$.

Figure 7.4 shows three curves (corresponding to $p = 1/3$, 0.5 and 2/3) of $\mathrm{DPP}_{\mathrm{yes}}(\pi, q, p)$ against q with a fixed π, where $\pi = 0.05$, 0.20, 0.50 and 0.95, respectively.

7.2 A Non-randomized Unrelated Question Model: The Parallel Model

In this section, we introduce a parallel model (Tian, 2012), which is a non-randomized version of the unrelated question model. Let $\{Y = 1\}$ denote the class of people who possess a sensitive characteristic (e.g., driving under influence) and $\{Y = 0\}$ denote the complementary class. The objective is to estimate the proportion $\pi = \Pr(Y = 1)$. Suppose that U and W are two non-sensitive dichotomous variates, and Y, U and W are mutually independent with known

$$q = \Pr(U = 1) \quad \text{and} \quad p = \Pr(W = 1).$$

Table 7.2 The parallel model and the corresponding cell probabilities

Category	$W = 0$	$W = 1$	Category	$W = 0$	$W = 1$	Marginal
$U = 0$	○		$U = 0$	$(1 - q)(1 - p)$		$1 - q$
$U = 1$	□		$U = 1$	$q(1 - p)$		q
$Y = 0$		○	$Y = 0$		$(1 - \pi)p$	$1 - \pi$
$Y = 1$		□	$Y = 1$		πp	π
			Marginal	$1 - p$	p	1

Note: Please connect the two circles by a straight line if you belong to one of the two circles or connect the two squares by a straight line if you belong to one of the two squares.

For example, we may define $U = 1$ if the respondent's birthday is in the second half of a month and $U = 0$ otherwise. Similarly, we define $W = 1$ if the respondent was born between July and December and $W = 0$ otherwise. Hence, it is reasonable to assume that $q \approx 0.5$ and $p \approx 0.5$. More discussions on the model assumptions and the choice of U and W are presented in Section 7.9.

7.2.1 The survey design for the parallel model

Interviewers may design a questionnaire in the format as shown at the left-hand side of Table 7.2 and ask each interviewee to connect the two circles by a straight line if he/she belongs to one of the two circles or connect the two squares by a straight line if he/she belongs to one of the two squares. Note that all $\{W = 0\}$, $\{W = 1\}$, $\{U = 0\}$, $\{U = 1\}$, and $\{Y = 0\}$ are non-sensitive classes, thus

$$\{U = 1,\ W = 0\} \cup \{Y = 1,\ W = 1\}$$

is also a non-sensitive subclass. Therefore, whether the interviewee belongs to the sensitive class is not known to the interviewers. Tian (2012) called this design the parallel model.

The corresponding cell probabilities are displayed at the right-hand side of Table 7.2. Since the three binary variables U, Y and W are independent, the joint probability is the product of two corresponding marginal probabilities.

Alternatively, the parallel model can be re-formulated as:

(1) If your birthday is in the first half of a year (i.e., $W = 0$), please answer 'yes' (i.e., $U = 1$) or 'no' (i.e., $U = 0$) to the question: *Is your birthday in the first half of a month?*

(2) If your birthday is in the second half of a year (i.e., $W = 1$), please answer 'yes' (i.e., $Y = 1$) or 'no' (i.e., $Y = 0$) to the question: *Have you ever driven a car although your blood alcohol was almost certainly over the legal limit?*

In general, interviewers do not know the information of birth date of a respondent. Hence, the interviewers receive the response (e.g., 'yes' or 'no') from a respondent without the knowledge of which question is being answered by the respondent. Since the birthday of a respondent cannot be controlled by the interviewers, we anticipate that all respondents may trust the survey.

7.2.2 Connection between the parallel model and the unrelated question model

Suppose there are a total of n straight lines (corresponding to n respondents in the survey) with $n - n'$ lines connecting the two circles and n' lines connecting the two squares (see Table 7.2). The observed data are denoted by $Y_{\mathrm{obs}} = \{n;\ n - n',\ n'\}$. The probability of drawing a line to connect the two squares is then $q(1 - p) + \pi p$. Defining it by a new parameter $\lambda = q(1 - p) + \pi p$, which is identical to (7.1), we have

$$\pi = \frac{\lambda - q(1 - p)}{p}.$$

The likelihood function is

$$L_{\mathrm{P}}(\pi|Y_{\mathrm{obs}}) = \binom{n}{n'} \lambda^{n'}(1 - \lambda)^{n-n'} \qquad (7.8)$$

so that the MLE of λ is given by $\hat{\lambda} = n'/n$. Therefore, the MLE of π is

$$\hat{\pi}_{\mathrm{P}} = \frac{\hat{\lambda} - q(1 - p)}{p}, \qquad (7.9)$$

provided that $\hat{\pi}_{\mathrm{P}} \in [0, 1]$, where the subscript 'P' refers to the 'parallel' model. When $\hat{\pi}_{\mathrm{P}} < 0$ or $\hat{\pi}_{\mathrm{T}} > 1$, we can use the EM algorithm (7.51)–(7.53) with $a = b = 1$ (i.e., if the uniform distribution on $[0, 1]$ is adopted as a prior of π, the posterior mode of π is equal to the MLE of π) to find the MLE of π. It is not difficult to verify that

$$\mathrm{Var}(\hat{\pi}_{\mathrm{P}}) = \mathrm{Var}(\hat{\pi}_{\mathrm{U}}) = \frac{\pi(1 - \pi)}{n} + \frac{(1 - p)f_{\mathrm{U}}(q|\pi, p)}{np^2}, \qquad (7.10)$$

where $f_{\mathrm{U}}(q|\pi, p)$ is defined by (7.4).

Since the MLE (7.9) is equal to (7.2) and the variances of $\hat{\pi}_{\mathrm{P}}$ and $\hat{\pi}_{\mathrm{U}}$ are the same, the parallel model can be regarded as a non-randomized version of the randomized unrelated question model.

7.2.3 Asymptotic properties of the MLE

In this subsection, first we derive an unbiased estimator of the variance of $\hat{\pi}_P$. Next, we provide two asymptotic confidence intervals and two bootstrap confidence intervals for π. Finally, we consider a modified MLE of π and theoretically investigate the asymptotic equivalence between this modified MLE and the MLE of π.

(a) An unbiased estimator of the variance of $\hat{\pi}_P$

Theorem 7.1 Let $f_U(q|\pi,p)$ be defined by (7.4) and

$$\overline{\mathrm{Var}}(\hat{\pi}_P) \triangleq \frac{\hat{\lambda}(1-\hat{\lambda})}{(n-1)p^2}.$$

Then, we have

$$\overline{\mathrm{Var}}(\hat{\pi}_P) = \frac{\hat{\pi}_P(1-\hat{\pi}_P)}{n-1} + \frac{(1-p)f_U(q|\hat{\pi}_P,p)}{(n-1)p^2}, \tag{7.11}$$

and it is an unbiased estimator of $\mathrm{Var}(\hat{\pi}_P) = \lambda(1-\lambda)/(np^2)$. ¶

Proof. We first verify (7.11), which is equivalent to

$$\hat{\lambda}(1-\hat{\lambda}) - p^2\hat{\pi}_P(1-\hat{\pi}_P)$$

$$= \quad (1-p)f_U(q|\hat{\pi}_P,p)$$

$$\overset{(7.4)}{=} \quad p(1-p)(1-2q)\hat{\pi}_P + q(1-p)(1-q+qp). \tag{7.12}$$

From (7.9), we obtain $\hat{\lambda} = q(1-p) + p\hat{\pi}_P$. Hence,

$$\hat{\lambda}(1-\hat{\lambda}) - p^2\hat{\pi}_P(1-\hat{\pi}_P)$$

$$= \quad \{q(1-p) + p\hat{\pi}_P\}(1-q+qp-p\hat{\pi}_P) - p^2\hat{\pi}_P(1-\hat{\pi}_P)$$

$$= \quad q(1-p)(1-q+qp) - q(1-p)p\hat{\pi}_P + p\hat{\pi}_P(1-q+qp) - p^2\hat{\pi}_P$$

$$= \quad q(1-p)(1-q+qp) + p(1-p)(1-2q)\hat{\pi}_P,$$

which implies (7.12).

Next, we prove the second part. Note that $n' \sim \mathrm{Binomial}(n,\lambda)$, we have $E(n') = n\lambda$ and $\mathrm{Var}(n') = n\lambda(1-\lambda)$ so that

$$E(\hat{\lambda}) = \frac{E(n')}{n} = \lambda \quad \text{and} \quad \mathrm{Var}(\hat{\lambda}) = \frac{\mathrm{Var}(n')}{n^2} = \frac{\lambda(1-\lambda)}{n}.$$

Therefore,

$$E\left\{\overline{\mathrm{Var}}(\hat{\pi}_{\mathrm{P}})\right\} = \frac{E(\hat{\lambda}) - E(\hat{\lambda}^2)}{(n-1)p^2}$$

$$= \frac{E(\hat{\lambda}) - \{E(\hat{\lambda})\}^2 - \mathrm{Var}(\hat{\lambda})}{(n-1)p^2}$$

$$= \frac{\lambda - \lambda^2 - \lambda(1-\lambda)/n}{(n-1)p^2}$$

$$= \frac{\lambda(1-\lambda)}{np^2}$$

$$= \mathrm{Var}(\hat{\pi}_{\mathrm{P}});$$

that is, $\overline{\mathrm{Var}}(\hat{\pi}_{\mathrm{P}})$ is an unbiased estimator of $\mathrm{Var}(\hat{\pi}_{\mathrm{P}})$.   □

(b) Wald and Wilson (score) confidence intervals for π

By the Central Limit Theorem, $\hat{\pi}_{\mathrm{P}}$ specified by (7.9) is asymptotically normally distributed, i.e.,

$$(\hat{\pi}_{\mathrm{P}} - \pi) \Big/ \sqrt{\mathrm{Var}(\hat{\pi}_{\mathrm{P}})} \sim N(0,1), \quad \text{as } n \to \infty.$$

The $(1-\alpha)100\%$ Wald confidence interval of π is given by

$$[\hat{\pi}_{\mathrm{P,WL}}, \ \hat{\pi}_{\mathrm{P,WU}}] = \left[\hat{\pi}_{\mathrm{P}} - z_{\alpha/2}\sqrt{\mathrm{Var}(\hat{\pi}_{\mathrm{P}})}, \ \hat{\pi}_{\mathrm{P}} + z_{\alpha/2}\sqrt{\mathrm{Var}(\hat{\pi}_{\mathrm{P}})}\right], \quad (7.13)$$

where z_α is the upper α-th quantile of the standard normal distribution.

The construction of the $(1-\alpha)100\%$ Wilson (score) confidence interval $[\hat{\pi}_{\mathrm{P,WSL}}, \ \hat{\pi}_{\mathrm{P,WSU}}]$ of π is based on

$$1-\alpha$$

$$= \mathrm{Pr}\left\{\left|\frac{\hat{\pi}_{\mathrm{P}} - \pi}{\sqrt{\mathrm{Var}(\hat{\pi}_{\mathrm{P}})}}\right| \leqslant z_{\alpha/2}\right\}$$

$$= \mathrm{Pr}\left\{(\hat{\pi}_{\mathrm{P}} - \pi)^2 \leqslant z_{\alpha/2}^2 \mathrm{Var}(\hat{\pi}_{\mathrm{P}})\right\}$$

$$\overset{(7.10)}{=} \mathrm{Pr}\left[(\hat{\pi}_{\mathrm{P}} - \pi)^2 \leqslant \frac{z_{\alpha/2}^2}{n}\left\{\pi(1-\pi) + \frac{(1-p)f_{\mathrm{U}}(q|\pi,p)}{p^2}\right\}\right]$$

$$\overset{(7.4)}{=} \quad \Pr\left\{\hat{\pi}_\text{P}^2 - 2\hat{\pi}_\text{P}\pi + \pi^2 \leqslant \frac{z_{\alpha/2}^2(-\pi^2 + \rho_1\pi + \rho_2)}{n}\right\}$$

$$= \quad \Pr\left\{(1 + z_*)\pi^2 - (2\hat{\pi}_\text{P} + z_*\rho_1)\pi + \hat{\pi}_\text{P}^2 - z_*\rho_2 \leqslant 0\right\}, \qquad (7.14)$$

where $z_* \doteq z_{\alpha/2}^2/n$,

$$\rho_1 \doteq \frac{1 - 2q(1 - p)}{p} \quad \text{and} \quad \rho_2 \doteq \frac{q(1 - p)(1 - q + qp)}{p^2}. \qquad (7.15)$$

Solving the quadratic inequality inside the probability in (7.14), we obtain the asymptotic confidence interval

$$[\hat{\pi}_{\text{P,WSL}}, \ \hat{\pi}_{\text{P,WSU}}] = \frac{2\hat{\pi}_\text{P} + z_*\rho_1 \pm \sqrt{(2\hat{\pi}_\text{P} + z_*\rho_1)^2 - 4(1 + z_*)(\hat{\pi}_\text{P}^2 - z_*\rho_2)}}{2(1 + z_*)},$$

$$(7.16)$$

which is, in general, within $[0, 1]$.

(c) Bootstrap confidence intervals of π

When the sample size n is small to moderate, asymptotic confidence intervals in (7.13) and (7.16) are not reliable. For such cases, the bootstrap approach can be used to obtain two bootstrap confidence intervals of π.

From the observed data $Y_{\text{obs}} = \{n; \ n - n', \ n'\}$, we first calculate the MLE of λ, i.e., $\hat{\lambda} = n'/n$. Then, we can generate

$$n'^* \sim \text{Binomial}(n, \hat{\lambda})$$

to produce $Y_{\text{obs}}^* = \{n; \ n - n'^*, \ n'^*\}$. Based on Y_{obs}^*, we can calculate a bootstrap replication $\hat{\pi}_\text{P}^*$ using the formula (7.9) or the EM algorithm (7.51)–(7.53) with $a = b = 1$. Independently repeating this process G times, we obtain G valid bootstrap replications $\{\hat{\pi}_\text{P}^*(g)\}_{g=1}^G$. Thus, the standard error, $\text{se}(\hat{\pi}_\text{P})$, of $\hat{\pi}_\text{P}$ can be estimated by the sample standard deviation of the G replications, i.e.,

$$\widehat{\text{se}}(\hat{\pi}_\text{P}) = \left[\frac{1}{G - 1}\sum_{g=1}^G \left\{\hat{\pi}_\text{P}^*(g) - \frac{\hat{\pi}_\text{P}^*(1) + \cdots + \hat{\pi}_\text{P}^*(G)}{G}\right\}^2\right]^{\frac{1}{2}}.$$

If $\{\hat{\pi}_\text{P}^*(g)\}_{g=1}^G$ is approximately normally distributed, a $(1 - \alpha)100\%$ bootstrap confidence interval for π is given by

$$\left[\hat{\pi}_\text{P} - z_{\alpha/2} \times \widehat{\text{se}}(\hat{\pi}_\text{P}), \ \hat{\pi}_\text{P} + z_{\alpha/2} \times \widehat{\text{se}}(\hat{\pi}_\text{P})\right]. \qquad (7.17)$$

If $\{\hat{\pi}_{\mathrm{P}}^*(g)\}_{g=1}^G$ is non-normally distributed, a $100(1-\alpha)\%$ bootstrap confidence interval for π can be obtained by

$$[\hat{\pi}_{\mathrm{P,BL}}, \hat{\pi}_{\mathrm{P,BU}}], \tag{7.18}$$

where $\hat{\pi}_{\mathrm{P,BL}}$ and $\hat{\pi}_{\mathrm{P,BU}}$ are the $100(\alpha/2)$ and $100(1-\alpha/2)$ percentiles of $\{\hat{\pi}_{\mathrm{P}}^*(g)\}_{g=1}^G$, respectively.

(d) A modified MLE of π

From (7.9), we have $0 \leqslant \hat{\pi}_{\mathrm{P}} \leqslant 1$ if and only if

$$q(1-p) \leqslant \hat{\lambda} \leqslant q(1-p) + p.$$

Therefore, a modified MLE of π is

$$\hat{\pi}_{\mathrm{PM}} = \mathrm{median}\{0, \hat{\pi}_{\mathrm{P}}, 1\} = \begin{cases} 0, & \text{if } \hat{\pi}_{\mathrm{P}} < 0, \\ \hat{\pi}_{\mathrm{P}}, & \text{if } 0 \leqslant \hat{\pi}_{\mathrm{P}} \leqslant 1, \\ 1, & \text{if } \hat{\pi}_{\mathrm{P}} > 1, \end{cases}$$

$$= \begin{cases} 0, & \text{if } 0 \leqslant \dfrac{n'}{n} < q(1-p), \\ \hat{\pi}_{\mathrm{P}}, & \text{if } q(1-p) \leqslant \dfrac{n'}{n} \leqslant q(1-p)+p, \\ 1, & \text{if } q(1-p)+p < \dfrac{n'}{n} \leqslant 1. \end{cases} \tag{7.19}$$

(e) The asymptotical equivalence between $\hat{\pi}_{\mathrm{PM}}$ and $\hat{\pi}_{\mathrm{P}}$

Theorem 7.2 If $0 < \pi < 1$, then $\sqrt{n}(\hat{\pi}_{\mathrm{PM}} - \pi)$ and $\sqrt{n}(\hat{\pi}_{\mathrm{P}} - \pi)$ have the same asymptotic distribution as $n \to \infty$.   ¶

Proof. It suffices to show that $\sqrt{n}(\hat{\pi}_{\mathrm{PM}} - \pi) - \sqrt{n}(\hat{\pi}_{\mathrm{P}} - \pi)$ converges to zero in probability as $n \to \infty$; i.e.,

$$\Pr\{|\sqrt{n}(\hat{\pi}_{\mathrm{PM}} - \hat{\pi}_{\mathrm{P}})| > 0\} \to 0, \quad \text{as } n \to \infty. \tag{7.20}$$

Noting that $\hat{\lambda}$ is the MLE of $\lambda = q(1-p) + \pi p$ and that $q(1-p) < \lambda < q(1-p) + p$ since $0 < \pi < 1$, we readily obtain $\Pr(|\hat{\lambda} - \lambda| > \varepsilon) \to 0$, as $n \to \infty$, for any $\varepsilon > 0$. We only need to prove

$$\Pr\{|\sqrt{n}(\hat{\pi}_{\mathrm{PM}} - \hat{\pi}_{\mathrm{P}})| > 0\} \leqslant \Pr(|\hat{\lambda} - \lambda| > \varepsilon),$$

or equivalently

$$\{|\sqrt{n}(\hat{\pi}_{\mathrm{PM}} - \hat{\pi}_{\mathrm{P}})| > 0\} \subseteq \{|\hat{\lambda} - \lambda| > \varepsilon\}, \tag{7.21}$$

for any $\varepsilon < \min\{q(1 - p) + p - \lambda, \ \lambda - q(1 - p)\}$. We consider three cases.

Case I: $q(1 - p) \leqslant \hat{\lambda} \leqslant q(1 - p) + p$. From (7.19), we obtain $\hat{\pi}_{\mathrm{PM}} = \hat{\pi}_{\mathrm{P}}$. Hence, (7.20) follows immediately.

Case II: $\hat{\lambda} < q(1 - p)$. Now $\hat{\pi}_{\mathrm{PM}} = 0$. If

$$|\sqrt{n}(\hat{\pi}_{\mathrm{PM}} - \hat{\pi}_{\mathrm{P}})| > 0$$

$$\Rightarrow \ \left|\sqrt{n}\left\{0 - \frac{\hat{\lambda} - q(1 - p)}{p}\right\}\right| > 0$$

$$\Rightarrow \ |\hat{\lambda} - q(1 - p)| > 0$$

$$\Rightarrow \ 0 < |\hat{\lambda} - q(1 - p)| = -\{\hat{\lambda} - q(1 - p)\}$$

$$= -(\hat{\lambda} - \lambda) - \{\lambda - q(1 - p)\}$$

$$\Rightarrow \ |\hat{\lambda} - \lambda| \geqslant -(\hat{\lambda} - \lambda) > \lambda - q(1 - p). \tag{7.22}$$

Noting that $\varepsilon < \min\{q(1 - p) + p - \lambda, \ \lambda - q(1 - p)\}$, we have

$$\{\lambda - q(1 - p)\} - \varepsilon > 0. \tag{7.23}$$

By combining (7.22) with (7.23), we obtain

$$|\hat{\lambda} - \lambda| - \varepsilon = |\hat{\lambda} - \lambda| - \{\lambda - q(1 - p)\} + \{\lambda - q(1 - p)\} - \varepsilon > 0$$

and hence (7.21) follows.

Case III: $\hat{\lambda} > q(1 - p) + p$. Now $\hat{\pi}_{\mathrm{PM}} = 1$. If

$$|\sqrt{n}(\hat{\pi}_{\mathrm{PM}} - \hat{\pi}_{\mathrm{P}})| > 0$$

$$\Rightarrow \ \left|\sqrt{n}\left\{1 - \frac{\hat{\lambda} - q(1 - p)}{p}\right\}\right| > 0$$

$$\Rightarrow \ |q(1 - p) + p - \hat{\lambda}| > 0$$

$$\Rightarrow \ 0 < |q(1 - p) + p - \hat{\lambda}| = -\{q(1 - p) + p - \hat{\lambda}\}$$

$$= -(\lambda - \hat{\lambda}) - \{q(1 - p) + p - \lambda\}$$

$$\Rightarrow \ |\lambda - \hat{\lambda}| \geqslant -(\lambda - \hat{\lambda}) > q(1 - p) + p - \lambda. \tag{7.24}$$

Noting that $\varepsilon < \min\{q(1-p)+p-\lambda,\ \lambda-q(1-p)\}$, we have

$$\{q(1-p)+p-\lambda\} - \varepsilon > 0. \tag{7.25}$$

By combining (7.24) and (7.25), we obtain

$$|\hat{\lambda}-\lambda| - \varepsilon = |\hat{\lambda}-\lambda| - \{q(1-p)+p-\lambda\} + \{q(1-p)+p-\lambda\} - \varepsilon > 0.$$

Hence, (7.21) follows.   □

7.3   Comparison with the Crosswise Model

One of the objectives of this section is to theoretically show why the unrelated question model is more popular than the Warner model in practice. Since the parallel model and the crosswise model are the non-randomized versions of the unrelated question model and the Warner model, respectively, it suffices to show that the parallel model is always more efficient than the crosswise model for most of the possible parameter values (see Corollary 7.1 below). To this end, we first present the following properties on a parabola by omitting the detailed proof. Figure 7.5 gives an illustration for these results.

Lemma 7.1   Let $a > 0$ and $D(f) = b^2 - 4ac$ denote the discriminant of a parabola $f(x) = ax^2 + bx + c$.

(1)   If $D(f) < 0$, then $f(x) > 0$ for all $x \in (-\infty, \infty)$.

(2)   If $D(f) = 0$, then $f(x) \geqslant 0$ for all $x \in (-\infty, \infty)$, and $f(x)$ reaches its minimum zero at $x = -b/(2a)$.

(3)   If $D(f) > 0$, then

$$f(x) \begin{cases} > 0, & \forall\ x \in (-\infty, x_1) \cup (x_2, \infty), \\ \leqslant 0, & \forall\ x \in [x_1, x_2], \end{cases}$$

   where

$$x_1 = \frac{-b - \sqrt{D(f)}}{2a} \quad \text{and} \quad x_2 = \frac{-b + \sqrt{D(f)}}{2a}. \qquad ¶$$

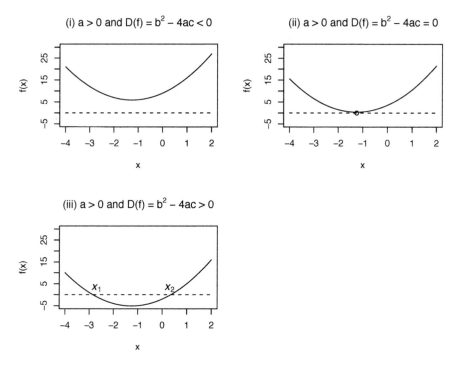

Figure 7.5 Three plots of a parabola $f(x) = ax^2 + bx + c$ with $a > 0$ for different values of the discriminant $D(f) = b^2 - 4ac$. (i) $f(x) > 0$ if $D(f) < 0$; (ii) $f(x) \geqslant 0$ and $f(x)$ has a unique root $x = -b/(2a)$ if $D(f) = 0$; (iii) $f(x)$ has two roots $x_1 = \{-b - \sqrt{D(f)}\}/(2a)$ and $x_2 = \{-b + \sqrt{D(f)}\}/(2a)$ if $D(f) > 0$.

7.3.1 The difference between variances

In the crosswise model there are only two parameters (i.e., π and p), while in the parallel model there is an additional parameter q besides π and p. By selecting q, we can control the value of the variance of $\hat{\pi}_{\mathrm{P}}$.

We first consider the difference between the variance of $\hat{\pi}_{\mathrm{C}}$ and the variance of $\hat{\pi}_{\mathrm{P}}$. From (2.9) and (7.10), we have

$$\mathrm{Var}(\hat{\pi}_{\mathrm{C}}) - \mathrm{Var}(\hat{\pi}_{\mathrm{P}}) = \frac{1 - p}{n} \left\{ \frac{p}{(2p - 1)^2} - \frac{f_{\mathrm{U}}(q|\pi, p)}{p^2} \right\},$$

$$= \frac{1 - p}{np^2(2p - 1)^2} \times h_{\mathrm{CP}}(q|\pi, p), \qquad (7.26)$$

where $p \neq 0.5$ and

$$h_{\mathrm{CP}}(q|\pi, p) = (1 - p)(2p - 1)^2 \times q^2 + (2\pi p - 1)(2p - 1)^2 \times q$$
$$+ \{p^3 - \pi p(2p - 1)^2\}$$

is a quadratic function of q with $(1-p)(2p-1)^2 > 0$ when both π and p are fixed. The discriminant of the parabola $h_{\mathrm{CP}}(q|\pi, p)$ is defined by

$$
\begin{aligned}
D(h_{\mathrm{CP}}) &= (2\pi p - 1)^2(2p-1)^4 - 4(1-p)(2p-1)^2\{p^3 - \pi p(2p-1)^2\} \\
&= 4p^2(2p-1)^4 \times \delta_{\mathrm{CP}}(\pi|p),
\end{aligned}
\tag{7.27}
$$

where

$$
\delta_{\mathrm{CP}}(\pi|p) = \pi^2 - \pi + \frac{1}{4p^2} - \frac{p(1-p)}{(2p-1)^2}.
\tag{7.28}
$$

We can see that both $D(h_{\mathrm{CP}})$ and $\delta_{\mathrm{CP}}(\pi|p)$ share signs. By applying Lemma 7.1, we immediately obtain the following results.

Theorem 7.3 Let $\pi \in (0,1)$, we have

(1) When

$$
p \in \{p:\ 0 < p < 1,\ p \neq 0.5 \text{ and } \delta_{\mathrm{CP}}(\pi|p) \leqslant 0\},
\tag{7.29}
$$

the parallel model is more efficient than the crosswise model for any $q \in (0,1)$, where $\delta_{\mathrm{CP}}(\pi|p)$ is defined by (7.28).

(2) When

$$
p \in \{p:\ 0 < p < 1,\ p \neq 0.5 \text{ and } \delta_{\mathrm{CP}}(\pi|p) > 0\},
\tag{7.30}
$$

the parallel model is always more efficient than the crosswise model for any q in $(0, q_{\mathrm{CP,L}})$ or in $(q_{\mathrm{CP,U}}, 1)$, where

$$
q_{\mathrm{CP,L}} = \max\left\{0,\ \frac{1 - 2\pi p - 2p\sqrt{\delta_{\mathrm{CP}}(\pi|p)}}{2(1-p)}\right\} \quad \text{and}
\tag{7.31}
$$

$$
q_{\mathrm{CP,U}} = \min\left\{1,\ \frac{1 - 2\pi p + 2p\sqrt{\delta_{\mathrm{CP}}(\pi|p)}}{2(1-p)}\right\},
\tag{7.32}
$$

respectively.   ¶

Next we investigate $\delta_{\mathrm{CP}}(\pi|p)$ for more detail.

Lemma 7.2 For the quadratic function $\delta_{\mathrm{CP}}(\pi|p)$ defined in (7.28), we have the following conclusions.

(1) If $p < 1/3$, then $\delta_{\mathrm{CP}}(\pi|p) > 0$ for all $\pi \in (0,1)$.

(2) If $p = 1/3$, then $\delta_{\mathrm{CP}}(\pi|p) = \delta_{\mathrm{CP}}(\pi|1/3) = (\pi - 0.5)^2 \geqslant 0$ for all $\pi \in (0,1)$, and $\delta_{\mathrm{CP}}(\pi|p)$ reaches its minimum zero at $\pi = 0.5$.

(3) If $p > 1/3$, then

$$\delta_{CP}(\pi|p) \begin{cases} > 0, & \forall\ \pi \in (0, \pi_{CP,L}) \cup (\pi_{CP,U}, 1), \\ \leqslant 0, & \forall\ \pi \in [\pi_{CP,L}, \pi_{CP,U}], \end{cases}$$

where

$$\pi_{CP,L} = \max\left\{0, \frac{1}{2} - \frac{1}{2}\sqrt{\frac{(1-p)(3p-1)}{p^2(2p-1)^2}}\right\} \quad \text{and} \quad (7.33)$$

$$\pi_{CP,U} = \min\left\{1, \frac{1}{2} + \frac{1}{2}\sqrt{\frac{(1-p)(3p-1)}{p^2(2p-1)^2}}\right\}, \quad (7.34)$$

respectively. ¶

Proof. The discriminant of the quadratic function $\delta_{CP}(\pi|p)$ is

$$D(\delta_{CP}) = 1 - 4\left\{\frac{1}{4p^2} - \frac{p(1-p)}{(2p-1)^2}\right\} = \frac{(1-p)(3p-1)}{p^2(2p-1)^2}.$$

Therefore, $D(\delta_{CP}) < 0$ if and only if $p < 1/3$. By using Lemma 7.1, we immediately obtain all three conclusions. □

By combining Theorem 7.3 with Lemma 7.2, we obtain the following results.

Corollary 7.1 Let $\pi \in (0, 1)$ and $p \neq 0.5$, we have

(1) When $p > 1/3$, the parallel model is more efficient than the crosswise model for any $q \in (0, 1)$ and $\pi \in [\pi_{CP,L}, \pi_{CP,U}]$, where $\pi_{CP,L}$ and $\pi_{CP,U}$ are given by (7.33) and (7.34), respectively.

(2) When $p < 1/3$, the parallel model is always more efficient than the crosswise model for any $\pi \in (0, 1)$ and $q \in (0, q_{CP,L}) \cup (q_{CP,U}, 1)$, where $q_{CP,L}$ and $q_{CP,U}$ are given by (7.31) and (7.32), respectively.

(3) When $p = 1/3$ and $\pi = 0.5$, the parallel model is more efficient than the crosswise model for any $q \in (0, 0.5) \cup (0.5, 1)$, while the two models are equivalent for $q = 0.5$.

(4) When $p = 1/3$ and $\pi \neq 0.5$, the parallel model is always more efficient than the crosswise model for any $q \in (0, q^*_{CP,L}) \cup (q^*_{CP,U}, 1)$, where

$$q^*_{CP,L} = \frac{3 - 2\pi - 2|\pi - 0.5|}{4} > 0 \quad \text{and} \quad (7.35)$$

$$q^*_{CP,U} = \frac{3 - 2\pi + 2|\pi - 0.5|}{4} < 1, \quad (7.36)$$

respectively.   ¶

Proof. (1) When $p > 1/3$ and $\pi \in [\pi_{\mathrm{CP,\,L}}, \pi_{\mathrm{CP,\,U}}]$, from Lemma 7.2 (3), we have $\delta_{\mathrm{CP}}(\pi|p) \leqslant 0$, which satisfies (7.29). Assertion (1) follows from Theorem 7.3 (1).

(2) When $p < 1/3$, from Lemma 7.2 (1), we have $\delta_{\mathrm{CP}}(\pi|p) > 0$, which satisfies (7.30). Assertion (2) follows from Theorem 7.3 (2).

(3) When $p = 1/3$ and $\pi = 0.5$, from Lemma 7.2 (2), we have $\delta_{\mathrm{CP}}(\pi|p) = 0$. Hence,

$$D(h_{\mathrm{CP}}) = 0 \quad \text{and} \quad h_{\mathrm{CP}}(q|\pi, p) = h_{\mathrm{CP}}\left(q\left|\frac{1}{2}, \frac{1}{3}\right.\right) = \frac{(2q-1)^2}{54}.$$

In this case, (7.26) becomes

$$\mathrm{Var}(\hat{\pi}_{\mathrm{C}}) - \mathrm{Var}(\hat{\pi}_{\mathrm{P}}) = \frac{(2q-1)^2}{n}\begin{cases} > 0, & \text{if } q \neq \dfrac{1}{2}, \\[2mm] = 0, & \text{if } q = \dfrac{1}{2}. \end{cases}$$

Assertion (3) follows immediately.

(4) When $p = 1/3$ and $\pi \neq 0.5$, from Lemma 7.2 (2), we have

$$\delta_{\mathrm{CP}}(\pi|p) = \left(\pi - \frac{1}{2}\right)^2 > 0,$$

which satisfies (7.30). Let $p = 1/3$ in (7.31) and (7.32), we obtain

$$q^*_{\mathrm{CP,\,L}} = \max\left(0, \frac{3 - 2\pi - 2|\pi - 0.5|}{4}\right) \quad \text{and}$$

$$q^*_{\mathrm{CP,\,U}} = \min\left(1, \frac{3 - 2\pi + 2|\pi - 0.5|}{4}\right).$$

On the one hand, from

$$\begin{aligned} 1 > \pi \quad &\Rightarrow \quad 8 > 8\pi \\ &\Rightarrow \quad 9 - 12\pi + 4\pi^2 > 4\pi^2 - 4\pi + 1 \\ &\Rightarrow \quad (3 - 2\pi)^2 > 4(\pi - 0.5)^2 \\ &\Rightarrow \quad 3 - 2\pi > 2|\pi - 0.5| \\ &\Rightarrow \quad 3 - 2\pi - 2|\pi - 0.5| > 0 \\ &\Rightarrow \quad \frac{3 - 2\pi - 2|\pi - 0.5|}{4} > 0, \end{aligned}$$

which implies (7.35). On the other hand, from

$$0 < \pi \Rightarrow 0 < 8\pi$$
$$\Rightarrow -4\pi + 1 < 1 + 4\pi$$
$$\Rightarrow 4(\pi^2 - \pi + 1/4) < 1 + 4\pi + 4\pi^2$$
$$\Rightarrow 4(\pi - 0.5)^2 < (1 + 2\pi)^2$$
$$\Rightarrow 2|\pi - 0.5| < 1 + 2\pi$$
$$\Rightarrow 3 - 2\pi + 2|\pi - 0.5| < 4$$
$$\Rightarrow \frac{3 - 2\pi + 2|\pi - 0.5|}{4} < 1,$$

which implies (7.36). Therefore, Assertion (4) follows from Theorem 7.3 (2).
□

Remark 7.1 For any $1/3 < p \leqslant 2/3$ $(p \neq 0.5)$, we have $\pi_{\mathrm{CP,\,L}} = 0$ and $\pi_{\mathrm{CP,\,U}} = 1$. Based on Corollary 7.1 (1), we assert that the parallel model is more efficient than the crosswise model for any $q \in (0, 1)$ and any $\pi \in (0, 1)$.
¶

7.3.2 Relative efficiency of the crosswise model to the parallel model

The criterion of variance difference considered in the previous subsection only concerns the sign instead of the magnitude. For example, let $\mathrm{Var}(\hat{\pi}_{\mathrm{C}}) = 3.0001$ and $\mathrm{Var}(\hat{\pi}_{\mathrm{P}}) = 3.0$. Although the sign of their difference is positive, it is very clear that the two models are approximately equivalent according to the relative efficiency.

Table 7.3 Relative efficiency $\mathrm{RE}_{\mathrm{C}\to\mathrm{P}}(\pi, q, p)$ for various combinations of π, q and p

π	q	p				
		0.34	0.40	0.51	0.60	2/3
0.05	1/3	1.4313	5.6387	1060.9	15.931	7.3636
0.10		1.3918	5.3421	965.12	14.057	6.3547
0.20		1.3256	4.8889	833.83	11.723	5.1979
0.30		1.2716	4.5662	751.63	10.390	4.5829
0.50		1.1863	4.1667	668.07	9.1629	4.0500
0.70		1.1183	3.9808	651.29	9.0453	4.0263
0.90		1.0578	3.9545	691.48	9.9675	4.5216
0.95		1.0430	3.9721	712.31	10.433	4.7833
0.05	0.50	1.1422	4.4467	823.54	12.293	5.6875
0.10		1.1392	4.3422	779.91	11.395	5.1925
0.20		1.1345	4.1834	717.32	10.191	4.5714
0.30		1.1313	4.0788	678.44	9.4890	4.2297
0.50		1.1289	4.0000	650.25	9.0000	4.0000
0.70		1.1313	4.0788	678.44	9.4890	4.2297
0.90		1.1392	4.3422	779.91	11.395	5.1925
0.95		1.1422	4.4467	823.54	12.293	5.6875
0.05	2/3	1.0430	3.9721	712.31	10.433	4.7833
0.10		1.0578	3.9545	691.48	9.9675	4.5216
0.20		1.0876	3.9487	663.66	9.3508	4.1897
0.30		1.1183	3.9808	651.29	9.0453	4.0263
0.50		1.1863	4.1667	668.07	9.1629	4.0500
0.70		1.2716	4.5662	751.63	10.390	4.5829
0.90		1.3918	5.3421	965.12	14.057	6.3547
0.95		1.4313	5.6387	1060.9	15.931	7.3636

Note: $\mathrm{RE}_{\mathrm{C}\to\mathrm{P}}(1-\pi, 1-q, p) = \mathrm{RE}_{\mathrm{C}\to\mathrm{P}}(\pi, q, p)$.

The RE of the crosswise model ($p \neq 0.5$) to the parallel model is

$$\mathrm{RE}_{\mathrm{C}\to\mathrm{P}}(\pi, q, p) = \frac{\mathrm{Var}(\hat{\pi}_{\mathrm{C}})}{\mathrm{Var}(\hat{\pi}_{\mathrm{P}})}$$

$$= \frac{\pi(1-\pi) + p(1-p)/(2p-1)^2}{\pi(1-\pi) + (1-p)f_{\mathrm{U}}(q|\pi, p)/p^2},$$

which is free of the sample size n and depends only on the parameters π, q and p. It is easy to verify that

$$\mathrm{RE}_{\mathrm{C}\to\mathrm{P}}(1-\pi, 1-q, p) = \mathrm{RE}_{\mathrm{C}\to\mathrm{P}}(\pi, q, p).$$

From Table 7.3, when $p = 2/3$ (or 0.60), the efficiency of the parallel model is 4–7 (or 9–16) times that of the crosswise model for any $q \in [1/3, 2/3]$. In particular, when $p = 0.51$, the efficiency of the parallel model is about 650–1060 times that of the crosswise model. Finally, when $p = 0.34$ the two models are approximately equivalent in terms of RE.

7.3.3 Degree of privacy protection

The goal of this subsection is to compare the DPP of the crosswise model with that of the parallel model. Because of the equivalence between the crosswise model and the Warner model in terms of DPP, we only need to compare the DPP of the Warner model with that of the parallel model. From (2.6) and (7.6), we have

$$
\begin{aligned}
\text{DPP}_{\text{yes}}(\pi, p) &= \frac{\pi p}{\pi p + (1 - \pi)(1 - p)} \\
&> \frac{\pi p}{\pi p + q(1 - p)} \\
&= \text{DPP}_{\text{yes}}(\pi, q, p) \quad (7.37)
\end{aligned}
$$

if and only if $q > 1 - \pi$ for any $\pi \in (0, 1)$ and any $p \in (0, 1)$. In addition, From (2.7) and (7.7), we always have

$$
\begin{aligned}
\text{DPP}_{\text{no}}(\pi, p) &= \frac{\pi(1 - p)}{\pi(1 - p) + (1 - \pi)p} \\
&> 0 \\
&= \text{DPP}_{\text{no}}(\pi, q, p).
\end{aligned}
$$

7.4 Comparison with the Triangular Model

7.4.1 The difference between variances

Let $p \in (0, 1)$. From (3.3) and (7.10), we have

$$
\begin{aligned}
\text{Var}(\hat{\pi}_{\text{T}}) - \text{Var}(\hat{\pi}_{\text{P}}) &= \frac{1}{n} \left\{ \frac{p(1 - \pi)}{1 - p} - \frac{(1 - p)f_{\text{U}}(q|\pi, p)}{p^2} \right\} \\
&= \frac{1}{np^2(1 - p)} \times h_{\text{TP}}(q|\pi, p), \quad (7.38)
\end{aligned}
$$

where

$$h_{\text{TP}}(q|\pi, p) = (1-p)^3 \times q^2 + (2\pi p - 1)(1-p)^2 \times q$$
$$+ \{p^3(1-\pi) - \pi p(1-p)^2\}$$

is a quadratic function of q with $(1-p)^3 > 0$ when both π and p are fixed. The discriminant of the parabola $h_{\text{TP}}(q|\pi, p)$ is

$$D(h_{\text{TP}}) = (2\pi p - 1)^2(1-p)^4 - 4(1-p)^3\{p^3(1-\pi) - \pi p(1-p)^2\}$$
$$= 4p^2(1-p)^4 \times \delta_{\text{TP}}(\pi|p), \tag{7.39}$$

where

$$\delta_{\text{TP}}(\pi|p) = \pi^2 - \frac{1-2p}{1-p} \cdot \pi + \frac{1}{4p^2} - \frac{p}{1-p}. \tag{7.40}$$

We can see that both $D(h_{\text{TP}})$ and $\delta_{\text{TP}}(\pi|p)$ share signs. By applying Lemma 7.1, we immediately obtain the following results.

Theorem 7.4 Let $\pi \in (0, 1)$, we have

(1) When

$$p \in \{p\colon 0 < p < 1 \text{ and } \delta_{\text{TP}}(\pi|p) \leqslant 0\}, \tag{7.41}$$

the parallel model is more efficient than the triangular model for any $q \in (0, 1)$, where $\delta_{\text{TP}}(\pi|p)$ is defined by (7.40).

(2) When

$$p \in \{p\colon 0 < p < 1 \text{ and } \delta_{\text{TP}}(\pi|p) > 0\}, \tag{7.42}$$

the parallel model is more efficient than the triangular model for any q in $(0, q_{\text{TP, L}})$ or in $(q_{\text{TP, U}}, 1)$, where

$$q_{\text{TP, L}} = \max\left\{0, \frac{1 - 2\pi p - 2p\sqrt{\delta_{\text{TP}}(\pi|p)}}{2(1-p)}\right\} \quad \text{and} \tag{7.43}$$

$$q_{\text{TP, U}} = \min\left\{1, \frac{1 - 2\pi p + 2p\sqrt{\delta_{\text{TP}}(\pi|p)}}{2(1-p)}\right\}, \tag{7.44}$$

respectively.

(3) When (7.42) holds, the triangular model is more efficient than the parallel model for any $q \in [q_{\text{TP, L}}, q_{\text{TP, U}}]$. ¶

Next we investigate $\delta_{\text{TP}}(\pi|p)$ for more detail.

Lemma 7.3 For the quadratic function $\delta_{\mathrm{TP}}(\pi|p)$ defined in (7.40), we have the following conclusions.

(1) If $p < 0.5$, then $\delta_{\mathrm{TP}}(\pi|p) > 0$ for all $\pi \in (0, 1)$.

(2) If $p = 0.5$, then $\delta_{\mathrm{TP}}(\pi|p) = \delta_{\mathrm{TP}}(\pi|0.5) = \pi^2 > 0$ for all $\pi \in (0, 1)$.

(3) If $p > 0.5$, then

$$\delta_{\mathrm{TP}}(\pi|p) \begin{cases} > 0, & \forall\ \pi \in (\pi_{\mathrm{TP,\,U}}, 1), \\ \leqslant 0, & \forall\ \pi \in (0, \pi_{\mathrm{TP,\,U}}], \end{cases}$$

where

$$\pi_{\mathrm{TP,\,U}} = \frac{1 - 2p + \sqrt{2p - 1}/p}{2(1 - p)}, \tag{7.45}$$

and $0 < \pi_{\mathrm{TP,\,U}} < 1$. ¶

Proof. (1) and (2). The discriminant of the quadratic function $\delta_{\mathrm{TP}}(\pi|p)$ is

$$D(\delta_{\mathrm{TP}}) = \left(\frac{1 - 2p}{1 - p}\right)^2 - 4\left(\frac{1}{4p^2} - \frac{p}{1 - p}\right) = \frac{2p - 1}{p^2(1 - p)^2}.$$

Therefore, $D(\delta_{\mathrm{TP}}) < 0$ if and only if $p < 0.5$. By using Lemma 7.1, we immediately obtain the first and second conclusions.

(3) If $p > 0.5$, then $\delta_{\mathrm{TP}}(\pi|p) = 0$ may have the following two roots:

$$\pi_{\mathrm{TP,\,L}} = \frac{1 - 2p - \sqrt{2p - 1}/p}{2(1 - p)} \quad \text{and}$$

$$\pi_{\mathrm{TP,\,U}} = \frac{1 - 2p + \sqrt{2p - 1}/p}{2(1 - p)}.$$

It is easy to verify that $\pi_{\mathrm{TP,\,L}} < 0$. We can also verify that $0 < \pi_{\mathrm{TP,\,U}} < 1$. On the one hand, from

$$\begin{aligned}
0.5 < p < 1 &\Rightarrow\ 0 < 2p - 1 < 1 \\
&\Rightarrow\ p\sqrt{2p - 1} < p < 1 \\
&\Rightarrow\ \sqrt{2p - 1} < 1/p \\
&\Rightarrow\ 2p - 1 < \sqrt{2p - 1}/p \\
&\Rightarrow\ 0 < 1 - 2p + \sqrt{2p - 1}/p \\
&\Rightarrow\ 0 < \frac{1 - 2p + \sqrt{2p - 1}/p}{2(1 - p)} = \pi_{\mathrm{TP,\,U}}.
\end{aligned}$$

On the other hand, from

$$(p-1)^2 > 0 \text{ and } p > 0.5 \quad \Rightarrow \quad p^2 > 2p - 1 > 0$$

$$\Rightarrow \quad p > \sqrt{2p-1}$$

$$\Rightarrow \quad 1 > \sqrt{2p-1}/p$$

$$\Rightarrow \quad 2 - 2p > 1 - 2p + \sqrt{2p-1}/p$$

$$\Rightarrow \quad 1 > \frac{1 - 2p + \sqrt{2p-1}/p}{2(1-p)} = \pi_{\text{TP, U}}.$$

Therefore, the third conclusion follows from Lemma 7.1 (3).   □

Combining Theorem 7.4 with Lemma 7.3, we obtain the following results.

Corollary 7.2 Let $\pi \in (0,1)$ and $p \in (0,1)$, we have

(1) The parallel model is always more efficient than the triangular model if one of the following three conditions is satisfied:

   (a1) $p \leqslant 0.5$, $\pi \in (0,1)$ and $q \in (0, q_{\text{TP, L}}) \cup (q_{\text{TP, U}}, 1)$, where $q_{\text{TP, L}}$ and $q_{\text{TP, U}}$ are given by (7.43) and (7.44), respectively.

   (b1) $p > 0.5$, $\pi \in (0, \pi_{\text{TP, U}}]$ and $q \in (0,1)$, where $\pi_{\text{TP, U}}$ is given by (7.45).

   (c1) $p > 0.5$, $\pi \in (\pi_{\text{TP, U}}, 1)$ and $q \in (0, q_{\text{TP, L}}) \cup (q_{\text{TP, U}}, 1)$.

(2) The triangular model is more efficient than the parallel model if one of the following two conditions is satisfied:

   (a2) $p \leqslant 0.5$, $\pi \in (0,1)$ and $q \in [q_{\text{TP, L}}, q_{\text{TP, U}}]$.

   (b2) $p > 0.5$, $\pi \in (\pi_{\text{TP, U}}, 1)$ and $q \in [q_{\text{TP, L}}, q_{\text{TP, U}}]$.   ¶

7.4.2   Relative efficiency of the triangular model to the parallel model

The RE of the triangular model to the parallel model is

$$\text{RE}_{\text{T} \to \text{P}}(\pi, q, p) = \frac{\text{Var}(\hat{\pi}_{\text{T}})}{\text{Var}(\hat{\pi}_{\text{P}})}$$

$$= \frac{\pi(1-\pi) + p(1-\pi)/(1-p)}{\pi(1-\pi) + (1-p)f_{\text{U}}(q|\pi, p)/p^2},$$

Table 7.4 Relative efficiency $\mathrm{RE}_{\mathrm{T}\to\mathrm{P}}(\pi, q, p)$ for various combinations of π, q and p

π	q	p				
		1/3	0.40	0.50	0.60	2/3
0.05	1/3	0.31930	0.63481	1.60959	3.87910	7.00400
0.10		0.31538	0.60526	1.45827	3.32402	5.74662
0.20		0.30288	0.55026	1.22727	2.58835	4.23529
0.30		0.28490	0.49754	1.05135	2.10824	3.33871
0.50		0.23376	0.38888	0.77143	1.46606	2.25000
0.70		0.16127	0.26282	0.51057	0.96134	1.47571
0.90		0.06234	0.10173	0.20094	0.39281	0.62740
0.95		0.03241	0.05309	0.10601	0.21135	0.34458
0.05	0.50	0.25518	0.50061	1.25078	2.99322	5.40972
0.10		0.25837	0.49197	1.17857	2.69439	4.69565
0.20		0.25925	0.47085	1.05495	2.25000	3.72487
0.30		0.25339	0.44444	0.94792	1.92530	3.08134
0.50		0.22222	0.37333	0.75000	1.44000	2.22222
0.70		0.16289	0.26929	0.53125	1.00849	1.55024
0.90		0.06698	0.11170	0.22619	0.44906	0.72050
0.95		0.03540	0.05943	0.12226	0.24901	0.40972
0.05	2/3	0.23363	0.44718	1.08457	2.54055	4.54964
0.10		0.24047	0.44805	1.04700	2.35684	4.08894
0.20		0.24901	0.44444	0.97738	2.06447	3.41379
0.30		0.25087	0.43376	0.91101	1.83528	2.93320
0.50		0.23376	0.38888	0.77143	1.46606	2.25000
0.70		0.18315	0.30147	0.58922	1.10432	1.67972
0.90		0.08176	0.13742	0.27987	0.55400	0.88176
0.95		0.04430	0.07536	0.15733	0.32271	0.53047

which is free of the sample size n and depends only on the parameters π, q and p.

Table 7.4 reports some values $\mathrm{RE}_{\mathrm{T}\to\mathrm{P}}(\pi, q, p)$ for various combinations of π, q and p. For example, when $\pi = 0.05$, $q = 0.50$, and $p = 0.60$, we have $\mathrm{RE}_{\mathrm{T}\to\mathrm{P}}(0.05, 0.50, 0.60) = 2.99322$, implying that the efficiency of the parallel model is about 3 times that of the triangular model. In particular, when $\pi = 0.90$, $q = 0.50$ and $p = 0.50$, the efficiency of the triangular model is about $1/0.22619 = 4.42$ times that of the parallel model. Finally, when

$\pi = 0.30$, $q = 1/3$, $p = 0.50$ the two models are approximately equivalent in terms of RE.

7.4.3 Degree of privacy protection

The goal of this subsection is to compare the DPP of the triangular model with that of the parallel model. From (3.6) and (7.7), we always have

$$\mathrm{DPP}_{\bigcirc}(\pi, p) = 0 = \mathrm{DPP}_{\mathrm{no}}(\pi, q, p).$$

On the other hand, from (3.7) and (7.6), we have

$$\mathrm{DPP}_{\triangle}(\pi, p) \;=\; \frac{\pi}{\pi + (1 - \pi)p} > \frac{\pi}{\pi + q(1 - p)/p}$$

$$=\; \frac{\pi p}{\pi p + q(1 - p)} = \mathrm{DPP}_{\mathrm{yes}}(\pi, q, p) \qquad (7.46)$$

if and only if $q > (1 - \pi)p^2/(1 - p)$ for any $\pi \in (0, 1)$ and any $p \in (0, 1)$.

7.5 Bayesian Inferences

In this section, we first derive the exact posterior distribution of π and its explicit posterior moments. We then derive the posterior mode via the EM algorithm (Dempster, Laird & Rubin, 1977) when the posterior distribution of π is highly skewed. Finally, we utilize the exact *inverse Bayes formulae* (IBF) sampler (Tian, Tan & Ng, 2007) to generate i.i.d. posterior samples.

7.5.1 Posterior moments in closed-form

From (7.8), the likelihood function of π based on the the observed data $Y_{\mathrm{obs}} = \{n; \, n - n', n'\}$ becomes

$$L_{\mathrm{p}}(\pi|Y_{\mathrm{obs}}) = \{q(1 - p) + \pi p\}^{n'}\{(1 - q)(1 - p) + (1 - \pi)p\}^{n - n'},$$

where $0 < \pi < 1$. If a beta distribution $\mathrm{Beta}(a, b)$ is adopted as the prior distribution of π, then the posterior distribution of π has the following explicit expression:

$$f(\pi|Y_{\mathrm{obs}}) = \frac{\pi^{a-1}(1 - \pi)^{b-1}L_{\mathrm{p}}(\pi|Y_{\mathrm{obs}})}{c_{\mathrm{p}}(a, b; \, n - n', n')}, \qquad (7.47)$$

where the normalizing constant is given by

$$c_{\mathrm{p}}(a, b; \, n - n', n') \;=\; p^{n} \sum_{j_1=0}^{n'} \sum_{j_2=0}^{n-n'} \binom{n'}{j_1}\binom{n - n'}{j_2}\left(\frac{1 - p}{p}\right)^{j_1 + j_2}$$

$$\times \, q^{j_1}(1 - q)^{j_2} B(a + n' - j_1, \, b + n - n' - j_2).$$

Therefore, the r-th posterior moment of π is given by

$$E(\pi^r | Y_{\text{obs}}) = \frac{c_{\text{p}}(a + r, b; \ n - n', n')}{c_{\text{p}}(a, b; \ n - n', n')}, \quad r \geqslant 1. \tag{7.48}$$

7.5.2 Calculation of the posterior mode via the EM algorithm

To derive the posterior mode, we introduce two latent variables Z_0 and Z_1, where Z_0 is the frequency of $\{U = 0, W = 0\}$ and Z_1 denotes the frequency of $\{U = 1, W = 0\}$ (cf. Table 7.2), respectively. In addition, let $\mathbf{z} = (Z_0, Z_1)^\top$ denote the random vector and $z = (z_0, z_1)^\top$ denote the realization of $\mathbf{z}$. Thus, the likelihood function of π for the complete-data is

$$
\begin{aligned}
L_{\text{p}}(\pi | Y_{\text{obs}}, z) \ &= \ \{q(1 - p)\}^{z_1} (\pi p)^{n' - z_1} \\
&\quad \times \{(1 - q)(1 - p)\}^{z_0} \{(1 - \pi)p\}^{n - n' - z_0} \\
&\propto \ \pi^{n' - z_1} (1 - \pi)^{n - n' - z_0}.
\end{aligned}
$$

If the prior distribution $\text{Beta}(a, b)$ is adopted, the complete-data posterior distribution and the conditional predictive distribution are given by

$$
\begin{aligned}
f(\pi | Y_{\text{obs}}, z) \ &= \ \text{Beta}(\pi | a + n' - z_1, b + n - n' - z_0), \quad \text{and} \tag{7.49} \\
f(z | Y_{\text{obs}}, \pi) \ &= \ f(z_0 | Y_{\text{obs}}, \pi) \times f(z_1 | Y_{\text{obs}}, \pi) \\
&= \ \text{Binomial}\left(z_0 \Big| n - n', \ \frac{(1 - q)(1 - p)}{(1 - q)(1 - p) + (1 - \pi)p}\right) \\
&\quad \times \text{Binomial}\left(z_1 \Big| n', \ \frac{q(1 - p)}{q(1 - p) + \pi p}\right), \tag{7.50}
\end{aligned}
$$

respectively. Using the EM algorithm, the M-step yields the following complete-data posterior mode

$$\tilde{\pi}_{\text{p}} = \frac{a + n' - z_1 - 1}{a + b + n - z_0 - z_1 - 2} \tag{7.51}$$

while the E-step replaces z_0 and z_1 by the conditional expectations

$$E(Z_0 | Y_{\text{obs}}, \pi) \ = \ \frac{(n - n')(1 - q)(1 - p)}{(1 - q)(1 - p) + (1 - \pi)p} \quad \text{and} \tag{7.52}$$

$$E(Z_1 | Y_{\text{obs}}, \pi) \ = \ \frac{n'q(1 - p)}{q(1 - p) + \pi p}, \tag{7.53}$$

respectively.

7.5.3   Generation of i.i.d. posterior samples via the exact IBF sampling

To apply the exact IBF sampling (cf. Appendix B) to the present model, we simply need to identify the conditional support of $z|(Y_{\text{obs}}, \pi)$. From (7.50), we have

$$\mathcal{S}_{(z|Y_{\text{obs}})} = \mathcal{S}_{(z|Y_{\text{obs}}, \pi)} = \{z_1, \ldots, z_K\}$$

$$= \left\{ \begin{array}{cccc} (0,0) & (0,1) & \cdots & (0, n') \\ (1,0) & (1,1) & \cdots & (1, n') \\ \vdots & \vdots & \ddots & \vdots \\ (n - n', 0) & (n - n', 1) & \cdots & (n - n', n') \end{array} \right\},$$

where $K = (n - n' + 1)(n' + 1)$. We then calculate $\{\omega_k\}_{k=1}^K$ according to (B.2) and (B.3) with $\pi_0 = 0.5$.

7.6   An Example: Induced Abortion in Mexico

In most countries, induced abortion is prohibited because of cultural and religious taboos against pregnancy termination. As in most of Latin America, abortion is prohibited under most circumstances; exceptions under Mexican law vary from one state to another. Lara *et al.* (2004) compared four methods (i.e., face-to-face interview, audio computer-assisted self-interview, self-administered questionnaire, and a random response technique) of gathering information on abortion in Mexico. The objective is to estimate the rates of attempted induced abortion in Mexico for three samples:

1°   hospital patients in Mexico City,

2°   rural women in Chiapas, and

3°   women randomly chosen as part of a house-to-house survey in Mexico City.

They used the unrelated question model with the unrelated question: "Were you born in April?" (i.e., $q = 1/12$). The following sensitive question is presented to the participant: "Did you ever interrupt a pregnancy?"

With $p = 0.5$, the survey received $n' = 56$ answers of 'yes' among $n = 370$ women for the first sample. By using (7.2) and (7.3), the rate of attempted induced abortion in the hospital sample of Mexico City is estimated to be

Table 7.5 Four 95% confidence intervals of π

Type of CIs	Confidence interval	Width
Wald CI (7.13)	[0.14623, 0.29250]	0.14627
Wilson CI (7.16)	[0.15352, 0.29955]	0.14602
Bootstrap CI (7.17)	[0.14631, 0.29233]	0.14602
Bootstrap CI (7.18)	[0.14910, 0.29505]	0.14595

$\hat{\pi}_{\mathrm{U}} = 0.21937$ with estimated variance being

$$\widehat{\mathrm{Var}}(\hat{\pi}_{\mathrm{U}}) = 0.0013886.$$

Since the parallel model is a non-randomized version of the unrelated question model, from (7.9) and (7.10), we have

$$\hat{\pi}_{\mathrm{P}} = \hat{\pi}_{\mathrm{U}} = 0.21937 \quad \text{and} \quad \widehat{\mathrm{Var}}(\hat{\pi}_{\mathrm{P}}) = \widehat{\mathrm{Var}}(\hat{\pi}_{\mathrm{U}}) = 0.0013886.$$

From (7.11), the unbiased estimate of $\hat{\pi}_{\mathrm{P}}$ is

$$\overline{\mathrm{Var}}(\hat{\pi}_{\mathrm{P}}) = 0.0013923.$$

From (7.13) and (7.16), the 95% Wald confidence interval and Wilson confidence interval of π are listed in Table 7.5. From (7.17) and (7.18), two 95% bootstrap confidence intervals for π with $G = 20{,}000$ bootstrap replications are also reported in Table 7.5. The corresponding standard error of $\hat{\pi}_{\mathrm{P}}$ is estimated by

$$\widehat{\mathrm{se}}(\hat{\pi}_{\mathrm{P}}) = 0.03725.$$

We noted that the width of the bootstrap confidence interval specified by (7.18) is the shortest among the four 95% confidence intervals in Table 7.5.

To illustrate the Bayesian methods presented in Section 7.5, we consider the uniform prior (i.e., $a = b = 1$). Note that $p = 0.5$, $q = 1/12$ and the observed data $Y_{\mathrm{obs}} = \{n;\ n - n',\ n'\} = \{370;\ 314,\ 56\}$. Using (7.48), we obtain $E(\pi|Y_{\mathrm{obs}}) = 0.22312$ and $E(\pi^2|Y_{\mathrm{obs}}) = 0.051173$. Thus, $\mathrm{Var}(\pi|Y_{\mathrm{obs}}) = 0.0013914$ and the 95% Bayesian credible interval for π based on normality approximation is $[0.15001, 0.29623]$.

Using $\pi^{(0)} = 0.5$ as an initial value, the EM algorithm based on (7.51)–(7.53) converges in 10 iterations. The posterior mode of π is $\tilde{\pi}_{\mathrm{P}} = 0.21937$, which is identical to the MLE $\hat{\pi}_{\mathrm{P}}$. Using the exact IBF sampling described in Section 7.5.3, we generate $L = 20{,}000$ i.i.d. posterior samples from $f(\pi|Y_{\mathrm{obs}})$. The histogram based on these samples is plotted in Figure 7.6 (ii). Figure 7.6 (i) shows that the exact IBF sampling recovers the density completely. The corresponding posterior mean, standard error, and 95% Bayesian credible interval for π are 0.22258, 0.037428, and $[0.14922, 0.29593]$.

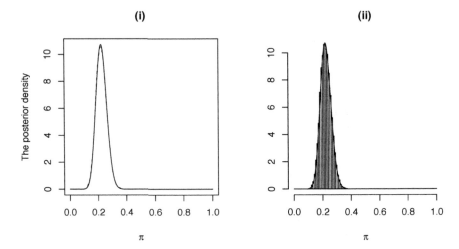

Figure 7.6   Posterior distribution of $\pi = \Pr(\text{interrupting a pregnancy})$ under the uniform prior (i.e., $a = b = 1$) with $p = \Pr(W = 1) = 0.5$ and $q = \Pr(U = 1) = 1/12$ for the parallel model for $n = 370$ and $n' = 56$. (i) The comparison between the posterior distribution (solid curve) exactly given by (7.47) with the dotted curve estimated by a kernel density smoother based on 20,000 i.i.d. posterior samples generated via the exact IBF sampling. (ii) The histogram based on 20,000 i.i.d. posterior samples generated via the exact IBF sampling.

7.7   A Case Study on College Students' Premarital Sexual Behavior at Wuhan

Because of the traditional restriction, premarital sexual activity is found to be rare in China before 1980. Since 1980, attitudes about sexual behavior have changed in China, and premarital sexual activity among adolescents has markedly increased. An earlier report stated that 15% of college students had participated in premarital sexual activity (Wu *et al.*, 1996). Similarly, an investigation of 50 universities in Shanghai showed that 16.8% of female students and 18.8% of male students had engaged in premarital sex (Li & Zhang, 1998).

   Due to sensitivity of this topic, under the supervision of the first author of this book, Miss Yin LIU used the non-randomized parallel model to conduct a survey in May 2011 among students at two universities in Wuhan, Hubei Province, P. R. China. The questions in the questionnaire are as follows:

(1)   If your birthday is in the first half of a year, please answer 'yes' or 'no' to the question: *Is the last digit of your cell phone number even?*

(2)   If your birthday is in the second half of a year, please answer 'yes' or 'no' to the question: *Have you ever had premarital sexual behavior?*

Table 7.6 Frequency distributions of 500 college students at two universities at Wuhan for sex

Sex	No. of 'yes' answer	No. of 'no' answer	No response	Total
Male	40	77	39	156
Female	64	164	29	257
Unknown	8	7	72	87
Total	112	248	140	500

Table 7.7 Frequency distributions of 500 college students at two universities at Wuhan for grade

Year of study	No. of 'yes' answer	No. of 'no' answer	No response	Total
Year 1	28	84	11	123
Year 2	45	108	39	192
Year 3	20	28	16	64
Year 4	9	19	1	29
Unknown	10	9	73	92
Total	112	248	140	500

At Central China Normal University, 250 students from School of Foreign Language, School of Mathematics and Statistics, School of Chinese Language and Literature, Department of Information and Management, School of Music, School of Chemistry, College of Physical Science and Technology, School of Fine Arts, and School of Sociology participated in the survey. At South-Central University for Nationalities, 250 students from School of Ethnology and Sociology, School of Economics, School of Management, School of Computer Science, School of Electronics and Information Engineering, School of Biological Science and Medical Engineering, and School of Life Sciences took part in the survey. Tables 7.6 and 7.7 report frequency distributions of 500 college students at the two universities based on sex and grade, where there are only 360 valid observations. Therefore, the observed data in Table 7.6 or Table 7.7 can be denoted by

$$Y_{\text{obs}} = \{n;\ n - n',\ n'\} = \{360;\ 248,\ 112\}$$

where $n - n' = 248$ implies 248 'no' answers (which is equivalent to 248 lines connecting the two circles in Table 7.2) and $n' = 112$ implies 112 'yes' answers (which is equivalent to 112 lines connecting the two squares in Table 7.2).

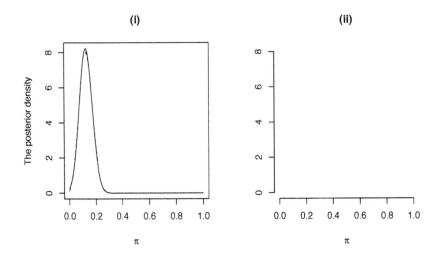

Figure 7.7 Posterior distribution of $\pi = \Pr(\text{having premarital sexual behavior})$ under the uniform prior (i.e., $a = b = 1$) and $p = \Pr(W = 1) \approx 0.5$ for the parallel model for $n = 360$ and $n' = 112$. (i) The comparison between the posterior distribution (solid curve) exactly given by (7.47) with the dotted curve estimated by a kernel density smoother based on 40,000 i.i.d. posterior samples generated via the exact IBF sampling. (ii) The histogram based on the 40,000 i.i.d. posterior samples generated via the exact IBF sampling.

In this survey, $p = \Pr(W = 1) = \Pr(\text{a respondent's birthday is in the second half of a year})$ is assumed to be 0.5, $q = \Pr(U = 1) = \Pr(\text{the last digit of a respondent's cell phone number is even})$ is also assumed to be 0.5. Let $\pi = \Pr(Y = 1) = \Pr(\text{having premarital sexual behavior})$. From (7.9) and (7.10), the MLE of π is $\hat{\pi}_{P} = 0.1222222$ and the estimated variance is 0.002381344. From (7.11), an unbiased estimate of $\text{Var}(\hat{\pi}_{P})$ is given by $\overline{\text{Var}}(\hat{\pi}_{P}) = 0.002387978$. From (7.13), the 95% Wald confidence interval of π is $[\hat{\pi}_{P,WL}, \hat{\pi}_{P,WU}] = [0.02644478, 0.2179997]$ with width being 0.1915549. From (7.16), the 95% Wilson confidence interval of π is $[\hat{\pi}_{P,WSL}, \hat{\pi}_{P,WSU}] = [0.03098918, 0.2214325]$ with width being 0.1904433, which is shorter than the width of the Wald confidence interval.

Alternatively, from (7.17) and (7.18), two 95% bootstrap confidence intervals of π with $G = 20,000$ bootstrap replications are $[0.026837, 0.217687]$ with width 0.19085 and $[0.02777778, 0.2166667]$ with width 0.1888889, respectively. The corresponding standard error of $\hat{\pi}_{P}$ is estimated by $\widehat{se}(\hat{\pi}_{P}) = 0.04868605$.

Now, we consider Bayesian inferences with the uniform prior (i.e., $a = b = 1$) on π. Using (7.48), we have $E(\pi|Y_{\text{obs}}) = 0.12487$ and $E(\pi^2|Y_{\text{obs}}) = 0.017890$. Thus, the 95% Bayesian credible interval for π based on normality approximation is given by $[0.030925, 0.218815]$.

Using $\pi^{(0)} = 0.5$ as an initial value, the EM algorithm (7.51)–(7.53) converges in 37 iterations. The posterior mode of π is $\tilde{\pi}_{\mathrm{P}} = 0.12222$, which is identical to the MLE $\hat{\pi}_{\mathrm{P}}$. Using the exact IBF sampling described in Section 7.5.3, we generate $L = 40{,}000$ i.i.d. posterior samples from $f(\pi|Y_{\mathrm{obs}})$. The histogram based on these samples is plotted in Figure 7.7 (ii). Figure 7.7 (i) shows that the exact IBF sampling can recover the density completely. The resultant posterior mean, standard error, and 95% Bayesian credible interval based on normality approximation for π are given by 0.12493, 0.047908, and [0.031031, 0.218829], respectively.

7.8 A Case Study on Plagiarism at The University of Hong Kong

Universities all over the world often define plagiarism in their regulations to prevent any misconduct in publications among their staffs and students. The University of Hong Kong (HKU) defines plagiarism as "the unacknowledged use, as one's own, of work of another person, whether or not such work has been published." If a student copies another person's work without acknowledgement, this student is committing plagiarism regardless of whether it is for an examination, assignment, project, thesis or dissertation.

Under the supervision of the first author of this book, a survey based on the non-randomized parallel model was conducted on plagiarism from September 17, 2012, to December 10, 2012, among students at HKU, Hong Kong, P. R. China.

Some questionnaires in printed form were distributed in the main campus of HKU and several lecture rooms. Lecture rooms were carefully chosen so that the data could be collected from students from various faculties and grades. Some questionnaires in electronic form were distributed via free online survey tools called Survey Monkey. Questions in the questionnaire include:

(1) If your birthday is in the first half of a year, please answer 'yes' or 'no' to the question: *Is the last digit of your mobile phone number even?*

(2) If your birthday is in the second half of a year, please answer 'yes' or 'no' to the question: *Have you ever committed plagiarism in assignments or class tests at HKU?*

Among 338 questionnaires, there were 185 'yes' answers and 153 'no' answers. Therefore, the observed data can be denoted by

$$Y_{\mathrm{obs}} = \{n;\ n - n',\ n'\} = \{338;\ 153,\ 185\}$$

where $n - n' = 153$ means 153 lines connecting the two circles in Table 7.2 and $n' = 185$ means 185 lines connecting the two squares in Table 7.2.

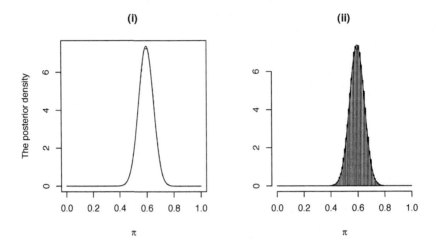

Figure 7.8 Posterior distribution of $\pi = \Pr(\text{having plagiarism behavior})$ under the uniform prior (i.e., $a = b = 1$) and $p = \Pr(W = 1) \approx 0.5$ for the parallel model for $n = 338$ and $n' = 185$. (i) The comparison between the posterior distribution (solid curve) exactly given by (7.47) with the dotted curve estimated by a kernel density smoother based on 40,000 i.i.d. posterior samples generated via the exact IBF sampling. (ii) The histogram based on the 40,000 i.i.d. posterior samples generated via the exact IBF sampling.

In this survey, $p = \Pr(W = 1) = \Pr(\text{a respondent's birthday is in the second half of a year})$ is assumed to be 0.5, $q = \Pr(U = 1) = \Pr(\text{the last digit of a respondent's cell phone number is even})$ is also assumed to be 0.5. Let $\pi = \Pr(Y = 1) = \Pr(\text{having plagiarism behavior})$. From (7.9) and (7.10), the MLE of π is $\hat{\pi}_{\text{P}} = 0.5946746$ and the estimated variance is

$$\widehat{\text{Var}}(\hat{\pi}_{\text{P}}) = 0.002932061.$$

From (7.11), an unbiased estimate of $\text{Var}(\hat{\pi}_{\text{P}})$ is $\overline{\text{Var}}(\hat{\pi}_{\text{P}}) = 0.002940762$. From (7.13), the 95% Wald confidence interval of π is $[\hat{\pi}_{\text{P,WL}}, \ \hat{\pi}_{\text{P,WU}}] = [0.4883881, \ 0.700961]$ with width being 0.2125730. From (7.16), the 95% Wilson confidence interval of π is $[\hat{\pi}_{\text{P,WSL}}, \ \hat{\pi}_{\text{P,WSU}}] = [0.4880742, \ 0.6991471]$ with width being 0.211073, which is shorter than the width of the Wald confidence interval.

Alternatively, from (7.17) and (7.18), two 95% bootstrap confidence intervals of π with $G = 20,000$ bootstrap replications are $[0.488845, \ 0.7008994]$ with width 0.2120544 and $[0.4881657, \ 0.7011834]$ with width 0.2130178, respectively. The corresponding standard error of $\hat{\pi}_{\text{P}}$ is estimated by $\widehat{\text{se}}(\hat{\pi}_{\text{P}}) = 0.05409552$.

For Bayesian inferences, we again adopt the uniform prior for π. Using (7.48), we have $E(\pi|Y_{\text{obs}}) = 0.5941176$ and $E(\pi^2|Y_{\text{obs}}) = 0.3558824$. Thus, the 95% Bayesian credible interval for π based on normality approximation is given by $[0.4884488, 0.6997865]$.

Using $\pi^{(0)} = 0.5$ as an initial value, the EM algorithm (7.51)–(7.53) converges in 25 iterations. The posterior mode of π is $\tilde{\pi}_{\mathrm{P}} = 0.5946746$, which is identical to the MLE $\hat{\pi}_{\mathrm{P}}$. Using the exact IBF sampling described in Section 7.5.3, we generate $L = 40{,}000$ i.i.d. posterior samples from $f(\pi|Y_{\mathrm{obs}})$. The histogram based on these samples is plotted in Figure 7.8 (ii). Figure 7.8 (i) shows that the exact IBF sampling can recover the density completely. The resultant posterior mean, standard error, and 95% Bayesian credible interval based on normality approximation for π are given by 0.5940246, 0.05375692, and $[0.4886611, 0.6993882]$, respectively.

7.9   Discussion

In this chapter, we introduce a general framework of design and analysis for the non-randomized parallel model. When comparing with the crosswise model and the triangular model, the parallel model has one excellent merit, that is, it can be applied to the case where both $\{Y = 0\}$ and $\{Y = 1\}$ are sensitive (cf. Table 7.2). For example, let $Y = 0$ if the number of sex partners is not more than three and $Y = 1$ if the number of sex partners is more than three. In addition, the parallel model is always more efficient than both the crosswise model and the triangular model for most of the possible parameter ranges. Finally, the parallel model is better than the two models from the viewpoint of privacy protection of respondents.

The availability of the proposed parallel model depends on the following four assumptions:

(1) *Existence* of two non-sensitive binary variates U and W.
    In fact, we could define $U = 1$ (or $W = 1$) if the respondent was born between July and December; or the respondent was born in an even numbered month; or the respondent's birthday is in the second half of the month; or the respondent's age is even numbered; or the respondent's house/apartment number is odd; or the last digit of the respondent's ID card/phone number is even; or the birthday of the respondent's mother is in the first six months (January–June).

(2) *Independence* of the three random variables Y, U and W.
    Suppose that we define $U = 1$ if the last digit of the respondent's phone number is even and define $W = 1$ if the respondent was born between July and December. Furthermore we let $Y = 1$ if the respondent is a drug user. Intuitively, the assumption that Y, U and W are mutually independent is reasonable, although we did not use real datesets to test the independency hypothesis statistically.

(3) *Known quantity* of both $q = \Pr(U = 1)$ and $p = \Pr(W = 1)$.

In some surveys, if it is difficult to choose an appropriate non-sensitive dichotomous variate U with known $q = \Pr(U = 1)$, we may define, say, $U = 1$ if the respondent lives in Hong Kong Island and $U = 0$ otherwise. In this case, q is unknown and is required to be estimated. More recently, Liu & Tian (2012b) developed a variant of the parallel model to address this issue (see Chapter 10 of this book).

(4) *Uniformity* of the two random variables U and W.

Petróczi *et al.* (2011) used one US life insurance application dataset ($n = 481{,}040$) and two UK universities registration datasets ($n = 495{,}870$ and $n = 11{,}157$) to illustrate the uniformity of birthday distribution. These results are summarized in Table 7.8. In addition, they also demonstrated the uniformity of the distribution of house and phone numbers.

Table 7.8 Birthday distributions for three real datasets

Dataset	Birthday	Count	Prob.	Count	Prob.
I	odd/even days	245,269	0.5099	235,771	0.4901
	1-st/2-nd half of the month	239,157	0.4972	241,883	0.5028
	1-st/2-nd half of the year	232,666	0.4837	248,374	0.5163
	odd/even numbered months	242,683	0.5045	238,357	0.4955
II	odd/even days	253,438	0.5111	242,432	0.4889
	1-st/2-nd half of the month	247,927	0.4999	247,943	0.5001
	1-st/2-nd half of the year	247,447	0.4990	248,423	0.5010
	odd/even numbered months	251,226	0.5066	244,644	0.4934
III	odd/even days	5,739	0.5143	5,418	0.4857
	1-st/2-nd half of the month	5,562	0.4985	5,595	0.5015
	1-st/2-nd half of the year	5,606	0.5024	5,551	0.4976
	odd/even numbered months	5,731	0.5137	5,426	0.4863

Source: Petróczi *et al.* (2011).
Note: Dataset I = US life insurance application data ($n = 481{,}040$); Dataset II = UK university registration data ($n = 495{,}870$); Dataset III = UK university registration data ($n = 11{,}157$).

CHAPTER 8

Sample Size Calculation
for the Parallel Model

The main objective of this chapter is to introduce sample size formulae for the parallel design by using the power analysis method for both the one- and two-sample problems. Numerical and theoretical comparisons of sample sizes needed for the parallel model with those required for the crosswise and triangular models are also provided.

8.1 Sample Sizes for One-sample Problem

Suppose that Y is a Bernoulli random variable corresponding to a sensitive question Q_Y (e.g., have you ever taken drugs?). Let $Y = 1$ if the answer to the question Q_Y is 'yes' and $Y = 0$ if the answer to the question Q_Y is 'no'. We are interested in estimating the unknown proportion $\pi = \Pr(Y = 1)$. To this end, we assume that there are two non-sensitive dichotomous variates W and U such that W, U and Y are mutually independent and $p = \Pr(W = 1)$ and $q = \Pr(U = 1)$ are known. For example, we may define $W = 1$ if a respondent was born in the first half of a year and $W = 0$ otherwise. Similarly, we could define $U = 1$ if the last digit of a respondent's cell phone number is even and $U = 0$ otherwise. Thus, it is reasonable to assume that $p = q \approx 0.5$. Table 7.2 shows the survey scheme for the parallel model.

For the parallel model described in Table 7.2, we define a Bernoulli random variable Y_P as

$$Y_P = \begin{cases} 1, & \text{if the two squares are connected,} \\ 0, & \text{if the two circles are connected,} \end{cases}$$

where the subscript 'P' refers to the 'parallel' model. From Table 7.2, the corresponding probabilities of $Y_P = 1$ and $Y_P = 0$ are given by

$$\Pr(Y_P = 1) = q(1 - p) + \pi p$$

and

$$\Pr(Y_P = 0) = (1 - q)(1 - p) + (1 - \pi)p.$$

Let $Y_{\text{obs}} = \{y_{i,\text{P}} : i = 1, \ldots, n\}$ denote the observed data for the n respondents. The likelihood function for π is

$$L_{\text{P}}(\pi|Y_{\text{obs}}) = \prod_{i=1}^{n} \left\{ q(1-p) + \pi p \right\}^{y_{i,\text{P}}} \left\{ (1-q)(1-p) + (1-\pi)p \right\}^{1-y_{i,\text{P}}}.$$

Consequently, the MLE of π and its variance are given by (7.9) and (7.10) (cf. (7.3)), which can be rewritten as

$$\hat{\pi}_{\text{P}} = \frac{\bar{y}_{\text{P}} - q(1-p)}{p} \quad \text{and} \quad \text{Var}(\hat{\pi}_{\text{P}}) = \frac{\delta(1-\delta)}{np^2}, \tag{8.1}$$

where

$$\bar{y}_{\text{P}} = \frac{1}{n} \sum_{i=1}^{n} y_{i,\text{P}} \quad \text{and} \quad \delta = q(1-p) + \pi p.$$

According to the Central Limit Theorem, $\hat{\pi}_{\text{P}}$ is asymptotically normally distributed as $n \to \infty$, i.e.,

$$\frac{\hat{\pi}_{\text{P}} - \pi}{\sqrt{\text{Var}(\hat{\pi}_{\text{P}})}} = \frac{n\hat{\pi}_{\text{P}} - n\pi}{\sqrt{n\delta(1-\delta)}/p} \mathrel{\dot\sim} N(0,1). \tag{8.2}$$

8.1.1   A one-sided test

In order to test whether the population proportion (π) with the sensitive characteristic is identical to a pre-specified value (π_0), the following hypotheses are often considered,

$$H_0: \pi = \pi_0 \quad \text{versus} \quad H_1: \pi < \pi_0. \tag{8.3}$$

If the null hypothesis H_0 is true, from (8.2), we have

$$\frac{n_{\text{P}}\hat{\pi}_{\text{P}} - n_{\text{P}}\pi_0}{\sqrt{n_{\text{P}}\delta_0(1-\delta_0)}/p} \mathrel{\dot\sim} N(0,1), \quad \text{as } n_{\text{P}} \to \infty,$$

where n_{P} denotes the sample size needed for the parallel design for above one-sided test, and

$$\delta_0 = q(1-p) + \pi_0 p.$$

Let z_α denote the upper α-th quantile of the standard normal distribution. When the event

$$\mathbb{E}_{\text{P}} = \left\{ n_{\text{P}}\hat{\pi}_{\text{P}} \leqslant n_{\text{P}}\pi_0 - z_\alpha \sqrt{n_{\text{P}}\delta_0(1-\delta_0)} \Big/ p \right\} \tag{8.4}$$

is observed, we should reject the null hypothesis H_0 at the α level of significance.

If H_1 is true, without loss of generality, we can assume that $\pi = \pi_1$ with $\pi_1 < \pi_0$. Thus, the power of the one-sided test can be calculated approximately by

$$\text{Power (at } \pi_1)$$

$$= \Pr(\text{rejecting } H_0|\pi = \pi_1)$$

$$= \Pr\left\{ \frac{n_\text{P}\hat{\pi}_\text{P} - E_{H_1}(n_\text{P}\hat{\pi}_\text{P})}{\sqrt{\text{Var}_{H_1}(n_\text{P}\hat{\pi}_\text{P})}} \leqslant \frac{n_\text{P}\pi_0 - z_\alpha\sqrt{n_\text{P}\delta_0(1-\delta_0)}/p - n_\text{P}\pi_1}{\sqrt{n_\text{P}\delta_1(1-\delta_1)}/p} \right\}$$

$$\approx \Phi\left(\frac{\sqrt{n_\text{P}}(\pi_0-\pi_1)p - z_\alpha\sqrt{\delta_0(1-\delta_0)}}{\sqrt{\delta_1(1-\delta_1)}} \right), \tag{8.5}$$

where

$$\delta_1 = q(1-p) + \pi_1 p$$

and $\Phi(\cdot)$ denotes the cumulative distribution function of the standard normal distribution. For a given power, say, $1 - \beta$, the required sample size n_P can be determined by solving the following equation

$$\sqrt{n_\text{P}}(\pi_0 - \pi_1)p - z_\alpha\sqrt{\delta_0(1-\delta_0)} = z_\beta\sqrt{\delta_1(1-\delta_1)},$$

which yields

$$n_\text{P} = \left\{ \frac{z_\alpha\sqrt{\delta_0(1-\delta_0)} + z_\beta\sqrt{\delta_1(1-\delta_1)}}{(\pi_0 - \pi_1)p} \right\}^2. \tag{8.6}$$

8.1.2 A two-sided test

For a two-sided test, the two-sided hypotheses are specified by

$$H_0: \pi = \pi_0 \quad \text{versus} \quad H_1: \pi \neq \pi_0.$$

Given a significance level α, we only consider the equal-tailed rejection region. Note that the relationship among the power, sample size, and effect size is approximately given by

$$\text{Power (at } \pi_1) \approx \Phi\left(\frac{\sqrt{n_{\text{P},2}}|\pi_0 - \pi_1|p - z_{\alpha/2}\sqrt{\delta_0(1-\delta_0)}}{\sqrt{\delta_1(1-\delta_1)}} \right).$$

The corresponding sample size formula is

$$n_{\text{P},2} = \left\{ \frac{z_{\alpha/2}\sqrt{\delta_0(1-\delta_0)} + z_\beta\sqrt{\delta_1(1-\delta_1)}}{(\pi_0 - \pi_1)p} \right\}^2. \tag{8.7}$$

8.1.3 Evaluation of the performance by comparing exact power with asymptotic power

The asymptotic power function for the one-sided test is given by (8.5). To derive the exact power formula, we define a new random variable

$$X_{\mathrm{P}} = \sum_{i=1}^{n_{\mathrm{P}}} Y_{i,\mathrm{P}}.$$

Then, we have $X_{\mathrm{P}} \sim \mathrm{Binomial}\,(n_{\mathrm{P}}, \delta)$ with $\delta = q(1-p) + \pi p$. The rejection region $\mathbb{E}_{\mathrm{P}}$ specified in (8.4) can be rewritten as

$$\mathbb{E}_{\mathrm{P}} = \left\{ X_{\mathrm{P}} \colon X_{\mathrm{P}} \leqslant n_{\mathrm{P}} \delta_0 - z_\alpha \sqrt{n_{\mathrm{P}} \delta_0 (1 - \delta_0)} \right\}.$$

The exact power (at π_1) for any particular sample size n is determined by the following formula

$$
\begin{aligned}
\text{Exact power (at } \pi_1) \;\; &= \;\; \sum_{x \in \mathbb{E}_{\mathrm{P}}} \mathrm{Binomial}\,(x|n_{\mathrm{P}},\, q(1-p) + \pi_1 p) \\
&= \;\; \sum_{x \in \mathbb{E}_{\mathrm{P}}} \binom{n_{\mathrm{P}}}{x} \delta_1^x (1 - \delta_1)^{n_{\mathrm{P}} - x}, \qquad (8.8)
\end{aligned}
$$

where

$$\delta_1 = q(1-p) + \pi_1 p.$$

To compare the accuracy of the approximate power formula given by (8.5), in Figure 8.1, we plot the exact and asymptotic powers against the sample size n_{P} for various combinations of (π_0, π_1) at $p = q = 0.5$ and $\alpha = 0.05$. Figure 8.1 shows that, in general, the asymptotic power function given by (8.5) is a satisfactory approximation to the exact power defined by (8.8). Especially for large sample sizes, the approximate power and the exact power are nearly the same (cf. Figure 8.1 (iv)).

8.1.4 Evaluation of the performance by calculating n_{P} and $n_{\mathrm{P}}/n_{\mathrm{D}}$

For a given pair of (π_0, π_1), we note that n_{P} is a decreasing function of p and an increasing function of q. It is clear that the parallel design reduces to the design of direct questioning when $p = 1$. Let n_{D} denote the sample size of the design of direct questioning. In (8.6), setting $p = 1$, we obtain

$$n_{\mathrm{D}} = \left\{ \frac{z_\alpha \sqrt{\pi_0 (1 - \pi_0)} + z_\beta \sqrt{\pi_1 (1 - \pi_1)}}{\pi_0 - \pi_1} \right\}^2. \qquad (8.9)$$

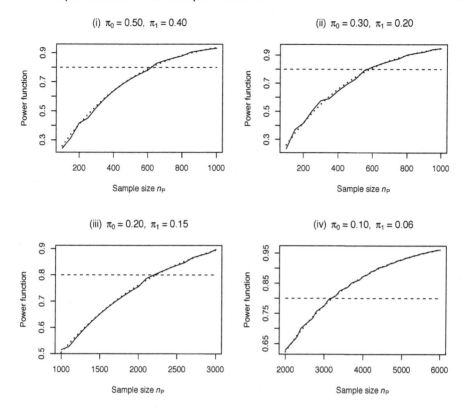

Figure 8.1   Comparisons of the exact power (8.8) (denoted by solid line) with the asymptotic power (8.5) (denoted by dotted line) against the sample size n_{P} for various combinations of (π_0, π_1) at $p = q = 0.5$ and $\alpha = 0.05$. (i) $(\pi_0, \pi_1) = (0.50, 0.40)$; (ii) $(\pi_0, \pi_1) = (0.30, 0.20)$; (iii) $(\pi_0, \pi_1) = (0.20, 0.15)$; (iv) $(\pi_0, \pi_1) = (0.10, 0.06)$.

Given $\alpha = 5\%$ and 80% power, Table 8.1 reports the sample size n_{P} defined by (8.6) and the corresponding ratio $n_{\mathrm{P}}/n_{\mathrm{D}}$ for various combinations of (π_0, π_1, q, p). For example, when $(\pi_0, \pi_1, q, p) = (0.40, 0.25, 1/3, 0.50)$, we have $n_{\mathrm{P}}/n_{\mathrm{D}} = 4.03$, indicating that the sample size required for the parallel design is about four times that required for the design of direct questioning in order to achieve the same power for the one-sided test.

In Table 8.1, we choose the non-sensitive dichotomous variate W to be a respondent's birthday and U to be the birthday of a respondent's mother. For example, $p = 0.42$ (i.e., $5/12$), 0.50 (i.e., $6/12$) and 0.58 (i.e., $7/12$) represent that we define $W = 1$ if a respondent was born between January to May, January to June, and January to July of a year, respectively. Similarly, $q = 1/3, 1/2$ and $2/3$ represent that we define $U = 1$ if a respondent's mother was born between the 1st to the 10th, the 1st to the 15th, and the 1st to the 20th of a month, respectively.

Table 8.1 Sample size n_{P} for testing H_0: $\pi = \pi_0$ versus H_1: $\pi = \pi_1 < \pi_0$ with $\alpha = 5\%$ and 80% power and the ratio $n_{\mathrm{P}}/n_{\mathrm{D}}$

π_0	π_1	q	$p=1$	$p=0.42$		$p=0.50$		$p=0.58$	
			n_{D}	n_{P}	$n_{\mathrm{P}}/n_{\mathrm{D}}$	n_{P}	$n_{\mathrm{P}}/n_{\mathrm{D}}$	n_{P}	$n_{\mathrm{P}}/n_{\mathrm{D}}$
0.50	0.40	1/3	153	832	5.44	592	3.87	444	2.90
	0.35		67	367	5.48	261	3.90	195	2.91
	0.30		37	204	5.51	145	3.92	108	2.92
0.40	0.35	1/3	583	3206	5.50	2273	3.90	1697	2.91
	0.30		142	793	5.58	561	3.95	418	2.94
	0.25		61	348	5.70	246	4.03	183	3.00
0.30	0.25	1/3	501	3009	6.01	2108	4.21	1555	3.10
	0.20		119	742	6.24	518	4.35	381	3.20
	0.18		81	512	6.32	357	4.41	262	3.23
0.20	0.16	1/3	584	4333	7.42	2972	5.09	2142	3.67
	0.13		181	1400	7.73	957	5.29	687	3.80
	0.10		83	678	8.17	462	5.57	330	3.98
0.10	0.08	1/3	1303	15634	12.00	10372	7.96	7185	5.51
	0.06		301	3874	12.87	2562	8.51	1767	5.87
	0.04		121	1706	14.10	1124	9.29	772	6.38
0.50	0.40	1/2	153	875	5.72	617	4.03	458	2.99
	0.35		67	388	5.79	273	4.07	203	3.03
	0.30		37	217	5.86	153	4.14	113	3.05
0.40	0.35	1/2	583	3470	5.95	2438	4.18	1803	3.09
	0.30		142	864	6.08	606	4.27	447	3.15
	0.25		61	382	6.26	268	4.39	197	3.23
0.30	0.25	1/2	501	3388	6.76	2356	4.70	1721	3.43
	0.20		119	841	7.07	583	4.90	425	3.57
	0.18		81	583	7.20	404	4.99	293	3.62
0.20	0.16	1/2	584	5095	8.72	3483	5.96	2491	4.27
	0.13		181	1655	9.14	1128	6.23	804	4.44
	0.10		83	806	9.71	548	6.60	389	4.69
0.10	0.08	1/2	1303	19347	14.85	12898	9.90	8928	6.85
	0.06		301	4814	15.99	3202	10.64	2210	7.34
	0.04		121	2129	17.60	1413	11.68	972	8.03
0.50	0.40	2/3	153	851	5.56	606	3.96	454	2.97
	0.35		67	380	5.67	270	4.03	202	3.01
	0.30		37	214	5.78	152	4.11	114	3.08
0.40	0.35	2/3	583	3472	5.96	2466	4.23	1837	3.15
	0.30		142	870	6.13	617	4.35	458	3.23
	0.25		61	387	6.34	274	4.49	203	3.33
0.30	0.25	2/3	501	3504	6.99	2466	4.92	1814	3.62
	0.20		119	875	7.35	615	5.17	451	3.79
	0.18		81	608	7.51	426	5.26	312	3.85
0.20	0.16	2/3	584	5449	9.33	3780	6.47	2727	4.67
	0.13		181	1776	9.81	1230	6.80	885	4.89
	0.10		83	869	10.47	600	7.23	430	5.18
0.10	0.08	2/3	1303	21422	16.44	14564	11.18	10221	7.84
	0.06		301	5345	17.76	3628	12.05	2539	8.44
	0.04		121	2371	19.60	1606	13.27	1121	9.26

Note: n_{D} is the sample size of the design of direct questioning, given by (8.9).

8.2 Comparison with the Crosswise Model

In this section, we first compare sample sizes required for the crosswise design with those needed for the parallel design numerically. Second, a theoretical justification is provided.

For the same one-sided hypotheses specified by (8.3), the sample size formula associated with the crosswise model is given by (4.11); that is,

$$n_C = \left\{ \frac{z_\alpha \sqrt{\lambda_0(1 - \lambda_0)} + z_\beta \sqrt{\lambda_1(1 - \lambda_1)}}{(\pi_0 - \pi_1)(2p - 1)} \right\}^2, \tag{8.10}$$

where $\lambda_i = (1 - \pi_i)(1 - p) + \pi_i p$, $i = 0, 1$, and $\pi_1 < \pi_0$.

8.2.1 Numerical comparisons

Intuitively, the optimal degree of privacy protection is attained at $p = 0.5$. When p is either too small or too large, the privacy of respondents cannot be protected sufficiently. Therefore, investigators should choose a p within some interval around $p = 0.5$ except for the point $p = 0.5$ at which the MLE of π does not exist.

In Table 8.2, we select several p's within $[0.42, 0.5) \cup (0.5, 0.65]$ and report the ratio n_C/n_P for testing $H_0: \pi = \pi_0$ versus $H_1: \pi = \pi_1 < \pi_0$ with 5% level of significance and 80% power. From Table 8.2, we can see that when p is near 0.5, the parallel model is far more efficient than the crosswise model. For example, when $p = 0.49$ or $p = 0.51$, the efficiency of the parallel model is about 601–909 or 651–1003 times that of the crosswise model.

8.2.2 A theoretical justification

The above observations are not surprising as we have the following theoretical results. Theorem 8.1 below identifies some conditions under which the parallel design is more efficient than the crosswise design. To Prove Theorem 8.1, we first present a lemma.

Lemma 8.1 Let $1/3 < p < 1$, $p \neq 1/2$, $0 < q < 1$ and $q \neq 1/2$. Define

$$H(p, q) = \frac{p - q + pq}{3p - 1}. \tag{8.11}$$

We have the following conclusions:

(1) If $(2p - 1)(2q - 1) > 0$, then $1 - q < H(p, q)$.

(2) If $(2p - 1)(2q - 1) < 0$, then $1 - q > H(p, q)$.   ¶

Table 8.2 The ratio n_C/n_P for testing H_0: $\pi = \pi_0$ versus H_1: $\pi = \pi_1 < \pi_0$ with $\alpha = 5\%$ and 80% power

π_0	π_1	q	p							
			0.42	0.45	0.49	0.51	0.55	0.58	0.6	0.65
0.50	0.40	1/3	7.25	21.26	627.29	677.91	31.41	13.59	6.88	4.83
	0.35		6.90	20.28	601.88	651.62	30.36	13.18	6.71	4.71
	0.30		7.09	20.74	612.37	662.79	30.66	13.30	6.74	4.73
0.40	0.35	1/3	7.31	21.46	633.71	684.21	31.65	13.75	6.96	4.88
	0.30		6.91	20.32	602.58	652.99	30.52	13.21	6.73	4.73
	0.25		7.06	20.69	611.16	660.52	30.66	13.28	6.73	4.73
0.30	0.25	1/3	7.39	21.70	639.74	690.01	32.19	13.96	7.04	5.08
	0.20		6.95	20.44	607.55	657.15	30.66	13.35	6.76	4.76
	0.18		7.05	20.66	611.40	661.65	30.66	13.23	6.76	4.76
0.20	0.16	1/3	7.52	22.08	653.53	707.05	32.80	14.21	7.19	5.04
	0.13		6.95	20.47	608.51	659.95	30.76	13.38	6.80	4.78
	0.10		6.95	20.37	602.34	651.88	30.26	13.13	6.66	4.68
0.10	0.08	1/3	7.60	22.32	661.64	715.55	33.21	14.42	7.28	5.10
	0.06		6.97	20.56	611.39	663.91	30.95	13.48	6.85	4.82
	0.04		6.93	20.35	601.87	651.69	30.28	13.16	6.66	4.68
0.50	0.40	1/2	7.69	22.57	670.82	727.67	33.80	14.62	7.41	5.18
	0.35		7.01	20.67	615.52	668.21	31.19	13.58	6.89	4.88
	0.30		6.91	20.30	602.56	652.97	30.36	13.18	6.69	4.70
0.40	0.35	1/2	7.99	23.61	703.47	763.60	35.59	15.46	7,83	5.49
	0.30		7.09	21.02	628.77	683.86	32.04	13.96	7.12	5.02
	0.25		6.86	20.22	601.60	652.67	30.44	13.25	6.74	4.75
0.30	0.25	1/2	8.09	23.95	715,51	777.41	36.29	15.76	8.00	5.61
	0.20		7.14	21.16	634.44	691.19	32.40	14.12	7.22	5.08
	0.18		6.86	20.24	602.77	654.87	30.60	13.31	6.79	4.78
0.20	0.16	1/2	8.14	24.12	721.27	782.26	36.55	15.90	8.06	5.67
	0.13		7.15	21.25	637.33	693.32	32.55	14.22	7.26	5.11
	0.10		6.85	20.24	604.31	656.02	30.68	13.36	6.81	4.80
0.10	0.08	1/2	8.62	25.76	777.42	848.45	39.93	17.44	8.89	6.25
	0.06		7.33	21.92	662.83	724.48	34.21	15.00	7.69	5.43
	0.04		6.86	20.37	611.78	666.67	31.33	13.70	7.01	4.95
0.50	0.40	2/3	8.71	26.07	787.67	860.76	40.58	17.74	9.06	6.37
	0.35		7.36	22.06	667.62	730.73	34.56	15.16	7.78	5.50
	0.30		6.86	20.41	614.06	669.32	31.47	13.77	7.05	4.99
0.40	0.35	2/3	8.80	26.39	799.88	874.07	41.25	18.08	9.24	6.51
	0.30		7.40	22.20	673.07	737.29	34.89	15.34	7.87	5.57
	0.25		6.87	20.46	616.17	671.90	31.66	13.88	7.10	5.02
0.30	0.25	2/3	9.49	28.82	887.00	976.81	46.74	20.65	10.7	7.57
	0.20		7.67	23.23	713.53	785.31	37.58	16.62	8.61	6.13
	0.18		6.93	20.81	633.25	694.05	32.97	14.52	7.49	5.32
0.20	0.16	2/3	9.57	29.10	897.55	989.10	47.43	20.98	10.9	7.71
	0.13		7.70	23.35	718.36	791.09	37.91	16.78	8.70	6.20
	0.10		6.94	20.86	635.37	696.80	33.13	14.60	7.54	5.36
0.10	0.08	2/3	9.65	29.40	908.60	1002.9	48.17	21.33	11.1	7.88
	0.06		7.73	23.47	723.08	797.22	38.23	16.94	8.81	6.28
	0.04		6.95	20.91	637.59	699.97	33.30	14.69	7.59	5.40

Proof. (1) If $(2p-1)(2q-1) > 0$, then we have $2p-1-2q(2p-1) < 0$, or

$$(1-q)(3p-1) < p-q+pq.$$

Since $p > 1/3$, i.e., $3p-1 > 0$, we immediately obtain

$$1-q < \frac{p-q+pq}{3p-1} = H(p,q).$$

Similarly, we can prove (2). □

Theorem 8.1 Let π, p, $q \in (0, 1)$. For the parallel model and the crosswise model, we have

(1) When $p = 1/3$, the parallel model is always more efficient than the crosswise model in the sense that $n_P \leqslant n_C$, if one of the following three conditions is satisfied:

    (a) $q = 1/2$ and $\pi \in (0, 1)$;
    (b) $q \in (0, \min(1/2, 1-\pi))$ and $\pi \in (0, 1)$;
    (c) $q \in (\max(1/2, 1-\pi), 1)$ and $\pi \in (0, 1)$.

(2) When $1/3 < p < 1$ and $p \neq 1/2$, the parallel model is always more efficient than the crosswise model in the sense that $n_P \leqslant n_C$, if one of the following five conditions is satisfied:

    (a) $q = 1/2$ and $\pi \in (0, 1)$;
    (b) $q > 1/2$, $p > 1/2$ and $\pi \in (0, 1-q) \cup (H(p,q), 1)$;
    (c) $q < 1/2$, $p < 1/2$ and $\pi \in (0, 1-q) \cup (\min\{1, H(p,q)\}, 1)$;
    (d) $q > 1/2$, $p < 1/2$ and $\pi \in (0, \max\{0, H(p,q)\}) \cup (1-q, 1)$;
    (e) $q < 1/2$, $p > 1/2$ and $\pi \in (0, H(p,q)) \cup (1-q, 1)$,

where $H(p,q)$ is given by (8.11). ¶

Proof. From (8.6) and (8.10), we have

$$\frac{n_P}{n_C} = \left(\frac{2p-1}{p}\right)^2 \times \left\{\frac{z_\alpha\sqrt{\delta_0(1-\delta_0)} + z_\beta\sqrt{\delta_1(1-\delta_1)}}{z_\alpha\sqrt{\lambda_0(1-\lambda_0)} + z_\beta\sqrt{\lambda_1(1-\lambda_1)}}\right\}^2, \qquad (8.12)$$

where

$$\delta_i = q(1-p) + \pi_i p \quad \text{and} \quad \lambda_i = (1-\pi_i)(1-p) + \pi_i p, \quad i = 0, 1.$$

Note that when $1/3 \leqslant p < 1$ and $p \neq 1/2$, we always have

$$p^2 \geqslant (2p-1)^2,$$

i.e., the first term on the right-hand side of (8.12) is less than or equal to 1. To obtain $n_P \leqslant n_C$, it suffices to show that

$$\delta(1-\delta) \leqslant \gamma(1-\gamma)$$

or equivalently

$$\left\{q(1-p) + \pi p\right\}\left\{(1-q)(1-p) + (1-\pi)p\right\}$$

$$\leqslant \left\{(1-\pi)(1-p) + \pi p\right\}\left\{(1-\pi)p + \pi(1-p)\right\}. \qquad (8.13)$$

After some simplifications, it can be showed that (8.13) is equivalent to

$$h_C(\pi|p,q) \hat{=} (3p-1)\pi^2 + (1-4p+2pq)\pi + (1-q)(p-q+pq) \geqslant 0. \quad (8.14)$$

(1) When $p = 1/3$, (8.14) reduces to

$$(2q-1)(q-1+\pi) \geqslant 0. \qquad (8.15)$$

(a)   If $q = 1/2$, then (8.15) is always true for any $\pi \in (0, 1)$.

(b)   If $0 < q < 1/2$, then (8.15) is equivalent to $q < 1 - \pi$. Therefore, (8.15) is always true for any $0 < q < \min(1/2, 1 - \pi)$ and $\pi \in (0, 1)$.

(c)   If $1/2 < q < 1$, then (8.15) is equivalent to $q > 1 - \pi$. Hence, (8.15) is always true for any $\max(1/2, 1 - \pi) < q < 1$ and $\pi \in (0, 1)$.

(2) When $1/3 < p < 1$ and $p \neq 1/2$, we always have $3p - 1 > 0$. Note that the discriminant for the quadratic function $h_C(\pi|p,q)$ defined in (8.14) is given by

$$\begin{aligned} \Delta_C &= (1-4p+2pq)^2 - 4(3p-1)(1-q)(p-q+pq) \\ &= (2p-1)^2(2q-1)^2. \end{aligned}$$

(a)   If $q = 1/2$, then $\Delta_C = 0$. Hence, $h_C(\pi|p,q) \geqslant 0$ (i.e., (8.14)) is true for all $\pi \in (0, 1)$.

If $q \neq 1/2$, then $\Delta_C > 0$. Hence, $h_C(\pi|p,q) > 0$ for any $\pi \in (0, \pi_{C,L}) \cup (\pi_{C,U}, 1)$, where

$$\pi_{C,L} = \max\left\{0, \frac{-(1-4p+2pq) - |(2p-1)(2q-1)|}{2(3p-1)}\right\} \qquad (8.16)$$

and

$$\pi_{C,U} = \min\left\{1,\ \frac{-(1-4p+2pq)+|(2p-1)(2q-1)|}{2(3p-1)}\right\}. \qquad (8.17)$$

(b) If $q > 1/2$ and $p > 1/2$, then (8.16) and (8.17) can be simplified as

$$\pi_{C,L} \quad = \quad \max\left\{0,\ \frac{-(1-4p+2pq)-(2p-1)(2q-1)}{2(3p-1)}\right\}$$

$$= \quad \max(0,\ 1-q)$$

$$= \quad 1-q, \quad \text{and}$$

$$\pi_{C,U} \quad = \quad \min\left\{1,\ \frac{-(1-4p+2pq)+(2p-1)(2q-1)}{2(3p-1)}\right\}$$

$$= \quad \min\left(1,\ \frac{p-q+pq}{3p-1}\right)$$

$$= \quad \frac{p-q+pq}{3p-1} \qquad\qquad (8.18)$$

$$\overset{(8.11)}{=} \quad H(p,q),$$

respectively, where (8.18) can be proved from

$$\frac{1}{2} < p < 1 \text{ and } q > 0 \ \Rightarrow\ q > 0 > \frac{1-2p}{1-p}$$

$$\Rightarrow\ q - pq > 1 - 2p$$

$$\Rightarrow\ 3p - 1 > p - q + pq$$

$$\Rightarrow\ 1 > \frac{p-q+pq}{3p-1}.$$

Finally, from Lemma 8.1 (1), we have $1 - q < H(p,q)$; that is $\pi_{C,L} < \pi_{C,U}$.

(c) If $q < 1/2$ and $p < 1/2$, then (8.16) and (8.17) become

$$\pi_{C,L} = 1-q \quad \text{and} \quad \pi_{C,U} = \min\{1,\ H(p,q)\},$$

respectively. However, we cannot simplify $\pi_{C,U}$. On the one hand, from Lemma 8.1 (1), we have $1 - q < H(p,q)$. On the other hand, $1 - q < 1$. Hence,

$$\pi_{C,L} = 1-q < \min\{1,\ H(p,q)\} = \pi_{C,U}.$$

(d)   If $q > 1/2$ and $p < 1/2$, then (8.16) and (8.17) become

$$\pi_{\mathrm{C,L}} = \max\{0,\ H(p,q)\} \quad \text{and} \quad \pi_{\mathrm{C,U}} = 1 - q,$$

respectively. However, we cannot simplify $\pi_{\mathrm{C,L}}$. On the one hand, from Lemma 8.1 (2), we have $1 - q > H(p,q)$. On the other hand, $1 - q > 0$. Hence,

$$\pi_{\mathrm{C,L}} = \max\{0,\ H(p,q)\} < 1 - q = \pi_{\mathrm{C,U}}.$$

(e)   If $q < 1/2$ and $p > 1/2$, then (8.16) and (8.17) become

$$\pi_{\mathrm{C,L}} = \max\{0,\ H(p,q)\} \quad \text{and} \quad \pi_{\mathrm{C,U}} = 1 - q,$$

respectively. Now, we have $\pi_{\mathrm{C,L}} = H(p,q)$, which can be proved from

$$p > \frac{1}{2} > q > 0 \quad \Rightarrow \quad p - q > 0$$

$$\Rightarrow \quad p - q + pq > 0$$

$$\Rightarrow \quad \frac{p - q + pq}{3p - 1} > 0 \quad (\text{as } p > 1/3)$$

$$\Rightarrow \quad H(p,q) > 0.$$

On the one hand, from Lemma 8.1 (2), we have $1 - q > H(p,q)$. On the other hand, $1 - q > 0$. Hence,

$$\pi_{\mathrm{C,L}} = \max\{0,\ H(p,q)\} < 1 - q = \pi_{\mathrm{C,U}}. \qquad \square$$

8.3   Comparison with the Triangular Model

For the same one-sided hypotheses specified by (8.3), the sample size formula associated with the triangular model is given by (4.6); that is,

$$n_{\mathrm{T}} = \left\{ \frac{z_\alpha \sqrt{\theta_0(1 - \theta_0)} + z_\beta \sqrt{\theta_1(1 - \theta_1)}}{(\pi_0 - \pi_1)(1 - p)} \right\}^2, \tag{8.19}$$

where $\theta_i = \pi_i + (1 - \pi_i)p$, $i = 0,\ 1$, and $\pi_1 < \pi_0$.

8.3.1 Numerical comparisons

In Table 8.3, we select several values of p within the interval $[0.48, 0.72]$ and report the ratio $n_{\mathrm{T}}/n_{\mathrm{P}}$ for testing H_0: $\pi = \pi_0$ against H_1: $\pi = \pi_1 < \pi_0$ with 5% level of significance and 80% power. From Table 8.3, we can see that when $p = 0.58 \approx 7/12$ or 0.72, the efficiency of the parallel model is about 1–3 or 3–10 times that of the triangular model. In particular, when $0.54 \leqslant p \leqslant 0.66$ (which is the optimal range such that the privacy of respondents is protected for the triangle model), the efficiency of the parallel design is about 1–6 times that of the triangular design.

8.3.2 A theoretical justification

The above observations are further confirmed by the following theoretical result. Theorem 8.2 below identifies the conditions under which the parallel design is more efficient than the triangular design.

Theorem 8.2 Let $\pi, p, q \in (0, 1)$. For the parallel model and the triangular model, we have

(1) When $p = 1/2$, the parallel model is always more efficient than the triangular model (in the sense that $n_{\mathrm{P}} \leqslant n_{\mathrm{T}}$) for any $q \in (0, 1 - 2\pi]$ and $\pi \in (0, 1/2)$.

(2) When $1/2 < p < 1$, the parallel model is always more efficient than the triangular model (in the sense that $n_{\mathrm{P}} \leqslant n_{\mathrm{T}}$) for any $q \in (0, 1)$ and $0 < \pi < (1 - p)(1 - q)$.

Proof. From (8.6) and (8.19),we have

$$\frac{n_{\mathrm{P}}}{n_{\mathrm{T}}} = \left(\frac{1-p}{p}\right)^2 \times \left\{\frac{z_\alpha\sqrt{\delta_0(1-\delta_0)} + z_\beta\sqrt{\delta_1(1-\delta_1)}}{z_\alpha\sqrt{\theta_0(1-\theta_0)} + z_\beta\sqrt{\theta_1(1-\theta_1)}}\right\}^2, \qquad (8.20)$$

where

$$\delta_i = q(1-p) + \pi_i p \quad \text{and} \quad \theta_i = \pi_i + (1-\pi_i)p, \quad i = 0, 1.$$

Note that when $1/2 \leqslant p < 1$, we always have

$$p^2 \geqslant (1-p)^2,$$

i.e., the first term on the right-hand side of (8.20) is less than or equal to 1. To obtain $n_{\mathrm{P}} \leqslant n_{\mathrm{T}}$, it suffices to show that

$$\delta(1-\delta) \leqslant \theta(1-\theta),$$

Table 8.3 The ratio n_T/n_P for testing H_0: $\pi = \pi_0$ versus H_1: $\pi = \pi_1 < \pi_0$ with $\alpha = 5\%$ and 80% power

π_0	π_1	q	0.48	0.50	0.54	0.58	0.62	0.66	0.70	0.72
0.50	0.40	1/3	0.71	0.82	1.06	1.37	1.77	2.30	2.99	3.44
	0.35		0.73	0.84	1.09	1.42	1.83	2.37	3.10	3.55
	0.30		0.75	0.86	1.12	1.46	1.89	2.46	3.21	3.69
0.40	0.35	1/3	0.81	0.93	1.22	1.59	2.06	2.69	3.53	4.06
	0.30		0.83	0.95	1.25	1.63	2.13	2.79	3.67	4.21
	0.25		0.85	0.98	1.28	1.68	2.21	2.89	3.79	4.38
0.30	0.25	1/3	0.93	1.08	1.43	1.90	2.50	3.31	4.40	5.08
	0.20		0.96	1.11	1.48	1.96	2.60	3.45	4.58	5.31
	0.18		0.96	1.12	1.50	1.99	2.63	3.51	4.68	5.42
0.20	0.16	1/3	1.07	1.25	1.70	2.30	3.10	4.18	5.67	6.62
	0.13		1.09	1.28	1.74	2.35	3.18	4.30	5.86	6.86
	0.10		1.11	1.30	1.77	2.42	3.27	4.45	6.07	7.12
0.10	0.08	1/3	1.25	1.48	2.07	2.88	4.00	5.57	7.82	9.30
	0.06		1.26	1.50	2.10	2.93	4.09	5.72	8.05	9.60
	0.04		1.28	1.52	2.13	2.99	4.19	5.87	8.31	9.93
0.50	0.40	1/2	0.68	0.78	1.02	1.33	1.73	2.25	2.94	3.37
	0.35		0.70	0.80	1.04	1.36	1.77	2.31	3.01	3.47
	0.30		0.71	0.82	1.07	1.40	1.82	2.38	3.13	3.59
0.40	0.35	1/2	0.75	0.87	1.14	1.49	1.95	2.56	3.38	3.89
	0.30		0.76	0.88	1.16	1.53	2.01	2.63	3.49	4.02
	0.25		0.78	0.90	1.19	1.56	2.05	2.72	3.59	4.17
0.30	0.25	1/2	0.83	0.97	1.29	1.71	2.28	3.03	4.05	4.71
	0.20		0.85	0.98	1.32	1.76	2.34	3.13	4.19	4.89
	0.18		0.85	0.99	1.33	1.78	2.37	3.17	4.26	4.97
0.20	0.16	1/2	0.91	1.07	1.46	1.98	2.68	3.64	4.98	5.85
	0.13		0.92	1.08	1.48	2.01	2.73	3.72	5.11	6.01
	0.10		0.93	1.09	1.50	2.05	2.79	3.81	5.25	6.18
0.10	0.08	1/2	1.00	1.19	1.66	2.31	3.22	4.52	6.38	7.63
	0.06		1.01	1.20	1.68	2.34	3.28	4.60	6.52	7.81
	0.04		1.01	1.21	1.70	2.37	3.33	4.69	6.68	8.01
0.50	0.40	2/3	0.70	0.80	1.03	1.34	1.74	2.25	2.94	3.37
	0.35		0.71	0.81	1.05	1.37	1.77	2.31	3.01	3.45
	0.30		0.72	0.82	1.07	1.39	1.82	2.35	3.09	3.54
0.40	0.35	2/3	0.75	0.86	1.12	1.46	1.91	2.50	3.30	3.80
	0.30		0.75	0.87	1.14	1.49	1.95	2.56	3.38	3.91
	0.25		0.76	0.88	1.15	1.52	1.99	2.62	3.46	4.01
0.30	0.25	2/3	0.80	0.92	1.23	1.63	2.15	2.86	3.84	4.46
	0.20		0.81	0.93	1.24	1.65	2.21	2.94	3.95	4.61
	0.18		0.81	0.94	1.25	1.67	2.22	2.97	4.00	4.66
0.20	0.16	2/3	0.85	0.99	1.34	1.81	2.44	3.32	4.54	5.34
	0.13		0.85	0.99	1.35	1.83	2.48	3.37	4.64	5.45
	0.10		0.86	1.00	1.36	1.85	2.52	3.43	4.73	5.60
0.10	0.08	2/3	0.89	1.05	1.46	2.02	2.81	3.92	5.53	6.62
	0.06		0.89	1.06	1.47	2.04	2.84	3.91	5.63	6.74
	0.04		0.90	1.06	1.48	2.06	2.87	4.03	5.73	6.87

or equivalently

$$\left\{q(1-p)+\pi p\right\}\left\{(1-q)(1-p)+(1-\pi)p\right\}$$

$$\leqslant \left\{\pi+(1-\pi)p\right\}(1-\pi)(1-p). \tag{8.21}$$

After some simplifications, we can show that (8.21) is equivalent to

$$h_{_\mathrm{T}}(\pi|p,q) \quad \hat{=} \quad (2p-1)\pi^2 + \{1-2p-2p(1-p)(1-q)\}\pi$$

$$+ (1-p)(1-q)(p-q+pq) \geqslant 0. \tag{8.22}$$

(1) When $p = 1/2$, (8.22) reduces to

$$\frac{(1-q)(-2\pi+1-q)}{4} \geqslant 0. \tag{8.23}$$

Hence, for any $q \in (0, 1-2\pi]$ and any $\pi \in (0, 1/2)$, (8.23) is true.

(2) When $1/2 < p < 1$, we always have $2p - 1 > 0$. Note that the discriminant for the quadratic function $h_{_\mathrm{T}}(\pi|p,q)$ defined in (8.22) is

$$\Delta_{_\mathrm{T}} \quad = \quad [(1-2p) - 2p(1-p)(1-q)]^2$$

$$- 4(2p-1)(1-p)(1-q)(p-q+pq)$$

$$= \quad \{p^2 + (1-p)^2(1-2q)\}^2.$$

We can show that $p^2 + (1-p)^2(1-2q) > 0$. In fact, from

$$\frac{1}{2} < p < 1 \text{ and } q < 1 \quad \Rightarrow \quad p^2 > (1-p)^2 \text{ and } q < 1$$

$$\Rightarrow \quad \frac{p^2 + (1-p)^2}{2(1-p)^2} > 1 > q$$

$$\Rightarrow \quad p^2 + (1-p)^2 > 2q(1-p)^2$$

$$\Rightarrow \quad p^2 + (1-p)^2(1-2q) > 0.$$

In other words, $\Delta_{_\mathrm{T}} > 0$ for any $q \in (0,1)$. Hence, $h_{_\mathrm{T}}(\pi|p,q) > 0$ for any $q \in (0,1)$ and $\pi \in (0, \pi_{_{\mathrm{T,L}}}) \cup (\pi_{_{\mathrm{T,U}}}, 1)$, where

$$\pi_{_{\mathrm{T,L}}} \quad = \quad \max\left\{0, \; \frac{-1+2p+2p(1-p)(1-q)-p^2-(1-p)^2(1-2q)}{2(2p-1)}\right\}$$

$$= \quad \max\{0, \; (1-p)(1-q)\}$$

$$= \quad (1-p)(1-q) < 1,$$

and

$$\pi_{\mathrm{T,U}} \;=\; \min\left\{1, \; \frac{-1 + 2p + 2p(1-p)(1-q) + p^2 + (1-p)^2(1-2q)}{2(2p-1)}\right\}$$

$$= \; \min\left(1, \; \frac{p - q + pq}{2p - 1}\right).$$

In the following equation, we show that $\pi_{\mathrm{T,U}} = 1$. From

$$\frac{1}{2} < p < 1 \text{ and } q < 1 \;\;\Rightarrow\;\; 2p - 1 > 0 \text{ and } (1-p)(1-q) > 0$$

$$\Rightarrow\;\; 2p - 1 > 0 \text{ and } 2p - 1 < p - q + pq$$

$$\Rightarrow\;\; 1 < \frac{p - q + pq}{2p - 1},$$

which implies $\pi_{\mathrm{T,U}} = 1$. In a summary, $h_{\mathrm{T}}(\pi|p,q) > 0$ for any $q \in (0,1)$ and $0 < \pi < (1-p)(1-q)$. □

8.4   Sample Size for Two-sample Problem

In this section, we consider two independent surveys on the same sensitive question in two different populations or regions (labeled as $k = 1$, 2) by using the parallel design. The purpose here is to determine the sample sizes in each survey in order to compare the proportions (π_k) of subjects with the sensitive characteristic. For a fixed k, we define a binary random variable $Y_{k,\mathrm{P}}$ as follows:

$$Y_{k,\mathrm{P}} = \begin{cases} 1, & \text{if the two squares are connected,} \\ 0, & \text{if the two circles are connected.} \end{cases}$$

Let π_k denote the proportion of subjects with the sensitive characteristic in population k ($k = 1$, 2). We have

$$\Pr(Y_{k,\mathrm{P}} = 1) = q_k(1 - p_k) + \pi_k p_k$$

and

$$\Pr(Y_{k,\mathrm{P}} = 0) = (1 - q_k)(1 - p_k) + (1 - \pi_k)p_k,$$

where $p_k = \Pr(W_k = 1)$ and $q_k = \Pr(U_k = 1)$ ($k = 1$, 2) are assumed to be known but neither p_1 and p_2 nor q_1 and q_2 are necessarily the same.

Suppose that there are a total of $n_1 + n_2$ individuals taking part in the survey, where n_1 respondents participating in the survey are from the first population and n_2 respondents are from the second population. Let $Y_{\text{obs}} = \{y_{ik,\text{P}}: i = 1, \ldots, n_k, k = 1, 2\}$ denote the observed data. The likelihood function for π_1 and π_2 is given by

$$
\begin{aligned}
L_{\text{P}}(\pi_1, \pi_2 | Y_{\text{obs}}) &= \prod_{k=1}^{2} \prod_{i=1}^{n_k} \left\{ q_k(1 - p_k) + \pi_k p_k \right\}^{y_{ik,\text{P}}} \\
&\quad \times \left\{ (1 - q_k)(1 - p_k) + (1 - \pi_k)p_k \right\}^{1 - y_{ik,\text{P}}} \\
&= \prod_{k=1}^{2} \left\{ q_k(1 - p_k) + \pi_k p_k \right\}^{n_k \bar{y}_{k,\text{P}}} \\
&\quad \times \left\{ (1 - q_k)(1 - p_k) + (1 - \pi_k)p_k \right\}^{n_k(1 - \bar{y}_{k,\text{P}})},
\end{aligned}
$$

where $\bar{y}_{k,\text{P}} = (1/n_k) \sum_{i=1}^{n_k} y_{ik,\text{P}}$ denote the average number of respondents connecting the two squares in the k-th population. The resulting MLE of π_k and its variance are given by

$$
\hat{\pi}_k = \frac{\bar{y}_{k,\text{P}} - q_k(1 - p_k)}{p_k} \quad \text{and} \quad \text{Var}(\hat{\pi}_k) = \frac{\Delta_k(1 - \Delta_k)}{n_k p_k^2},
$$

where $\Delta_k = q_k(1 - p_k) + \pi_k p_k$. Thus,

$$
\widehat{\text{Var}}(\hat{\pi}_k) = \frac{\bar{y}_{k,\text{P}}(1 - \bar{y}_{k,\text{P}})}{n_k p_k^2}
$$

is the MLE of $\text{Var}(\hat{\pi}_k)$.

Now, we consider the following two-sided hypotheses

$$
H_0: \pi_1 = \pi_2 \quad \text{versus} \quad H_1: \pi_1 \neq \pi_2.
$$

Let $\text{SE}_+ = [\sum_{k=1}^{2} \text{Var}(\hat{\pi}_k)]^{1/2}$ and $\widehat{\text{SE}}_+ = [\sum_{k=1}^{2} \widehat{\text{Var}}(\hat{\pi}_k)]^{1/2}$ denote the MLE of SE_+. Then the null hypothesis H_0 will be rejected at the α level of significance if

$$
\left| \frac{\hat{\pi}_1 - \hat{\pi}_2}{\widehat{\text{SE}}_+} \right| > z_{\alpha/2}.
$$

Under the alternative hypothesis H_1, i.e., $\pi_1 - \pi_2 \neq 0$, the power of the two-sided test is approximately given by

$$
\Phi\left(\frac{|\pi_1 - \pi_2| - z_{\alpha/2} \times \widehat{\text{SE}}_+}{\text{SE}_+} \right),
$$

which can be further approximated by (Chow, Shao & Wang, 2003)

$$\Phi\left(\frac{|\pi_1 - \pi_2|}{\mathrm{SE}_+} - z_{\alpha/2}\right).$$

Consequently, to achieve a desired power of $1 - \beta$, we need to solve the following equation:

$$\frac{|\pi_1 - \pi_2|}{\mathrm{SE}_+} - z_{\alpha/2} = z_\beta. \tag{8.24}$$

Let $\rho = n_1/n_2$ be known. Then, from (8.24), we have

$$n_1 = \rho n_2, \tag{8.25}$$

and

$$n_2 = \frac{(z_{\alpha/2} + z_\beta)^2}{(\pi_1 - \pi_2)^2}\left\{\frac{\Delta_1(1 - \Delta_1)}{\rho p_1^2} + \frac{\Delta_2(1 - \Delta_2)}{p_2^2}\right\}. \tag{8.26}$$

8.5   An Example

Monto (2001) reported a sexual practice study carried in three Western cities (San Francisco, Las Vegas, and Portland, Oregon) of the United States. In this investigation, there are 343 individuals graduating at most from some high school and 927 individuals receiving at least some college training. In addition, it was also observed that 593 respondents have no more than one sexual partner and 668 respondents have no fewer than two sexual partners. The investigators would like to estimate the proportion of persons with more than one sexual partner in a population.

We first define $W = 1$ if the birthday of the respondent is between May to December and $W = 0$ if the birthday of the respondent is between January to April, and let

$$p = \Pr(W = 1) \approx \frac{8}{12} = \frac{2}{3}.$$

We then define $U = 1$ if the respondent receives at least some college training and $U = 0$ if the respondent graduates at most from some high school, and let

$$q = \Pr(U = 1) = \frac{927}{343 + 927} \approx 0.73.$$

Finally, we define $Y = 1$ if the respondent has at least two sexual partners and $Y = 0$ otherwise. For the purpose of illustration, we assume that W, U and Y are mutually independent although we have noted the possible association between U and Y.

With $p = 2/3$ and $q = 0.73$, the survey with the parallel design will yield

$$n\bar{y}_P = \sum_{i=1}^{n} y_{i,P} = 927 \times (1 - p) + 668 \times p \approx 754$$

lines connecting the two squares and

$$n - n\bar{y}_P = 343 \times (1 - p) + 593 \times p \approx 509$$

lines connecting the two circles, where $n = 1263$. If the crosswise design is employed, it can be observed that

$$n\bar{y}_C = \sum_{i=1}^{n} y_{i,C} = 593 \times (1 - p) + 668 \times p \approx 643$$

ticks will be put in the upper circle and

$$n - n\bar{y}_C = 593 \times p + 668 \times (1 - p) \approx 618$$

ticks will be put in the upper square, where $n = 1261$. Finally, for the survey with the triangular design, there are

$$n\bar{y}_T = \sum_{i=1}^{n} y_{i,T} = 593 \times p + 668 = 1063$$

ticks in the circle and

$$n - n\bar{y}_T = 593 \times (1 - p) \approx 198$$

ticks in the upper square, where $n = 1261$.

Table 8.4 reports MLEs of π based on (8.1), (2.8) and (3.3), estimated standard errors and 95% Wald confidence intervals of π for the three models. From Table 8.4, we can see that the width of the 95% confidence interval of π for the parallel model is the shortest among the three models.

Table 8.4 MLEs of π, estimated standard errors and 95% Wald confidence intervals of π for three models for the sexual practice data

Model	$\hat{\pi}$	$\widehat{SE}(\hat{\pi})$	95% Wald CI of π	Width of the Wald CI
Parallel	0.53049	0.020703	[0.48991, 0.57106]	0.081155
Crosswise	0.52974	0.042233	[0.44696, 0.61251]	0.165552
Triangular	0.52895	0.030736	[0.46870, 0.58919]	0.120485

To illustrate the proposed methods, we now determine the sample sizes required in order to guarantee 80% power with 0.05 level of significance by using the one-sided test for testing $\pi_0 = 0.65$ against $\pi_1 = 0.55$. Using the sample size formulae (8.6), (8.10) and (8.19), we obtain $n_P = 314$, $n_C = 1382$ and $n_T = 618$, which are required sample sizes for the parallel, crosswise and triangular designs, respectively.

Finally, we estimate how many subjects are required for comparing the proportions that people having more than one sexual partner in a population between two regions with 80% power and 0.05 level of significance using the two-sided test for testing $\pi_1 = \pi_2$ against $\pi_1 \neq \pi_2$. Assume that true proportions with sensitive character in the two regions are $\pi_1 = 0.68$ and $\pi_2 = 0.75$, respectively. Using the parallel design with $p_1 = 0.55$, $p_2 = 0.6$, $q_1 = 0.5$ and $q_2 = 0.4$, the sample sizes with $\rho = 1$ (equal allocation) are given by $n_1 = n_2 = 2306$ via (8.25) and (8.26) while the desired sample sizes are $n_1 = n_2 = 50054$ for the crosswise model and $n_1 = n_2 = 2849$ for the triangular model.

CHAPTER 9

The Multi-category Parallel Model

In Chapter 5, we introduced the multi-category triangular model for a sensitive question with multiple answers based on an assumption that one subclass (e.g., $\{Y = 1\}$) is non-sensitive and the other subclasses (denoted by $\{Y = i\}$ for $i = 2, \ldots, m$) are sensitive. Thus, the multi-category triangular model cannot be applied to such situations where each subclass is sensitive. Next, the multi-category triangular model still has a lower efficiency for some cases. Third, the parallel model introduced in Chapter 7 can only deal with the case of $m = 2$ groups where both $\{Y = 0\}$ and $\{Y = 1\}$ could be sensitive. These limitations motivate us to introduce a non-randomized multi-category parallel model, which is an extension of the parallel model (Liu & Tian, 2013a).

9.1 The Survey Design

Consider a sensitive question Q_Y (e.g., how many sex partners do you have within a certain period?) with m possible answers (e.g., 0–3, 4–6 or $\geqslant 7$), which classify the target population into m mutually exclusive categories and each category has a certain degree of a sensitive attribute. Let Y denote a categorical random variable associated with the question Q_Y and $\{Y = i\}$ denote that a person in the target population belongs to the i-th category $(i = 1, \ldots, m)$. The purpose is to estimate proportions $\pi_i = \Pr(Y = i)$, $i = 1, \ldots, m$. Let $\boldsymbol{\pi} = (\pi_1, \ldots, \pi_m)^{\top}$, we have $\boldsymbol{\pi} \in \mathbb{T}_m$, where $\mathbb{T}_m$ is defined by (5.1).

To carry out a survey in which each category includes sensitive attribute, we need to introduce a non-sensitive dichotomous variate W and another non-sensitive multichotomous variate U so that the three variables W, U and Y are mutually independent with known proportions

$$q = \Pr(W = 1) \quad \text{and} \quad p_i = \Pr(U = i), \quad i = 1, \ldots, m.$$

For example, when $m = 4$, we may define $W = 0$ if a respondent's birthday is in the first half of a month and $W = 1$ otherwise. Similarly, we define $U = i$ if a respondent was born in the i-th quarter of a year $(i = 1, \ldots, 4)$. Hence, it is reasonable to assume that $q \approx 0.5$ and $p_i \approx 0.25$ for each i.

Table 9.1 The multi-category parallel model and the corresponding cell probabilities

Category	$W = 0$	$W = 1$	Category	$W = 0$	$W = 1$	Marginal
$U = 1$	○		$U = 1$	$p_1(1 - q)$		p_1
$U = 2$	△		$U = 2$	$p_2(1 - q)$		p_2
$\vdots$	$\vdots$		$\vdots$	$\vdots$		$\vdots$
$U = m$	●		$U = m$	$p_m(1 - q)$		p_m
$Y = 1$		○	$Y = 1$		$\pi_1 q$	π_1
$Y = 2$		△	$Y = 2$		$\pi_2 q$	π_2
$\vdots$		$\vdots$	$\vdots$		$\vdots$	$\vdots$
$Y = m$		●	$Y = m$		$\pi_m q$	π_m
			Marginal	$1 - q$	q	1

Note: Please truthfully link the two circles by a straight line if you belong to $\{U = 1,\ W = 0\} \cup \{Y = 1,\ W = 1\}$, or link the two triangles by a straight line if you belong to $\{U = 2,\ W = 0\} \cup \{Y = 2,\ W = 1\}, \ldots$, or link the two dots by a straight line if you belong to $\{U = m,\ W = 0\} \cup \{Y = m,\ W = 1\}$.

Some practical guidelines to choose the two non-sensitive variates W and U are given in Section 9.7.

An interviewer may design a questionnaire in the format as shown at the left-hand side of Table 9.1 and ask an interviewee to truthfully link the two circles by a straight line if he/she belongs to one of the two circles (i.e., $\{U = 1,\ W = 0\}$ or $\{Y = 1,\ W = 1\}$); or to connect the two triangles by a straight line if he/she belongs one of the two triangles (i.e., $\{U = 2,\ W = 0\}$ or $\{Y = 2,\ W = 1\}$); ...; or to connect the two dots by a straight line if he/she belongs to one of the two dots (i.e., $\{U = m,\ W = 0\}$ or $\{Y = m,\ W = 1\}$). Note that all $\{W = 0\}$, $\{W = 1\}$ and $\{U = i\}$ are non-sensitive subclasses, thus

$$\{U = i,\ W = 0\} \cup \{Y = i,\ W = 1\}, \quad i = 1, \ldots, m,$$

are also non-sensitive subclasses. Therefore, each interviewee's privacy is well protected, and the interviewer does not have information on whether the interviewee belongs to a sensitive subclass or not. Liu & Tian (2013a) called this the multi-category parallel model. The right-hand side of Table 9.1 shows the corresponding cell probabilities. Since the three random variables W, U and Y are independent, the joint probability is the product of two corresponding marginal probabilities.

For those who may not completely understand the questionnaire shown in Table 9.1, we can formulate the survey design of the multi-category parallel model in verbal format. For example, let $m = 4$ and define $Y = 1$, 2, 3 or 4 if the number of days of the illegal drug usage last month for a respondent is 0–1, 2, 3 or $\geqslant 4$. Thus, the 4-category parallel model can be re-formulated in the following way:

(1) If your birthday is in the first half of a month (i.e., $W = 0$), please answer '1' (i.e., $U = 1$), or '2' (i.e., $U = 2$), or '3' (i.e., $U = 3$), or '4' (i.e., $U = 4$) to the question: *Which quarter of the year is your birthday?*

(2) If your birthday is in the second half of a month (i.e., $W = 1$), please answer '1' (i.e., $Y = 1$), or '2' (i.e., $Y = 2$), or '3' (i.e., $Y = 3$), or '4' (i.e., $Y = 4$) to the question: *How many days have you used illegal drugs in last month?*

9.2 Likelihood-based Inferences

9.2.1 MLEs via the EM algorithm

Suppose that a survey with n respondents is conducted and we observe n_1 respondents connecting the two circles, n_2 respondents connecting the two triangles, $\ldots$, and n_m respondents connecting the two dots (see, Table 9.1). Let $Y_{\mathrm{obs}} = \{n; n_1, \ldots, n_m\}$ denote the observed data with $n = n_1 + \cdots + n_m$. Hence, the observed-data likelihood function for $\boldsymbol{\pi} = (\pi_1, \ldots, \pi_m)^{\top}$ is

$$L_{\mathrm{MP}}(\boldsymbol{\pi}|Y_{\mathrm{obs}}) = \binom{n}{n_1, \ldots, n_m} \prod_{i=1}^{m} \{p_i(1-q) + \pi_i q\}^{n_i}, \qquad (9.1)$$

where the subscript 'MP' refers to the 'multi-category parallel' model.

We employ the EM algorithm to calculate the MLEs of $\{\pi_i\}_{i=1}^{m}$ by introducing a latent vector $\mathbf{z} = (Z_1, \ldots, Z_m)^{\top}$, where Z_i denotes the number of respondents belonging to the sensitive subclass $\{Y = i, W = 1\}$. Let $\mathbf{z} = (z_1, \ldots, z_m)^{\top}$ denote the realization of $\mathbf{z}$. We denote the complete data by $Y_{\mathrm{com}} = \{Y_{\mathrm{obs}}, \mathbf{z}\}$. Note that $\{p_i\}_{i=1}^{m}$ and q are known. The complete-data likelihood function for $\boldsymbol{\pi}$ is

$$L_{\mathrm{MP}}(\boldsymbol{\pi}|Y_{\mathrm{obs}}, \mathbf{z}) \propto \prod_{i=1}^{m} \{p_i(1-q)\}^{n_i - z_i} (\pi_i q)^{z_i}$$

$$\propto \prod_{i=1}^{m} \pi_i^{z_i}. \qquad (9.2)$$

Therefore, the M-step is to calculate the complete-data MLEs of $\{\pi_i\}_{i=1}^m$, which are given by

$$\pi_i = \frac{z_i}{z_1 + \cdots + z_m}, \quad i = 1, \ldots, m. \tag{9.3}$$

Since the conditional predictive density is

$$
\begin{aligned}
f(\boldsymbol{z}|Y_{\text{obs}}, \boldsymbol{\pi}) &= \prod_{i=1}^m f(z_i|Y_{\text{obs}}, \pi_i) \\
&= \prod_{i=1}^m \text{Binomial}\left(z_i \middle| n_i, \frac{\pi_i q}{p_i(1-q) + \pi_i q}\right), \tag{9.4}
\end{aligned}
$$

the E-step is to replace $\{z_i\}_{i=1}^m$ in (9.3) by the conditional expectations

$$E(Z_i|Y_{\text{obs}}, \pi_i) = \frac{n_i \pi_i q}{p_i(1-q) + \pi_i q}, \quad i = 1, \ldots, m. \tag{9.5}$$

9.2.2   Two bootstrap confidence intervals

We utilize the bootstrap method to derive the corresponding confidence intervals of $\{\pi_i\}_{i=1}^m$. Based on the obtained MLE $\hat{\boldsymbol{\pi}}_{\text{MP}} = (\hat{\pi}_{\text{MP}1}, \ldots, \hat{\pi}_{\text{MP}m})^{\top}$ of $\boldsymbol{\pi}$, we can generate

$$(n_1^*, \ldots, n_m^*)^{\top} \sim \text{Multinomial}_m\left(n, \ (1-q)\boldsymbol{p} + q\hat{\boldsymbol{\pi}}_{\text{MP}}\right).$$

where $\boldsymbol{p} = (p_1, \ldots, p_m)^{\top}$. For the bootstrap sample $\{n_1^*, \ldots, n_m^*\}$, we can compute the bootstrap replication $\hat{\pi}_{\text{MP}i}^*$ via the EM algorithm (9.3) and (9.5) by replacing $\{n_1, \ldots, n_m\}$ with $\{n_1^*, \ldots, n_m^*\}$. Independently repeating this process G times, we obtain G bootstrap replications $\{\hat{\pi}_{\text{MP}i}^*(g)\}_{g=1}^G$. Therefore, the standard error, $\text{se}(\hat{\pi}_{\text{MP}i})$, of $\hat{\pi}_{\text{MP}i}$ can be estimated by the sample standard deviation of the G replications, i.e.,

$$\widehat{\text{se}}(\hat{\pi}_{\text{MP}i}) = \left[\frac{1}{G-1}\sum_{g=1}^G \left\{\hat{\pi}_{\text{MP}i}^*(g) - \frac{\hat{\pi}_{\text{MP}i}^*(1) + \cdots + \hat{\pi}_{\text{MP}i}^*(G)}{G}\right\}^2\right]^{\frac{1}{2}}. \tag{9.6}$$

If $\{\hat{\pi}_{\text{MP}i}^*(g)\}_{g=1}^G$ is approximately normally distributed, a $(1-\alpha)100\%$ bootstrap confidence interval for π_i is given by

$$\left[\hat{\pi}_{\text{MP}i} - z_{\alpha/2} \times \widehat{\text{se}}(\hat{\pi}_{\text{MP}i}), \ \hat{\pi}_{\text{MP}i} + z_{\alpha/2} \times \widehat{\text{se}}(\hat{\pi}_{\text{MP}i})\right], \tag{9.7}$$

where z_{α} is the upper α-th quantile of the standard normal distribution. Alternatively, if $\{\hat{\pi}_{\text{MP}i}^*(g)\}_{g=1}^G$ is non-normally distributed or the bootstrap confidence interval (9.7) is beyond the unit interval $(0, 1)$, a $(1 - \alpha)100\%$ bootstrap confidence interval of π_i can be obtained by

$$[\hat{\pi}_{\mathrm{MP}i,\mathrm{BL}}, \ \hat{\pi}_{\mathrm{MP}i,\mathrm{BU}}], \tag{9.8}$$

where $\hat{\pi}_{\mathrm{MP}i,\mathrm{BL}}$ and $\hat{\pi}_{\mathrm{MP}i,\mathrm{BU}}$ are the $100(\alpha/2)$ and $100(1 - \alpha/2)$ percentiles of $\{\hat{\pi}^*_{\mathrm{MP}i}(g)\}_{g=1}^{G}$, respectively.

9.2.3   Explicit solutions to the valid estimators

Although the resulting MLE $\hat{\boldsymbol{\pi}}_{\mathrm{MP}}$ via the EM algorithm (9.3) and (9.5) definitely belongs to $\mathbb{T}_m$, we can only obtain numerical solution to $\hat{\boldsymbol{\pi}}_{\mathrm{MP}}$. In addition, the variance-covariance matrix of $\hat{\boldsymbol{\pi}}_{\mathrm{MP}}$ does not have a closed-form expression. However, for some cases, we can obtain explicit solutions to $\hat{\boldsymbol{\pi}}_{\mathrm{MP}}$ and its variance-covariance matrix.

From (9.1), the log-likelihood function is given by

$$\ell_{\mathrm{MP}}(\boldsymbol{\pi}|Y_{\mathrm{obs}}) = c + \sum_{i=1}^{m} n_i \log\{p_i(1 - q) + \pi_i q\},$$

where c is a constant not depending on $\boldsymbol{\pi}$. Let

$$\frac{\partial \ell_{\mathrm{MP}}(\boldsymbol{\pi}|Y_{\mathrm{obs}})}{\partial \pi_i} = 0, \quad i = 1, \dots, m - 1,$$

an alternative estimator of $\boldsymbol{\pi}$ is given by

$$\begin{aligned}
\hat{\boldsymbol{\pi}}_v &= (\hat{\pi}_{v1}, \dots, \hat{\pi}_{vm})^{\top} \\
&= \left(\frac{n_1/n - p_1(1 - q)}{q}, \dots, \frac{n_m/n - p_m(1 - q)}{q}\right)^{\top}.
\end{aligned} \tag{9.9}$$

Although $\hat{\pi}_{vi}$ is an unbiased estimator of the true proportion π_i, $\hat{\boldsymbol{\pi}}_v$ may not belong to $\mathbb{T}_m$. For example, let $m = 4$, $p_1 = \dots = p_4 = 0.25$, $q = 1/3$ and $(n_1, \dots, n_4)^{\top} = (15, 19, 7, 9)^{\top}$, then

$$\hat{\boldsymbol{\pi}}_v = (0.40, \ 0.64, \ -0.08, \ 0.04)^{\top} \notin \mathbb{T}_4.$$

In this chapter, the estimator $\hat{\boldsymbol{\pi}}_v$ given by (9.9) is said to be *valid* if $\hat{\boldsymbol{\pi}}_v \in \mathbb{T}_m$. Clearly, if $\hat{\boldsymbol{\pi}}_v$ specified by (9.9) is a valid estimator of $\boldsymbol{\pi}$, then $\hat{\boldsymbol{\pi}}_v = \hat{\boldsymbol{\pi}}_{\mathrm{MP}}$. In the following discussion, we only consider the case of valid estimators.

Note that

$$(n_1, \dots, n_m)^{\top} \sim \mathrm{Multinomial}_m(n, \boldsymbol{\lambda}),$$

where $\boldsymbol{\lambda} = (\lambda_1, \dots, \lambda_m)^{\top} \in \mathbb{T}_m$,

$$\lambda_i = p_i(1 - q) + \pi_i q, \quad i = 1, \dots, m. \tag{9.10}$$

Let $\hat{\boldsymbol{\lambda}} = (\hat{\lambda}_1, \ldots, \hat{\lambda}_m)^\top = (n_1/n, \ldots, n_m/n)^\top$ denote the MLE of $\boldsymbol{\lambda}$. In the matrix form, then (9.9) can be rewritten as

$$\hat{\boldsymbol{\pi}}_{\mathrm{MP}} = (\hat{\pi}_{\mathrm{MP}1}, \ldots, \hat{\pi}_{\mathrm{MP}m})^\top = \frac{\hat{\boldsymbol{\lambda}} - (1-q)\boldsymbol{p}}{q}, \qquad (9.11)$$

where $\boldsymbol{p} = (p_1, \ldots, p_m)^\top$. Thus, the variance-covariance matrix of $\hat{\boldsymbol{\pi}}_{\mathrm{MP}}$ is given by

$$\begin{aligned}
&\mathrm{Var}(\hat{\boldsymbol{\pi}}_{\mathrm{MP}}) \\
&= \frac{1}{q^2}\mathrm{Var}(\hat{\boldsymbol{\lambda}}) \\
&= \frac{1}{nq^2}\begin{pmatrix}
\lambda_1(1-\lambda_1) & -\lambda_1\lambda_2 & \cdots & -\lambda_1\lambda_m \\
-\lambda_2\lambda_1 & \lambda_2(1-\lambda_2) & \cdots & -\lambda_2\lambda_m \\
\vdots & \vdots & \ddots & \vdots \\
-\lambda_m\lambda_1 & -\lambda_m\lambda_2 & \cdots & \lambda_m(1-\lambda_m)
\end{pmatrix}.
\end{aligned} \qquad (9.12)$$

9.2.4 Three asymptotic confidence intervals

(a) Wald confidence intervals

From (9.12), it is not difficult to show the following result.

Theorem 9.1 Let

$$\overline{\mathrm{Var}}(\hat{\pi}_{\mathrm{MP}i}) = \frac{\hat{\lambda}_i(1-\hat{\lambda}_i)}{(n-1)q^2}, \quad i = 1, \ldots, m. \qquad (9.13)$$

We have

$$\overline{\mathrm{Var}}(\hat{\pi}_{\mathrm{MP}i}) = \frac{\hat{\pi}_{\mathrm{MP}i}(1-\hat{\pi}_{\mathrm{MP}i})}{n-1} + \frac{(1-q)f(\hat{\pi}_{\mathrm{MP}i}, p_i, q)}{(n-1)q^2},$$

where

$$f(\pi_i, p_i, q) \mathrel{\hat{=}} q(1-2p_i)\pi_i + p_i(1-p_i+qp_i),$$

and $\overline{\mathrm{Var}}(\hat{\pi}_{\mathrm{MP}i})$ is an unbiased estimator of

$$\mathrm{Var}(\hat{\pi}_{\mathrm{MP}i}) = \frac{\lambda_i(1-\lambda_i)}{nq^2}, \quad i = 1, \ldots, m. \qquad \P$$

Based on the large-sample property of MLE, we have

$$(\hat{\pi}_{\text{MP}i} - \pi_i)\Big/\sqrt{\text{Var}(\hat{\pi}_{\text{MP}i})} \sim N(0,1), \quad \text{as } n \to \infty, \quad i = 1,\ldots,m.$$

Thus, the $(1-\alpha)100\%$ Wald confidence interval of π_i is given by

$$\left[\hat{\pi}_{\text{MP}i} - z_{\alpha/2}\sqrt{\text{Var}(\hat{\pi}_{\text{MP}i})}, \ \hat{\pi}_{\text{MP}i} + z_{\alpha/2}\sqrt{\text{Var}(\hat{\pi}_{\text{MP}i})}\right]. \tag{9.14}$$

(b) Wilson (score) confidence intervals

The $(1-\alpha)100\%$ Wilson (score) confidence interval of π_i can be constructed based on

$$
\begin{aligned}
& 1 - \alpha \\
=\ & \Pr\left\{\left|\frac{\hat{\pi}_{\text{MP}i} - \pi_i}{\sqrt{\text{Var}(\hat{\pi}_{\text{MP}i})}}\right| \leqslant z_{\alpha/2}\right\} \\
=\ & \Pr\left\{(\hat{\pi}_{\text{MP}i} - \pi_i)^2 \leqslant z_{\alpha/2}^2 \text{Var}(\hat{\pi}_{\text{MP}i})\right\} \\
\overset{(9.12)}{=}\ & \Pr\left[(\hat{\pi}_{\text{MP}i} - \pi_i)^2 \leqslant \frac{z_{\alpha/2}^2}{n}\left\{\pi_i(1-\pi_i) + \frac{(1-q)f(\pi_i, p_i, q)}{q^2}\right\}\right] \\
=\ & \Pr\left\{\hat{\pi}_{\text{MP}i}^2 - 2\hat{\pi}_{\text{MP}i}\pi_i + \pi_i^2 \leqslant \frac{z_{\alpha/2}^2(-\pi_i^2 + \rho_1\pi_i + \rho_2)}{n}\right\} \\
=\ & \Pr\left\{(1+z_*)\pi_i^2 - (2\hat{\pi}_{\text{MP}i} + z_*\rho_1)\pi_i + \hat{\pi}_{\text{MP}i}^2 - z_*\rho_2 \leqslant 0\right\}, \tag{9.15}
\end{aligned}
$$

where $z_* \overset{\circ}{=} z_{\alpha/2}^2/n$,

$$\rho_1 \overset{\circ}{=} \frac{1 - 2p_i(1-q)}{q} \quad \text{and} \quad \rho_2 \overset{\circ}{=} \frac{p_i(1-q)(1-p_i+qp_i)}{q^2}.$$

Solving the quadratic inequality inside the probability in (9.15), we obtain the Wilson (score) confidence interval of π_i by

$$\frac{2\hat{\pi}_{\text{MP}i} + z_*\rho_1 \pm \sqrt{(2\hat{\pi}_{\text{MP}i} + z_*\rho_1)^2 - 4(1+z_*)(\hat{\pi}_{\text{MP}i}^2 - z_*\rho_2)}}{2(1+z_*)}, \tag{9.16}$$

which is, in general, within $[0,1]$.

The Wilson confidence interval has been shown to have better performance than the Wald confidence interval and the exact (Clopper–Pearson) confidence interval, see Clopper & Pearson (1934), Agresti & Coull (1998), Newcombe (1998), and Brown, Cai & DasGupta (2001) for more detail.

(c) Likelihood ratio confidence intervals

For sensitive responses where some of the true values $\{\pi_i\}$ are often small, *likelihood ratio confidence intervals* (LRCIs) could provide better performance than other alternatives. To construct the LRCI of π_i $(i = 1, \ldots, m)$, we consider to test the following hypothses

$$H_0: \pi_i = \pi_{i0} \quad \text{against} \quad H_1: H_0 \text{ is not true.}$$

Let $\hat{\boldsymbol{\pi}}_{\mathrm{R}} = (\hat{\pi}_{1,\mathrm{R}}, \ldots, \hat{\pi}_{m,\mathrm{R}})^{\top}$ denote the restricted MLE of $\boldsymbol{\pi}$ under H_0. It can be verified that

$$\begin{cases} \hat{\pi}_{i,\mathrm{R}} = \pi_{i0}, \\ \hat{\pi}_{j,\mathrm{R}} = \dfrac{\{1 - p_i(1 - q) - q\pi_{i0}\}n_j/(n - n_i) - p_j(1 - q)}{q}, \end{cases}$$

where $j = 1, \ldots, m$ and $j \neq i$. When $n \to \infty$, it is well known that

$$\Lambda(\pi_{i0}) = -2\{\ell_{\mathrm{MP}}(\hat{\boldsymbol{\pi}}_{\mathrm{R}}|Y_{\mathrm{obs}}) - \ell_{\mathrm{MP}}(\hat{\boldsymbol{\pi}}_v|Y_{\mathrm{obs}})\} \stackrel{\cdot}{\sim} \chi^2(1),$$

where $\hat{\boldsymbol{\pi}}_v$ denotes the unrestricted MLE of $\boldsymbol{\pi}$ specified by (9.9). Since

$$\begin{aligned} \Lambda(\pi_{i0}) = -2\Bigg[& n_i \log\{p_i(1 - q) + q\pi_{i0}\} + \sum_{j=1, j\neq i}^{m} n_j \log\{p_j(1 - q) + q\hat{\pi}_{j,\mathrm{R}}\} \\ & - \sum_{k=1}^{m} n_k \log\{p_k(1 - q) + q\hat{\pi}_{vk}\} \Bigg], \end{aligned} \tag{9.17}$$

it is easy to verify that $\Lambda(\pi_{i0})$ is a decreasing function of π_{i0} when

$$\pi_{i0} \in \left[0, \; \frac{n_i/n - p_i(1 - q)}{q}\right]$$

and an increasing function of π_{i0} when

$$\pi_{i0} \in \left[\frac{n_i/n - p_i(1 - q)}{q}, \; 1\right].$$

Therefore, for a given significance level α, the $(1 - \alpha)100\%$ LRCI for π_i is

$$[\hat{\pi}_{\text{MP}i,\,\text{LRL}}, \ \hat{\pi}_{\text{MP}i,\,\text{LRU}}], \tag{9.18}$$

where $\hat{\pi}_{\text{MP}i,\,\text{LRL}}$ and $\hat{\pi}_{\text{MP}i,\,\text{LRU}}$ are two roots of π_{i0} to the following equation

$$\Lambda(\pi_{i0}) = \chi^2(\alpha, 1), \tag{9.19}$$

where $\chi^2(\alpha, 1)$ denotes the upper α-th quantile of the chi-squared distribution with one degree of freedom.

The asymptotic confidence intervals (9.14),(9.16) and (9.18) are appropriate for the cases of large sample sizes. When n is small to moderate, we could use the bootstrap confidence intervals (9.7) and/or (9.8).

9.3 Bayesian Inferences

In many applications, a certain prior knowledge on the proportions $\{\pi_i\}_{i=1}^m$ is available before data are collected. For example, Kim, Tebbs & An (2006) reported that the prevalence of homosexuality among American males is between 1% and 10%. Such prior information should be included in obtaining the estimate of $\boldsymbol{\pi}$. In this section, we first utilize the Bayesian approach to derive the exact posterior distribution of $\boldsymbol{\pi}$ and its mixed posterior moments. We then employ the EM algorithm to find the posterior mode when the posterior distribution of $\boldsymbol{\pi}$ is highly skewed. Finally, we generate posterior samples of $\boldsymbol{\pi}$ via the data augmentation algorithm of Tanner & Wong (1987).

9.3.1 Posterior moments

The likelihood function of $\boldsymbol{\pi}$ is given by (9.1). If the Dirichlet distribution $\text{Dirichlet}_m(\boldsymbol{a})$ with $\boldsymbol{a} = (a_1, \ldots, a_m)^\top$ is adopted as the prior distribution of $\boldsymbol{\pi}$, then the posterior distribution of $\boldsymbol{\pi}$ is given by

$$
\begin{aligned}
f(\boldsymbol{\pi}|Y_{\text{obs}}) &= \frac{l_{\text{MP}}(\boldsymbol{\pi}|Y_{\text{obs}}) \times \text{Dirichlet}_m(\boldsymbol{\pi}|\boldsymbol{a})}{\displaystyle\int_{\mathbb{T}_m} l_{\text{MP}}(\boldsymbol{\pi}|Y_{\text{obs}}) \times \text{Dirichlet}_m(\boldsymbol{\pi}|\boldsymbol{a}) \, d\boldsymbol{\pi}} \\[2ex]
&= \frac{l_{\text{MP}}(\boldsymbol{\pi}|Y_{\text{obs}}) \times \prod_{i=1}^m \pi_i^{a_i-1}}{c_{\text{MP}}(\boldsymbol{a}; \boldsymbol{n})},
\end{aligned} \tag{9.20}
$$

where

$$l_{\text{MP}}(\boldsymbol{\pi}|Y_{\text{obs}}) \hat{=} \prod_{j=1}^m \{p_j(1-q) + \pi_j q\}^{n_j}$$

and $\boldsymbol{n} = (n_1, \ldots, n_m)^{\mathsf{T}}$. Since

$$l_{\mathrm{MP}}(\boldsymbol{\pi}|Y_{\mathrm{obs}}) \times \prod_{i=1}^{m} \pi_i^{a_i-1}$$

$$= \prod_{j=1}^{m} \sum_{k_j=0}^{n_j} \binom{n_j}{k_j} \{p_j(1-q)\}^{k_j} (\pi_j q)^{n_j-k_j} \times \prod_{i=1}^{m} \pi_i^{a_i-1}$$

$$= \prod_{j=1}^{m} \sum_{k_j=0}^{n_j} \binom{n_j}{k_j} q^{n_j} p_j^{k_j} \left(\frac{1-q}{q}\right)^{k_j} \pi_j^{a_j+n_j-k_j-1}$$

$$= q^n \sum_{k_1=0}^{n_1} \cdots \sum_{k_m=0}^{n_m} \left(\frac{1-q}{q}\right)^{k_+} \left\{\prod_{j=1}^{m} \binom{n_j}{k_j} p_j^{k_j}\right\} \prod_{i=1}^{m} \pi_i^{a_i+n_i-k_i-1},$$

where $k_+ = \sum_{i=1}^{m} k_i$, we obtain

$$c_{\mathrm{MP}}(\boldsymbol{a};\boldsymbol{n})$$

$$= \int_{\mathbb{T}_m} l_{\mathrm{MP}}(\boldsymbol{\pi}|Y_{\mathrm{obs}}) \times \prod_{i=1}^{m} \pi_i^{a_i-1} \, \mathrm{d}\boldsymbol{\pi}$$

$$= q^n \sum_{k_1=0}^{n_1} \cdots \sum_{k_m=0}^{n_m} \left(\frac{1-q}{q}\right)^{k_+} \left\{\prod_{j=1}^{m} \binom{n_j}{k_j} p_j^{k_j}\right\} \int_{\mathbb{T}_m} \prod_{i=1}^{m} \pi_i^{a_i+n_i-k_i-1} \, \mathrm{d}\boldsymbol{\pi}$$

$$= q^n \sum_{k_1=0}^{n_1} \cdots \sum_{k_m=0}^{n_m} \left(\frac{1-q}{q}\right)^{k_+} \frac{\displaystyle\prod_{i=1}^{m} \binom{n_i}{k_i} p_i^{k_i} \Gamma(a_i + n_i - k_i)}{\Gamma(a_+ + n - k_+)}, \qquad (9.21)$$

where $a_+ = \sum_{i=1}^{m} a_i$. Therefore, For any $r_1, \ldots, r_m \geqslant 0$, the mixed posterior moment of $\boldsymbol{\pi}$ is given by

$$E\left(\prod_{i=1}^{m} \pi_i^{r_i} \middle| Y_{\mathrm{obs}}\right) = \frac{c_{\mathrm{MP}}(\boldsymbol{a}+\boldsymbol{r};\boldsymbol{n})}{c_{\mathrm{MP}}(\boldsymbol{a};\boldsymbol{n})},$$

where $\boldsymbol{r} = (r_1, \ldots, r_m)^{\mathsf{T}}$.

9.3.2   Calculation of the posterior mode via the EM algorithm

The complete-data likelihood function of $\boldsymbol{\pi}$ is given by (9.2). Again, we consider Dirichlet$_m(\boldsymbol{a})$ as the prior distribution of $\boldsymbol{\pi}$. The complete-data posterior distribution is

$$f(\boldsymbol{\pi}|Y_{\text{obs}}, \boldsymbol{z}) = \text{Dirichlet}\,(\boldsymbol{\pi}|a_1 + z_1, \ldots, a_m + z_m) \qquad (9.22)$$

and the conditional predictive density is given by (9.4). Let $\boldsymbol{\pi}^{(t)}$ be the t-th approximation to the mode $\tilde{\boldsymbol{\pi}}$ of the observed-data posterior density $f(\boldsymbol{\pi}|Y_{\text{obs}})$. The E-step of the EM algorithm is to calculate the Q function:

$$\begin{aligned}
Q(\boldsymbol{\pi}|\boldsymbol{\pi}^{(t)}) &= E\left\{\log f(\boldsymbol{\pi}|Y_{\text{obs}}, \boldsymbol{z})\,\middle|\,Y_{\text{obs}}, \boldsymbol{\pi}^{(t)}\right\} \\[2mm]
&= \int \log f(\boldsymbol{\pi}|Y_{\text{obs}}, \boldsymbol{z}) \times f(\boldsymbol{z}|Y_{\text{obs}}, \boldsymbol{\pi}^{(t)})\,\mathrm{d}\boldsymbol{z} \\[2mm]
&= \int \left\{c(\boldsymbol{z}) + \sum_{i=1}^{m}(a_i + z_i - 1)\log \pi_i\right\} \times f(\boldsymbol{z}|Y_{\text{obs}}, \boldsymbol{\pi}^{(t)})\,\mathrm{d}\boldsymbol{z} \\[2mm]
&= E\left\{c(\boldsymbol{z})\,\middle|\,Y_{\text{obs}}, \boldsymbol{\pi}^{(t)}\right\} + \sum_{i=1}^{m}\left\{a_i + E(Z_i|Y_{\text{obs}}, \boldsymbol{\pi}^{(t)}) - 1\right\}\log \pi_i,
\end{aligned}$$

where $c(\boldsymbol{z}) = -\log B(a_1 + Z_1, \ldots, a_m + Z_m)$. The M-step is to find

$$\boldsymbol{\pi}^{(t+1)} = \arg\max_{\boldsymbol{\pi} \in \mathbb{T}_m} Q(\boldsymbol{\pi}|\boldsymbol{\pi}^{(t)}),$$

which results in

$$\pi_i^{(t+1)} = \frac{a_i + E(Z_i|Y_{\text{obs}}, \boldsymbol{\pi}^{(t)}) - 1}{a_+ + E(Z_+|Y_{\text{obs}}, \boldsymbol{\pi}^{(t)}) - m}, \quad i = 1, \ldots, m, \qquad (9.23)$$

where $Z_+ = \sum_{i=1}^{m} Z_i$ and

$$E(Z_i|Y_{\text{obs}}, \boldsymbol{\pi}^{(t)}) = \frac{n_i \pi_i^{(t)} q}{p_i(1-q) + \pi_i^{(t)} q}, \quad i = 1, \ldots, m. \qquad (9.24)$$

9.3.3 Generation of posterior samples via the data augmentation algorithm

To make Bayesian inferences on the parameter vector $\boldsymbol{\pi}$, we need to generate posterior samples from the observed posterior distribution $f(\boldsymbol{\pi}|Y_{\text{obs}})$ by using the data augmentation algorithm. The I-step of the data augmentation algorithm is to draw the missing values of $\boldsymbol{z}$ from (9.4) for any given Y_{obs} and $\boldsymbol{\pi}$, and the P-step is to draw $\boldsymbol{\pi}$ from (9.22) for given Y_{obs} and updated $\boldsymbol{z}$.

9.4   A Special Case of the Multi-category Parallel Model

In this section, we consider a special case of the multi-category parallel model with four categories, which can be utilized to investigate the association of two binary sensitive variates. Some simulation studies are conducted to assess the performance of the likelihood ratio test and the chi-squared test by comparing their empirical Type I error rates (or the actual significance levels) and powers.

9.4.1   A four-category parallel model

Let X and Y be two dichotomous variates associated with two sensitive questions. For example, X represents whether or not a respondent is an illegal drug user and Y denotes whether a respondent is with AIDS or not. Let $X = 1$ and $Y = 1$ denote the sensitive characteristics of a respondent (e.g., $X = 1$ if the respondent is a drug user), and $X = 0$ and $Y = 0$ denote the non-sensitive characteristics of a respondent (e.g., $Y = 0$ if a respondent is without AIDS). Define

$$
\begin{aligned}
\pi_1 &= \Pr(X = 0,\ Y = 0), \\
\pi_2 &= \Pr(X = 0,\ Y = 1), \\
\pi_3 &= \Pr(X = 1,\ Y = 0), \quad \text{and} \\
\pi_4 &= \Pr(X = 1,\ Y = 1).
\end{aligned}
$$

Obviously, we have

$$
\boldsymbol{\pi} = (\pi_1, \ldots, \pi_4)^{\top} \in \mathbb{T}_4.
$$

From Table 9.1, the survey design for the four-category parallel model is displayed in Table 9.2. Two major objectives here are to collect sensitive data and to test whether or not the association exists between the two binary variates X and Y.

9.4.2   Testing hypotheses for association

A commonly used index for measuring the association of two binary variates is the odds ratio

$$
\psi = \frac{\pi_1 \pi_4}{\pi_2 \pi_3}.
$$

Assume that we want to test

$$
H_0\colon \psi = 1 \quad \text{against} \quad H_1\colon \psi \neq 1.
$$

Table 9.2 The four-category parallel model and the corresponding cell probabilities

Category	$W = 0$	$W = 1$	Category	$W = 0$	$W = 1$	Marginal
$U = 1$	○		$U = 1$	$p_1(1-q)$		p_1
$U = 2$	△		$U = 2$	$p_2(1-q)$		p_2
$U = 3$	□		$U = 3$	$p_3(1-q)$		p_3
$U = 4$	●		$U = 4$	$p_4(1-q)$		p_4
$X = 0, Y = 0$		○	$X = 0, Y = 0$		$\pi_1 q$	π_1
$X = 0, Y = 1$		△	$X = 0, Y = 1$		$\pi_2 q$	π_2
$X = 1, Y = 0$		□	$X = 1, Y = 0$		$\pi_3 q$	π_3
$X = 1, Y = 1$		●	$X = 1, Y = 1$		$\pi_4 q$	π_4
			Marginal	$1 - q$	q	1

(a) The likelihood ratio test

The likelihood ratio statistic is defined by

$$\Lambda_1 = -2\left\{\ell_{\mathrm{MP}}(\hat{\boldsymbol{\pi}}_0|Y_{\mathrm{obs}}) - \ell_{\mathrm{MP}}(\hat{\boldsymbol{\pi}}_{\mathrm{MP}}|Y_{\mathrm{obs}})\right\} \overset{\cdot}{\sim} \chi^2(1), \quad \text{as } n \to \infty, \quad (9.25)$$

where $\hat{\boldsymbol{\pi}}_0$ denotes the restricted MLE of $\boldsymbol{\pi}$ under H_0 and $\hat{\boldsymbol{\pi}}_{\mathrm{MP}}$ denotes the MLE of $\boldsymbol{\pi}$ given by (9.11). To calculate $\hat{\boldsymbol{\pi}}_0$, let

$$\pi_x = \Pr(X = 1) = \pi_3 + \pi_4$$

and

$$\pi_y = \Pr(Y = 1) = \pi_2 + \pi_4.$$

If H_0 is true, i.e., X and Y are mutually independent, we have

$$\begin{cases} \pi_1 &= (1 - \pi_x)(1 - \pi_y), \\ \pi_2 &= (1 - \pi_x)\pi_y, \\ \pi_3 &= \pi_x(1 - \pi_y), \\ \pi_4 &= \pi_x\pi_y. \end{cases} \quad (9.26)$$

If we could obtain the restricted MLEs $\hat{\pi}_{0x}$ of π_x and $\hat{\pi}_{0y}$ of π_y, from (9.26) the restricted MLEs $\hat{\boldsymbol{\pi}}_0 = (\hat{\pi}_{01}, \ldots, \hat{\pi}_{04})^\top$ can be calculated as

$$
\begin{cases}
\hat{\pi}_{01} & = & (1 - \hat{\pi}_{0x})(1 - \hat{\pi}_{0y}), \\
\hat{\pi}_{02} & = & (1 - \hat{\pi}_{0x})\hat{\pi}_{0y}, \\
\hat{\pi}_{03} & = & \hat{\pi}_{0x}(1 - \hat{\pi}_{0y}), \\
\hat{\pi}_{04} & = & \hat{\pi}_{0x}\hat{\pi}_{0y}.
\end{cases}
\tag{9.27}
$$

Recall that the number of the respondents belonging to the subclass $\{Y = i,\ W = 1\}$ is denoted by Z_i and the frequencies $\{Z_i\}$ are unobservable. From (9.2), the complete-data likelihood function for $\boldsymbol{\pi}$ under H_0 becomes

$$
\begin{aligned}
& L_{\mathrm{MP}}(\pi_x, \pi_y | Y_{\mathrm{obs}}, \boldsymbol{z}, H_0) \\
\propto\ & \{(1 - \pi_x)(1 - \pi_y)\}^{z_1} \{(1 - \pi_x)\pi_y\}^{z_2} \{\pi_x(1 - \pi_y)\}^{z_3} (\pi_x \pi_y)^{z_4} \\
=\ & \pi_x^{z_3 + z_4}(1 - \pi_x)^{z_1 + z_2} \times \pi_y^{z_2 + z_4}(1 - \pi_y)^{z_1 + z_3}.
\end{aligned}
$$

Thus, the M-step is to calculate the complete-data MLEs of π_x and π_y as follows:

$$
\hat{\pi}_{0x} = \frac{z_3 + z_4}{z_+} \quad \text{and} \quad \hat{\pi}_{0y} = \frac{z_2 + z_4}{z_+},
\tag{9.28}
$$

respectively. From (9.4), the E-step is to find the conditional expectations:

$$
E(Z_i | Y_{\mathrm{obs}}, \hat{\boldsymbol{\pi}}_0) = \frac{n_i \hat{\pi}_{0i} q}{p_i(1 - q) + \hat{\pi}_{0i} q}, \quad i = 1, \ldots, m,
\tag{9.29}
$$

where $\{\hat{\pi}_{0i}\}_{i=1}^4$ are defined by (9.27).

(b) The chi-squared test

Alternatively, the chi-squared statistic can be utilized to test the H_0 against the H_1. Let $\boldsymbol{p} = (p_1, \ldots, p_4)^\top$ and $\boldsymbol{\lambda} = (\lambda_1, \ldots, \lambda_4)^\top$, where

$$
\lambda_i = (1 - q)p_i + q\pi_i, \quad i = 1, \ldots, 4.
$$

Therefore, we have

$$
\boldsymbol{\lambda} = (1 - q)\boldsymbol{p} + q\boldsymbol{\pi}.
$$

Note that the restricted MLE $\hat{\boldsymbol{\pi}}_0 = (\hat{\pi}_{01}, \ldots, \hat{\pi}_{04})^\top$ of $\boldsymbol{\pi}$ under H_0 is given by (9.27). We have

$$
\hat{\boldsymbol{\lambda}}_0 = (\hat{\lambda}_{01}, \ldots, \hat{\lambda}_{04})^\top = (1 - q)\boldsymbol{p} + q\hat{\boldsymbol{\pi}}_0
$$

is the restricted MLE of λ under H_0. Therefore, under H_0, the chi-squared statistic

$$\Lambda_2 = \sum_{i=1}^{4} \frac{(n_i - n\hat{\lambda}_{0i})^2}{n\hat{\lambda}_{0i}} \overset{\cdot}{\sim} \chi^2(1), \quad \text{as } n \to \infty. \tag{9.30}$$

9.4.3 Comparison of the likelihood ratio test with the chi-squared test

For any given π_1, we consider the following three combinations of π_2, π_3 and π_4 such that $\sum_{i=1}^{4} \pi_i = 1$:

Scenario 1: $(\pi_2, \pi_3, \pi_4) = (3, 4, 1)\dfrac{1 - \pi_1}{8}, \quad \psi = \dfrac{2\pi_1}{3(1 - \pi_1)};$

Scenario 2: $(\pi_2, \pi_3, \pi_4) = (4, 10, 1)\dfrac{1 - \pi_1}{15}, \quad \psi = \dfrac{3\pi_1}{8(1 - \pi_1)};$ $\qquad$ (9.31)

Scenario 3: $(\pi_2, \pi_3, \pi_4) = (6, 13, 1)\dfrac{1 - \pi_1}{20}, \quad \psi = \dfrac{10\pi_1}{39(1 - \pi_1)}.$

The sample sizes are chosen to be $n = 50(50)500$. To compare the Type I error rates (i.e., $\pi_1\pi_4/(\pi_2\pi_3) = \psi = 1$), we take $\pi_1 = 3/5$ for scenario 1, $\pi_1 = 8/11$ for scenario 2, and $\pi_1 = 39/49$ for scenario 3. For the comparison of powers (i.e., $\psi \neq 1$), the chosen π_1 and the corresponding ψ are listed in Table 9.3.

Table 9.3 Various values of π_1 and ψ for the three scenarios specified by (9.31)

Scenario	π_1					
	0.200	0.300	0.500	0.700	0.800	0.900
Scenario 1: ψ	0.167	0.286	0.667	1.556	2.667	6.000
Scenario 2: ψ	0.094	0.161	0.375	0.875	1.500	3.375
Scenario 3: ψ	0.064	0.110	0.256	0.598	1.026	2.308

For a given pair (n, π_1), we independently generate

$$\left(n_1^{(l)}, \ldots, n_4^{(l)}\right)^{\top} \sim \text{Multinomial}_4\left(n, \frac{1}{8} + \frac{1}{2}\pi\right) \tag{9.32}$$

for $l = 1, \ldots, L$ $(L = 1000)$, where only $p_i = 1/4$ $(i = 1, \ldots, 4)$ and $q = 1/2$ are considered. All hypothesis testings are conducted at the level 0.05. Let

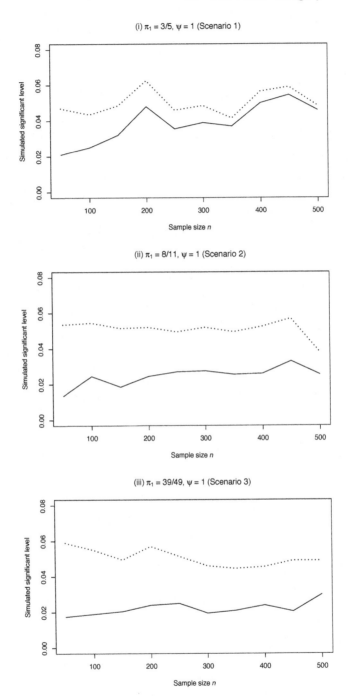

Figure 9.1 Comparisons of Type I error rates between the likelihood ratio test (solid line) and the chi-squared test (dotted line): (i) $\pi_1 = 3/5$, $\psi = 1$ (Scenario 1); (ii) $\pi_1 = 8/11$, $\psi = 1$ (Scenario 2); (iii) $\pi_1 = 39/49$, $\psi = 1$ (Scenario 3).

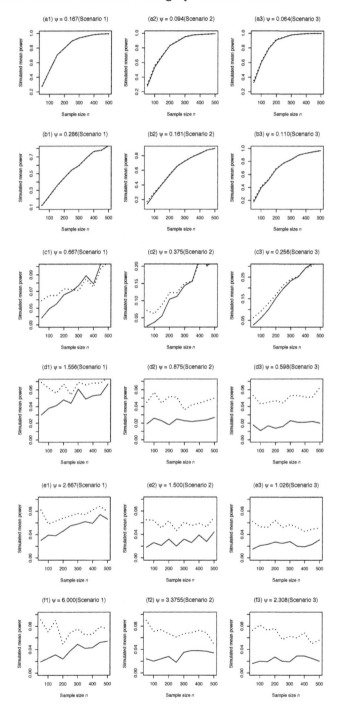

Figure 9.2 Comparisons of powers between the likelihood ratio test (solid line) and the chi-squared test (dotted line).

r_j denote the number of rejecting the null hypothesis (i.e., H_0: $\psi = 1$) by the statistics Λ_j $(j = 1, 2)$. Hence, the actual significance level can be estimated by r_j/L with $\psi = 1$ and the power of the test statistic Λ_j $(j = 1, 2)$ can be estimated by r_j/L with $\psi \neq 1$.

Figure 9.1 shows comparisons of Type I error rates between the likelihood ratio and chi-squared tests for three scenarios. In general, we find that the chi-squared test can well control its Type I error rates around the pre-chosen nominal level while the likelihood ratio test may be relatively conservative (see, Figures 9.1 (ii) and 9.1 (iii)).

Figure 9.2 gives the comparisons of powers between the likelihood ratio and chi-squared tests for different cases with $\psi \neq 1$. It is not difficult to find that there is no significant difference between the powers of the two test when ψ is small (i.e., < 0.40). When $0.60 < \psi < 1$, we always have that the chi-squared test is more powerful than the likelihood ratio test, no matter whether the sample size is large or small.

9.5 Comparison with the Multi-category Triangular Model

In this section, we theoretically compare the efficiency of the multi-category parallel model with the multi-category triangular model by comparing the two variance-covariance matrices of the MLEs of parameters based on the trace criterion. We also consider the comparison of the degree of privacy protection for the two models.

9.5.1 The difference between the trace of two variance-covariance matrices

(a) The variance-covariance matrix of $\hat{\boldsymbol{\pi}}_{\mathrm{MT}}$

For the multi-category triangular model, let $\boldsymbol{\theta} = (\theta_1, \ldots, \theta_m)^\top$, where

$$\theta_1 = p_1\pi_1 \quad \text{and} \quad \theta_j = p_j\pi_1 + \pi_j \ (j = 2, \ldots, m)$$

represent the proportions that the respondents belonging to Block j. In matrix notation, we have

$$\boldsymbol{\theta} = \mathbf{P}\boldsymbol{\pi} = \begin{pmatrix} p_1 & \mathbf{0}_{m-1}^\top \\ \boldsymbol{p}_{-1} & \mathbf{I}_{m-1} \end{pmatrix} \begin{pmatrix} \pi_1 \\ \vdots \\ \pi_m \end{pmatrix}, \tag{9.33}$$

where $\boldsymbol{p}_{-1} = (p_2, \ldots, p_m)^\top$, $\mathbf{0}_{m-1}$ is the $(m-1) \times 1$ vector of zeros and $\mathbf{I}_{m-1}$ denotes the $(m-1) \times (m-1)$ identity matrix. Since

$$(n_1, \ldots, n_m)^\top \sim \text{Multinomial}(n; \, \theta_1, \ldots, \theta_m),$$

the MLE of θ_j is $\hat{\theta}_j = n_j/n$. Thus, the variance-covariance matrix of $\hat{\boldsymbol{\theta}} = (\hat{\theta}_1, \ldots, \hat{\theta}_m)^\top$ is

$$\text{Var}(\hat{\boldsymbol{\theta}}) = \frac{1}{n}\Big\{\text{diag}(\boldsymbol{\theta}) - \boldsymbol{\theta}\boldsymbol{\theta}^\top\Big\}. \tag{9.34}$$

In general, the MLE $\hat{\boldsymbol{\pi}}_{\text{MT}}$ of $\boldsymbol{\pi}$ for the multi-category triangular model can be obtained by using the EM algorithm (Tang *et al.*, 2009). However, for some cases, we can obtain a closed-form solution to $\hat{\boldsymbol{\pi}}_{\text{MT}}$. In fact, from (9.33), we have $\boldsymbol{\pi} = \mathbf{P}^{-1}\boldsymbol{\theta}$. Since the MLE of $\boldsymbol{\theta}$ is $\hat{\boldsymbol{\theta}} = (n_1/n, \ldots, n_m/n)^\top$ so that an alternative estimator of $\boldsymbol{\pi}$ for the multi-category triangular model is given by

$$\hat{\boldsymbol{\pi}}_v = \mathbf{P}^{-1}\hat{\boldsymbol{\theta}} = \begin{pmatrix} 1/p_1 & \mathbf{0}_{m-1}^\top \\ -\boldsymbol{p}_{-1}/p_1 & \mathbf{I}_{m-1} \end{pmatrix} \begin{pmatrix} n_1/n \\ \vdots \\ n_m/n \end{pmatrix}. \tag{9.35}$$

It should be noted that it is possible that $\hat{\boldsymbol{\pi}}_v \notin \mathbb{T}_m$. For example, if $m = 4$, $p_1 = \cdots = p_4 = 0.25$ and $(n_1, \ldots, n_4)^\top = (12, 8, 6, 19)^\top$, then

$$\hat{\boldsymbol{\pi}}_v = (1.066667, \; -0.088889, \; -0.133333, \; 0.155556)^\top \notin \mathbb{T}_4.$$

In this chapter, the estimator $\hat{\boldsymbol{\pi}}_v$ given by (9.35) is said to be *valid* if $\hat{\boldsymbol{\pi}}_v \in \mathbb{T}_m$. Clearly, if $\hat{\boldsymbol{\pi}}_v$ specified by (9.35) is a valid estimator of $\boldsymbol{\pi}$ then $\hat{\boldsymbol{\pi}}_v = \hat{\boldsymbol{\pi}}_{\text{MT}}$. In the following discussion, we only consider the case of valid estimators.

Hence, from (9.35), (9.34) and (9.33), the variance-covariance matrix of $\hat{\boldsymbol{\pi}}_{\text{MT}}$ is

$$\begin{aligned}
\text{Var}(\hat{\boldsymbol{\pi}}_{\text{MT}}) &= \text{Var}(\hat{\boldsymbol{\pi}}_v) \\
&= \mathbf{P}^{-1}\text{Var}(\hat{\boldsymbol{\theta}})(\mathbf{P}^{-1})^\top \\
&= \frac{1}{n}\Big\{\mathbf{P}^{-1}\text{diag}(\mathbf{P}\boldsymbol{\pi})(\mathbf{P}^{-1})^\top - \boldsymbol{\pi}\boldsymbol{\pi}^\top\Big\},
\end{aligned} \tag{9.36}$$

or equivalently

$$\text{Var}(\hat{\pi}_{\text{MT1}}) = \frac{1}{n}\left(\frac{\pi_1}{p_1} - \pi_1^2\right),$$

$$\text{Var}(\hat{\pi}_{\text{MT}j}) = \frac{1}{n}\left(\frac{p_j^2}{p_1}\pi_1 + p_j\pi_1 + \pi_j - \pi_j^2\right), \quad 2 \leqslant j \leqslant m, \quad (9.37)$$

$$\text{Cov}(\hat{\pi}_{\text{MT}1}, \hat{\pi}_{\text{MT}j}) = \frac{1}{n}\left(-\frac{p_j}{p_1}\pi_1 - \pi_1\pi_j\right), \quad 2 \leqslant j \leqslant m, \quad \text{and}$$

$$\text{Cov}(\hat{\pi}_{\text{MT}i}, \hat{\pi}_{\text{MT}j}) = \frac{1}{n}\left(\frac{p_i p_j}{p_1}\pi_1 - \pi_i\pi_j\right), \quad i \neq j; \quad 2 \leqslant i, j \leqslant m.$$

(b) Comparison between Var($\hat{\boldsymbol{\pi}}_{\text{MT}}$) and Var($\hat{\boldsymbol{\pi}}_{\text{MP}}$)

In the multi-category triangular model, there are only two parameter vectors (i.e., $\boldsymbol{\pi}$ and $\boldsymbol{p}$), while in the multi-category parallel model, besides $\boldsymbol{\pi}$ and $\boldsymbol{p}$, there is an additional parameter q. By controlling q within a certain subset of the unit interval, we may have Var($\hat{\boldsymbol{\pi}}_{\text{MP}}$) being 'smaller' than Var($\hat{\boldsymbol{\pi}}_{\text{MT}}$). In what follows, we only apply the trace criterion in the comparison between Var($\hat{\boldsymbol{\pi}}_{\text{MT}}$) and Var($\hat{\boldsymbol{\pi}}_{\text{MP}}$).

First, from (9.36) and (9.37), we have

$$\text{tr}\left\{\text{Var}(\hat{\boldsymbol{\pi}}_{\text{MT}})\right\}$$

$$= \frac{1}{n}\left[\frac{\pi_1}{p_1} - \pi_1 + \pi_1(1-\pi_1) + \sum_{j=2}^{m}\left\{\frac{p_j^2}{p_1}\pi_1 + p_j\pi_1 + \pi_j(1-\pi_j)\right\}\right]$$

$$= \frac{\pi_1}{n}\left(\frac{1}{p_1} - p_1 + \sum_{j=2}^{m}\frac{p_j^2}{p_1}\right) + \frac{1}{n}\sum_{j=1}^{m}\pi_j(1-\pi_j).$$

Next, from (9.12) and (9.10), we obtain

$$\text{tr}\left\{\text{Var}(\hat{\boldsymbol{\pi}}_{\text{MP}})\right\}$$

$$= \frac{1}{nq^2}\sum_{j=1}^{m}\lambda_j(1-\lambda_j)$$

$$= \frac{1}{nq^2}\sum_{j=1}^{m}\{p_j(1-q)+\pi_j q\}\{1 - p_j(1-q) - \pi_j q\}$$

$$= \frac{1-q}{nq^2}\left\{1+q-2q\sum_{j=1}^{m}\pi_j p_j - (1-q)\sum_{j=1}^{m}p_j^2\right\} + \frac{1}{n}\sum_{j=1}^{m}\pi_j(1-\pi_j).$$

Thus, the difference between them is

$$
\operatorname{tr}\left\{\operatorname{Var}(\hat{\boldsymbol{\pi}}_{\mathrm{MT}})\right\} - \operatorname{tr}\left\{\operatorname{Var}(\hat{\boldsymbol{\pi}}_{\mathrm{MP}})\right\}
$$

$$
= \frac{\pi_1}{n}\left(\frac{1}{p_1} - p_1 + \sum_{j=2}^{m}\frac{p_j^2}{p_1}\right)
$$

$$
- \frac{1-q}{nq^2}\left\{1 + q - 2q\sum_{j=1}^{m}\pi_j p_j - (1-q)\sum_{j=1}^{m}p_j^2\right\}
$$

$$
= \frac{1}{nq^2} \times h(q|\boldsymbol{\pi},\boldsymbol{p}),
$$

where

$$
h(q|\boldsymbol{\pi},\boldsymbol{p}) = \left\{\pi_1\left(\frac{1}{p_1} - p_1 + \sum_{j=2}^{m}\frac{p_j^2}{p_1}\right) + \left(1 - 2\sum_{j=1}^{m}\pi_j p_j + \sum_{j=1}^{m}p_j^2\right)\right\}q^2
$$

$$
+ 2\left(\sum_{j=1}^{m}\pi_j p_j - \sum_{j=1}^{m}p_j^2\right)q - 1 + \sum_{j=1}^{m}p_j^2
$$

$$
\stackrel{\triangle}{=} aq^2 + bq + c \tag{9.38}
$$

is a quadratic function of q for given $\boldsymbol{\pi}$ and $\boldsymbol{p}$. In both survey designs (see Table 9.1 and Table 5.1), we require $p_1 \in (0,1)$ so that $1 - p_1^2 > 0$. In addition,

$$
0 \leqslant \sum_{j=1}^{m}\pi_j^2 \leqslant \sum_{j=1}^{m}\pi_j = 1.
$$

Thus,

$$
a = \pi_1\left(\frac{1}{p_1} - p_1 + \sum_{j=2}^{m}\frac{p_j^2}{p_1}\right) + \left(1 - 2\sum_{j=1}^{m}\pi_j p_j + \sum_{j=1}^{m}p_j^2\right)
$$

$$
= \frac{\pi_1}{p_1}\left(1 - p_1^2 + \sum_{j=2}^{m}p_j^2\right) + 1 - \sum_{j=1}^{m}\pi_j^2 + \sum_{j=1}^{m}(\pi_j - p_j)^2 > 0.
$$

Now, the discriminant of the quadratic function $h(q|\boldsymbol{\pi},\boldsymbol{p})$ is

$$
D(h) = b^2 - 4ac
$$

$$
= 4\left(\sum_{j=1}^{m}\pi_j p_j - \sum_{j=1}^{m}p_j^2\right)^2 + 4\pi_1\left(\frac{1}{p_1} - p_1 + \sum_{j=2}^{m}\frac{p_j^2}{p_1}\right)\left(1 - \sum_{j=1}^{m}p_j^2\right)
$$

$$
+ 4\left(1 - 2\sum_{j=1}^{m}\pi_j p_j + \sum_{j=1}^{m}p_j^2\right)\left(1 - \sum_{j=1}^{m}p_j^2\right)
$$

$$
= 4\left(1 - \sum_{j=1}^{m}\pi_j p_j\right)^2 + \frac{4\pi_1}{p_1}\left(1 - p_1^2 + \sum_{j=2}^{m}p_j^2\right)\left(1 - \sum_{j=1}^{m}p_j^2\right)
$$

$$
> 0.
$$

By applying Lemma 7.1 (3), we immediately obtain the following result.

Theorem 9.2 Let $\boldsymbol{\pi} \in \mathbb{T}_m$ and $\boldsymbol{p} \in \mathbb{T}_m$, we always have

$$
\mathrm{tr}\left\{\mathrm{Var}(\hat{\boldsymbol{\pi}}_{\mathrm{MT}})\right\} > \mathrm{tr}\left\{\mathrm{Var}(\hat{\boldsymbol{\pi}}_{\mathrm{MP}})\right\}
$$

for any $q \in (0, q_{\mathrm{L}}) \cup (q_{\mathrm{U}}, 1)$, where

$$
q_{\mathrm{L}} = \max\left(0, \; \frac{-b - \sqrt{b^2 - 4ac}}{2a}\right)
$$

and

$$
q_{\mathrm{U}} = \min\left(1, \; \frac{-b + \sqrt{b^2 - 4ac}}{2a}\right),
$$

where a, b and c are defined in (9.38).   ¶

9.5.2 Degree of privacy protection

In this subsection, we compare *degrees of privacy protection* (DPP) of the multi-category parallel model with those of the multi-category triangular model. For the multi-category parallel model (see Table 9.1), we define

$$
\mathrm{DPP}_{\mathrm{MP}}(\pi_1, p_1, q) \quad = \quad \Pr(Y = 1 \,|\, \text{two circles are connected}),
$$

$$
\mathrm{DPP}_{\mathrm{MP}}(\pi_2, p_2, q) \quad = \quad \Pr(Y = 2 \,|\, \text{two triangles are connected}),
$$

$$
\vdots
$$

$$
\mathrm{DPP}_{\mathrm{MP}}(\pi_m, p_m, q) \quad = \quad \Pr(Y = m \,|\, \text{two dots are connected}),
$$

where, for example, $\mathrm{DPP}_{\mathrm{MP}}(\pi_1, p_1, q)$ denotes the conditional probability that the respondent belongs to the subclass $\{Y = 1\}$ given that he/she connected the two circles. For the multi-category triangular model (see Table 5.1), we can similarly define

$$\mathrm{DPP}_{\mathrm{MT}}(\pi_1, p_1) = \Pr(Y = 1 \,|\, \text{a tick is put in Block 1}),$$

$$\mathrm{DPP}_{\mathrm{MT}}(\pi_2, p_2) = \Pr(Y = 2 \,|\, \text{a tick is put in Block 2}),$$

$$\vdots$$

$$\mathrm{DPP}_{\mathrm{MT}}(\pi_m, p_m) = \Pr(Y = m \,|\, \text{a tick is put in Block } m).$$

First, for any $q \in (0, 1)$, $\boldsymbol{\pi} \in \mathbb{T}_m$ and $\boldsymbol{p} \in \mathbb{T}_m$, we always have

$$\mathrm{DPP}_{\mathrm{MT}}(\pi_1, p_1) = 1$$

$$> \frac{\pi_1 q}{p_1(1 - q) + \pi_1 q}$$

$$= \mathrm{DPP}_{\mathrm{MP}}(\pi_1, p_1, q). \tag{9.39}$$

Next, when $0 < q < 1/(1 + \pi_1)$, we obtain

$$\mathrm{DPP}_{\mathrm{MT}}(\pi_j, p_j) = \frac{\pi_j}{p_j \pi_1 + \pi_j}$$

$$> \frac{\pi_j q}{p_j(1 - q) + \pi_j q}$$

$$= \mathrm{DPP}_{\mathrm{MP}}(\pi_j, p_j, q), \quad j = 2, \ldots, m, \tag{9.40}$$

for any $\boldsymbol{\pi} \in \mathbb{T}_m$ and $\boldsymbol{p} \in \mathbb{T}_m$. Inequalities (9.39) and (9.40) show that if we choose q within the open interval $(0, \; 1/(1 + \pi_1))$, the multi-category parallel model is more efficient than the multi-category triangular model in protecting the individual's privacy for any $\boldsymbol{\pi} \in \mathbb{T}_m$ and $\boldsymbol{p} \in \mathbb{T}_m$.

9.6 An Example

9.6.1 The income and sexual partner data

Williamson & Haber (1994) reported a study aimed to examine the relationship among disease status of cervical cancer, the number of sexual partners, and income. Respondents were women 20–79 years old in Fulton or Dekalb counties in Atlanta, Georgia. Table 9.4 displays the the cross-classification of income (low or high, denoted by $Y = 0$ or $Y = 1$) and number of sex partners ('few' (0–3) or 'many' ($\geqslant 4$), denoted by $X = 0$ or $X = 1$). Since all four questions (i.e., the combination of the number of sex partners and

Table 9.4 Survey data from Williamson & Haber (1994)

Number of	Income		
sex partners	$Y = 0$ (low)	$Y = 1$ (high)	Missing
$X = 0$ (0–3)	144 (m_1, π_1)	123 (m_2, π_2)	17
$X = 1$ ($\geqslant 4$)	237 (m_3, π_3)	148 (m_4, π_4)	17
Missing	68	34	26

income status) are highly sensitive to respondents, a sizable proportion (19.9% in this example) of the responses would be missing because of 'unknown' or 'refused to answer' in a telephone interview. The major objective is to examine if an association exists between the number of sex partners and income.

The multi-category triangular model and the corresponding statistical methods presented in Chapter 5 cannot be applied to the present study since each of the four subclasses

$$\{X = 0, \ Y = 0\},$$

$$\{X = 0, \ Y = 1\},$$

$$\{X = 1, \ Y = 0\}, \quad \text{and}$$

$$\{X = 1, \ Y = 1\}$$

is sensitive to respondents. To demonstrate the multi-category parallel design in Tables 9.1 and 9.2 and the estimation methods in Sections 9.2 and 9.4, we let $m = 4$ and define $W = 0$ if the respondent's birthday is in the first half of a month and $W = 1$ otherwise. Similarly, we define $U = i$ if the respondent was born in the i-th quarter of a year $(i = 1, \ldots, 4)$. Thus, it is reasonable to assume that

$$q = \Pr(W = 1) = 0.5 \quad \text{and} \quad p_i = \Pr(U = i) = 0.25$$

for each i.

To obtain the observed data $Y_{\text{obs}} = \{n; n_1, \ldots, n_4\}$ in the four-category parallel model (see Table 9.2), we only consider the complete observations in Table 9.4 and discard the associated missing data and obtain

$$n = m_1 + \cdots + m_4 = 144 + 123 + 237 + 148 = 652.$$

Note that n_1 denotes the number of respondents connecting the two circles in Table 9.2, n_2 is the number of respondents connecting the two triangles,

n_3 is the number of respondents connecting the two rectangles, and n_4 is the number of respondents connecting the two dots. Let z_1, z_2, z_3 and z_4 denote the numbers of respondents belonging to

$$\{X = 0,\ Y = 0\} \cap \{W = 1\},$$

$$\{X = 0,\ Y = 1\} \cap \{W = 1\},$$

$$\{X = 1,\ Y = 0\} \cap \{W = 1\}, \quad \text{and}$$

$$\{X = 1,\ Y = 1\} \cap \{W = 1\}$$

in Table 9.2 respectively. Since $q = 1/2$, we have $z_i = m_i/2$ for the ideal situation, i.e.,

$$(z_1,\ z_2,\ z_3,\ z_4)^\top \approx (72,\ 62,\ 118,\ 74)^\top.$$

Furthermore, let n_i' denote the number of respondents belonging to $\{U = i\} \cap \{W = 0\}$ for $i = 1, \ldots, 4$ in Table 9.2. To obtain these $\{n_i'\}_{i=1}^4$ by considering the sampling error, we first generate 50 i.i.d. samples from

$$\text{Multinomial}\left(n - \textstyle\sum_{i=1}^4 z_i;\ p_1,\ p_2,\ p_3,\ p_4\right)$$

$$= \text{Multinomial}_4(326,\ 0.25 \times \mathbf{1}_4)$$

and then average these counts for each component, yielding

$$(n_1',\ n_2',\ n_3',\ n_4')^\top = (81,\ 82,\ 81,\ 82)^\top.$$

Therefore, we obtain the following observed counts

$$(n_1,\ n_2,\ n_3,\ n_4)^\top = (z_1 + n_1',\ z_2 + n_2',\ z_3 + n_3',\ z_4 + n_4')^\top$$

$$= (153,\ 144,\ 199,\ 156)^\top.$$

9.6.2 Likelihood-based analysis

Using $\pi^{(0)} = 0.25 \times \mathbf{1}_4$ as the initial values, the EM algorithm in (9.3) and (9.5) converges in 25 iterations. The resultant MLEs for $\pi = (\pi_1, \ldots, \pi_4)^\top$ and the odds ratio ψ are listed in the second column of Table 9.5. Based on (9.6), we generate $G = 10{,}000$ bootstrap samples to estimate the standard errors of $\{\hat{\pi}_{\text{MP}i}\}_{i=1}^4$ and $\hat{\psi}$, which are given in the third column of Table 9.5. The corresponding 95% normal-based bootstrap confidence intervals and non-normal-based bootstrap confidence intervals are displayed in the fourth and the fifth columns of Table 9.5. Since the two bootstrap confidence intervals of ψ include 1, we do not have reason to believe that there exists association between the number of sex partners and income status.

Table 9.5 MLEs and two bootstrap confidence intervals of parameters for the observed counts $(n_1, n_2, n_3, n_4)^\top = (153, 144, 199, 156)^\top$

Parameter	MLE	std	95% bootstrap CI†	95% bootstrap CI‡
π_1	0.2196	0.0328	[0.1553, 0.2839]	[0.1549, 0.2837]
π_2	0.1919	0.0327	[0.1277, 0.2561]	[0.1273, 0.2561]
π_3	0.3604	0.0359	[0.2900, 0.4308]	[0.2929, 0.4310]
π_4	0.2281	0.0338	[0.1619, 0.2943]	[0.1641, 0.2960]
ψ	0.7661	0.2712	[0.2344, 1.2977]	[0.3690, 1.4099]

Note: CI† = Normal-based bootstrap confidence intervals, cf. (9.7). CI‡ = Non-normal-based bootstrap confidence intervals, cf. (9.8).

According to (9.9), we obtain

$$\hat{\boldsymbol{\pi}}_v = (0.2193252, 0.1917178, 0.3604294, 0.2285276)^\top.$$

Since $\hat{\boldsymbol{\pi}}_v \in \mathbb{T}_4$, we know that $\hat{\boldsymbol{\pi}}_v$ is a valid estimator of $\boldsymbol{\pi}$ and $\hat{\boldsymbol{\pi}}_v = \hat{\boldsymbol{\pi}}_{\mathrm{MP}}$. Based on (9.12), the estimated variance-covariance matrix of $\hat{\boldsymbol{\pi}}_{\mathrm{MP}}$ is

$$\widehat{\mathrm{Var}}(\hat{\boldsymbol{\pi}}_{\mathrm{MP}}) = \begin{pmatrix} 0.00110 & -0.00032 & -0.00044 & -0.00034 \\ -0.00032 & 0.00106 & -0.00041 & -0.00032 \\ -0.00044 & -0.00041 & 0.00130 & -0.00045 \\ -0.00034 & -0.00032 & -0.00045 & 0.00112 \end{pmatrix}$$

so that the unbiased estimates of $\{\mathrm{Var}(\hat{\pi}_{\mathrm{MP}i})\}_{i=1}^4$, from (9.13), are given by

$$\left(\widehat{\mathrm{Var}}(\hat{\pi}_{\mathrm{MP}1}), \ldots, \widehat{\mathrm{Var}}(\hat{\pi}_{\mathrm{MP}4})\right)^\top = (0.00110, 0.00106, 0.00130, 0.00112)^\top.$$

Therefore, from (9.14), (9.16) and (9.18), the 95% Wald, Wilson and likelihood ratio confidence intervals of $\{\pi_i\}_{i=1}^4$ can be calculated and are given in Table 9.6. We note that the width of the 95% Wilson confidence interval of π_i is slightly shorter than those of the 95% Wald confidence intervals and LRCIs of π_i.

9.6.3 Bayesian analysis

If we adopt Dirichlet$(\mathbf{1}_4)$ as the prior distribution of $\boldsymbol{\pi}$, the EM algorithm specified by (9.23) and (9.24) is identical to the EM algorithm in (9.3) and (9.5). In other words, the posterior modes of $\{\pi_i\}_{i=1}^4$ are equal to their MLEs, which are listed in the second column of Table 9.7.

Table 9.6 Three 95% confidence intervals of parameters for large sample sizes

Three	Parameter			
CIs	π_1	π_2	π_3	π_4
Wald	[0.1542, 0.2844]	[0.1280, 0.2554]	[0.2897, 0.4312]	[0.1630, 0.2941]
Width	0.1302	0.1274	0.1415	0.1311
Wilson	[0.1575, 0.2874]	[0.1314, 0.2586]	[0.2922, 0.4332]	[0.1662, 0.2970]
Width	0.1299	0.1272	0.1410	0.1308
LR	[0.1564, 0.2864]	[0.1303, 0.2575]	[0.2914, 0.4326]	[0.1652, 0.2960]
Width	0.1300	0.1272	0.1412	0.1308

Table 9.7 Posterior estimates of parameters

Parameter	Posterior mode	Posterior mean	Posterior std	95% Bayesian credible interval
π_1	0.2193252	0.21943	0.0201	[0.1788, 0.2587]
π_2	0.1917178	0.19065	0.0193	[0.1529, 0.2280]
π_3	0.3604294	0.36162	0.0204	[0.3217, 0.4024]
π_4	0.2285276	0.22830	0.0197	[0.1894, 0.2668]
ψ	0.7253447	0.74015	0.1484	[0.4910, 1.0739]

Based on (9.4) and (9.22), we use the data augmentation algorithm to generate 20,000 posterior samples of π. By discarding the first half of the samples, we can calculate the posterior means, the posterior standard deviations and the 95% Bayesian credible intervals of π, which are given in the third, fourth, and fifth columns of Table 9.7.

Figures 9.3 and 9.4 show the corresponding posterior densities of $\{\pi_i\}_{i=1}^4$ and the odds ratio ψ via a kernel density smoother based on the second half posterior samples generated by the data augmentation algorithm. Since the lower bound of the Bayesian credible interval of ψ is less than 1, we have reason to believe that there exists association between the number of sexual partners and income status.

9.7 Discussion

How to choose the two non-sensitive variates W and U in Table 9.1 is an important issue in practice. On the one hand, since W is a binary variate,

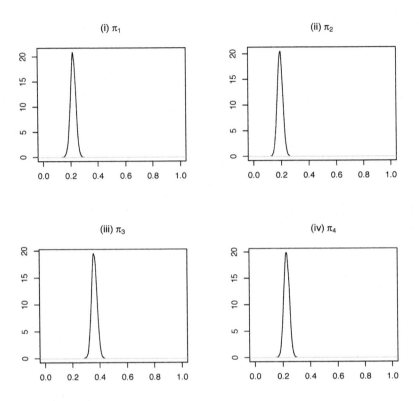

Figure 9.3 Posterior densities of $\{\pi_i\}_{i=1}^4$ via a kernel density smoother based on the second half posterior samples generated by the data augmentation algorithm in (9.4) and (9.22) with Dirichlet $(\mathbf{1}_4)$ as the prior distribution of $\boldsymbol{\pi}$. (i) The posterior density of π_1; (ii) The posterior density of π_2; (iii) The posterior density of π_3; (iv) The posterior density of π_4.

we could define $W = 0$ if a respondent was born between January and June; or a respondent was born in an odd numbered month; or a respondent's birthday is in the first half of the month; or a respondent's age is odd numbered; or a respondent's house/apartment number is even. On the other hand, since U is an m-category variate, for example, when $m = 3$, we may let

$U = 1$ if the respondent's mother was born in January–April;

$U = 2$ if the respondent's mother was born in May–August; and

$U = 3$ if the respondent's mother was born in September–December.

In this case, it is reasonable to assume that each $p_i = \Pr(U = i)$ is approx-

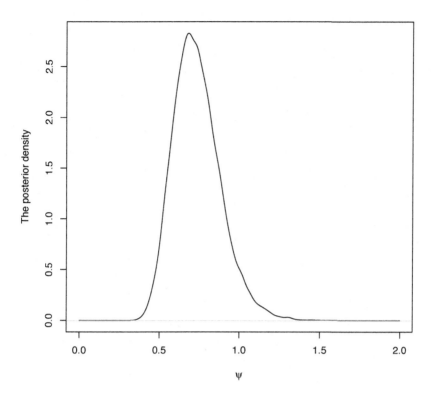

Figure 9.4 The posterior density of the odds ratio ψ via a kernel density smoother based on the second half posterior samples generated by the data augmentation algorithm in (9.4) and (9.22) with Dirichlet $(\mathbf{1}_4)$ as the prior distribution of $\boldsymbol{\pi}$.

imately equal to $1/3$. Similarly, when $m = 5$, we may define

$U = 1$ if the last digit of the respondent's ID/phone is 1 or 2;

$U = 2$ if the last digit of the respondent's ID/phone is 3 or 4;

$U = 3$ if the last digit of the respondent's ID/phone is 5 or 6;

$U = 4$ if the last digit of the respondent's ID/phone is 7 or 8; and

$U = 5$ if the last digit of a respondent's ID card/phone is 9 or 0.

When $m = 7$, we might define

$U = 1$ if the respondent's mother was born on Monday;

$U = 2$ if the respondent's mother was born on Tuesday;

$U = 3$ if the respondent's mother was born on Wednesday;

$U = 4$ if the respondent's mother was born on Thursday;

$U = 5$ if the respondent's mother was born on Friday;

$U = 6$ if the respondent's mother was born on Saturday; and

$U = 7$ if the respondent's mother was born on Sunday.

CHAPTER 10

A Variant of the Parallel Model

In Chapter 7, we introduced a general framework of design and analysis for the parallel model to estimate the unknown proportion, $\pi = \Pr(Y = 1)$, of individuals with a sensitive characteristic in a specific population. The presented survey design is based on the introduction of two non-sensitive dichotomous variates U and W such that Y, U and W are mutually independent. The presented statistical analysis methods are based on the assumption of known proportions $\theta = \Pr(U = 1)$ and $p = \Pr(W = 1)$. However, in survey practice, it is usually difficult to choose an appropriate non-sensitive dichotomous variate U with known $\theta = \Pr(U = 1)$. Even if such a binary variate U can be found and a constant θ_0 is assumed to be equal to the true value of the θ, testing the hypothesis H_0: $\theta = \theta_0$ is still not available for the parallel design. The purpose of this chapter is to present a variant of the parallel model with $\theta = \Pr(U = 1)$ being unknown.

10.1   The Survey Design and Basic Properties

10.1.1   The survey design

Let $\{Y = 1\}$ denote the population class with a sensitive characteristic and $\{Y = 0\}$ denote the complementary class. The objective is to estimate the proportion $\pi = \Pr(Y = 1)$. Suppose that U and W are two non-sensitive dichotomous variates, and Y, U and W are mutually independent with unknown $\theta = \Pr(U = 1)$ and known $p = \Pr(W = 1)$. For example, we may define $U = 1$ if the respondent lives at Hong Kong Island (or likes watching football/soccer on TV, or likes fishing/singing/shopping/traveling, or receives education at high school level or above) and $U = 0$ otherwise. Similarly, we could define $W = 1$ if the last digit of the respondent's identity card/cell phone number is odd (or the respondent's birthday is in the second half of a year/month) and $W = 0$ otherwise. Hence, it is reasonable to assume that $p \approx 0.5$.

An interviewer may design a questionnaire in the format as shown at the left-hand side of Table 10.1 and ask an interviewee to truthfully put a tick in

(a)   the circle if he/she belongs to $\{U = 0,\ W = 0\}$; or

Table 10.1 A survey design for the variant of the parallel model with unknown θ and known p

Category	$W = 0$	$W = 1$		Category	$W = 0$	$W = 1$	Marginal
$U = 0$	○			$U = 0$	$(1 - \theta)(1 - p)$		$1 - \theta$
$U = 1$	□			$U = 1$	$\theta(1 - p)$		θ
$Y = 0$		△		$Y = 0$		$(1 - \pi)p$	$1 - \pi$
$Y = 1$		□		$Y = 1$		πp	π
				Marginal	$1 - p$	p	1

Note: Please truthfully put a tick in the circle if you belong to $\{U = 0, W = 0\}$ or put a tick in the triangle if you belong to $\{Y = 0, W = 1\}$ or put a tick in the upper square if you belong to $\{U = 1, W = 0\} \cup \{Y = 1, W = 1\}$.

(b)   the triangle if he/she belongs to $\{Y = 0, W = 1\}$; or

(c)   the upper square if he/she belongs to $\{U = 1, W = 0\} \cup \{Y = 1, W = 1\}$.

Note that all $\{W = 0\}$, $\{W = 1\}$, $\{U = 0\}$, $\{U = 1\}$ and $\{Y = 0\}$ are non-sensitive classes. Thus,

$$\{U = 1, W = 0\} \cup \{Y = 1, W = 1\}$$

is also a non-sensitive subclass. Therefore, whether an interviewee belongs to the sensitive class $\{Y = 1, W = 1\}$ is not revealed. Since θ is unknown, Liu & Tian (2012b) called this a variant of the parallel model. The corresponding cell probabilities are displayed at the right-hand side of Table 10.1. Since the three binary variables U, Y and W are independent, the joint probability is the product of two corresponding marginal probabilities.

Let $Y = 1$ if a respondent is a drug user and $Y = 0$ otherwise. For those respondents not completely understanding the questionnaire shown in Table 10.1, investigators may formulate the questionnaire of the variant of the parallel model as follows:

(1)   If your birthday is in the first half of a year (i.e., $W = 0$), please answer '0' (i.e., $U = 0$), or '2' (i.e., $U = 1$) to the question: *Do you like shopping?*

(2)   If your birthday is in the second half of a month (i.e., $W = 1$), please answer '1' (i.e., $Y = 0$), or '2' (i.e., $Y = 1$) to the question: *Are you a drug user?*

10.1.2 Estimation

Suppose that a sample survey with n respondents is conducted. Let $Y_{\text{obs}} = \{n; n_1, n_2, n_3\}$ denote the observed data. Here, $n = n_1 + n_2 + n_3$, n_1 represents the number of respondents putting a tick in the circle, n_2 represents the number of respondents putting a tick in the triangle, and n_3 represents the number of individuals who put a tick in the upper square (see Table 10.1). The likelihood function of π and θ for the observed data Y_{obs} is

$$L_{\text{V}}(\pi, \theta | Y_{\text{obs}}) = \binom{n}{n_1, n_2, n_3} \{(1 - \theta)(1 - p)\}^{n_1}$$
$$\times \{(1 - \pi)p\}^{n_2} \{\theta(1 - p) + \pi p\}^{n_3}, \qquad (10.1)$$

where the subscript 'V' refers to the 'variant' of the parallel model. Hence, the corresponding log-likelihood function is given by

$$\ell_{\text{V}}(\pi, \theta | Y_{\text{obs}}) = c + n_1 \log(1 - \theta) + n_2 \log(1 - \pi) + n_3 \log\{\theta(1 - p) + \pi p\},$$

where c is a constant independent of π and θ. Letting

$$\frac{\partial \ell_{\text{V}}(\pi, \theta | Y_{\text{obs}})}{\partial \pi} = 0 \quad \text{and} \quad \frac{\partial \ell_{\text{V}}(\pi, \theta | Y_{\text{obs}})}{\partial \theta} = 0,$$

we obtain

$$\begin{cases} -\dfrac{n_2}{1 - \pi} + \dfrac{n_3 p}{\theta(1 - p) + \pi p} = 0 \quad \text{and} \\[3mm] -\dfrac{n_1}{1 - \theta} + \dfrac{n_3(1 - p)}{\theta(1 - p) + \pi p} = 0. \end{cases}$$

Hence, the MLEs of π and θ are given by

$$\hat{\pi}_{\text{V}} = 1 - \frac{n_2}{np} \quad \text{and} \quad \hat{\theta} = 1 - \frac{n_1}{n(1 - p)}, \qquad (10.2)$$

provided that both $\hat{\pi}_{\text{V}} \in [0, 1]$ and $\hat{\theta} \in [0, 1]$. When $\hat{\pi}_{\text{V}} \notin [0, 1]$ or $\hat{\theta} \notin [0, 1]$, we can use the EM algorithm (10.36)–(10.38) with $a_1 = b_1 = a_2 = b_2 = 1$ to find the MLEs of π and θ. Here, we note that if two independent uniform distributions on $[0, 1]$ are adopted as the priors for π and θ, then the posterior mode of π (or θ) is identical to the MLE of π (or θ).

To derive the expectation and variance of the $\hat{\pi}_{\text{V}}$, we define

$$\begin{aligned} \lambda_1 &= \Pr(U = 0, W = 0) = (1 - \theta)(1 - p), \\ \lambda_2 &= \Pr(Y = 0, W = 1) = (1 - \pi)p, \quad \text{and} \qquad (10.3) \\ \lambda_3 &= \Pr(U = 1, W = 0) + \Pr(Y = 1, W = 1) \\ &= \theta(1 - p) + \pi p. \end{aligned}$$

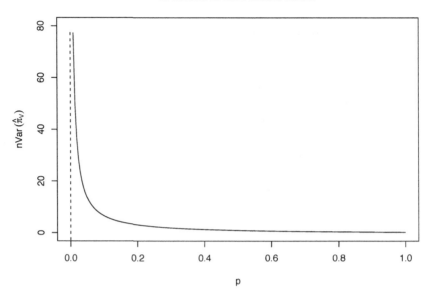

Figure 10.1 Plot of $n\mathrm{Var}(\hat{\pi}_{\mathrm{V}})$ defined by (10.5) against p with $\pi = 0.3$ for the variant of the parallel model.

Obviously, we have

$$(n_1,\, n_2,\, n_3)^{\top} \sim \mathrm{Multinomial}\,(n;\, \lambda_1,\, \lambda_2,\, \lambda_3).$$

Note that the MLEs of $\{\lambda_i\}_{i=1}^{3}$ are given by $\hat{\lambda}_i = n_i/n$ and $E(n_i) = n\lambda_i$, $i = 1,\, 2,\, 3$. It is easy to verify that $\hat{\pi}_{\mathrm{V}}$ is an unbiased estimator of π and the variance of $\hat{\pi}_{\mathrm{V}}$ is given by

$$\mathrm{Var}(\hat{\pi}_{\mathrm{V}}) = \frac{\lambda_2(1-\lambda_2)}{np^2} \stackrel{(10.3)}{=} \mathrm{Var}(\hat{\pi}_{\mathrm{D}}) + \frac{(1-p)(1-\pi)}{np}, \qquad (10.4)$$

where

$$\mathrm{Var}(\hat{\pi}_{\mathrm{D}}) = \frac{\pi(1-\pi)}{n}$$

denotes the variance of $\hat{\pi}_{\mathrm{D}}$ in the design of direct questioning. It is clear that when $p = 1$ the variant of the parallel design will reduce to the design of direct questioning. In addition, we observed that $\mathrm{Var}(\hat{\pi}_{\mathrm{V}})$ does not depend on the unknown parameter θ. Hence, for any fixed π,

$$n\mathrm{Var}(\hat{\pi}_{\mathrm{V}}) = \pi(1-\pi) + \frac{(1-p)(1-\pi)}{p} \qquad (10.5)$$

is a decreasing function of p as shown in Figure 10.1. We can see that $n\mathrm{Var}(\hat{\pi}_{\mathrm{V}}) \to \infty$ as $p \to 0$.

Table 10.2 Relative efficiency $\text{RE}_{\text{V}\to\text{D}}(\pi, p)$ for various combinations of π and p

π	p				
	1/3	0.40	0.50	0.60	2/3
0.05	41.000	31.000	21.000	14.333	11.000
0.10	21.000	16.000	11.000	7.6667	6.0000
0.20	11.000	8.5000	6.0000	4.3333	3.5000
0.30	7.6667	6.0000	4.3333	3.2222	2.6667
0.40	6.0000	4.7500	3.5000	2.6667	2.2500
0.50	5.0000	4.0000	3.0000	2.3333	2.0000
0.60	4.3333	3.5000	2.6667	2.1111	1.8333
0.70	3.8571	3.1429	2.4286	1.9524	1.7143
0.80	3.5000	2.8750	2.2500	1.8333	1.6250
0.90	3.2222	2.6667	2.1111	1.7407	1.5556
0.95	3.1053	2.5789	2.0526	1.7018	1.5263

10.1.3   Relative efficiency

The *relative efficiency* (RE) is a useful tool to compare two survey designs. The RE of the variant of the parallel design to the design of direct questioning is defined by

$$\text{RE}_{\text{V}\to\text{D}}(\pi, p) = \frac{\text{Var}(\hat{\pi}_{\text{V}})}{\text{Var}(\hat{\pi}_{\text{D}})} = 1 + \frac{1-p}{\pi p}.$$

It is noted that $\text{RE}_{\text{V}\to\text{D}}(\pi, p)$ does not depend on the unknown parameter θ and the sample size n. When p is fixed, $\text{RE}_{\text{V}\to\text{D}}(\pi, p)$ is a decreasing function of π. Similarly, when π is fixed, $\text{RE}_{\text{V}\to\text{D}}(\pi, p)$ is also a decreasing function of p. Table 10.2 lists the relative efficiency $\text{RE}_{\text{V}\to\text{D}}(\pi, p)$ for various combinations of π and p. For example, when $\pi = 0.10$ and $p = 2/3$, we have $\text{RE}_{\text{V}\to\text{D}}(0.10, 2/3) = 6$, which implies that the sample size needed for the variant of the parallel design is about 6 times that needed for the DDQ in order to achieve the same estimation precision. When $\pi = 0.10$ and $p = 0.50$, we have $\text{RE}_{\text{V}\to\text{D}}(0.10, 0.50) = 11$.

10.1.4   Degree of privacy protection

To evaluate how the respondent's privacy is protected, we investigate the *degree of privacy protection* (DDP) for the variant of the parallel model.

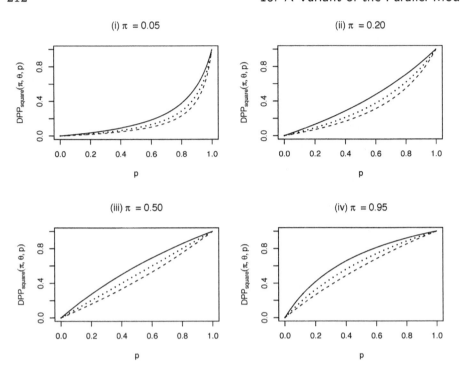

Figure 10.2 Plots of $\text{DPP}_\square(\pi, \theta, p)$ defined by (10.6) against p for the variant of the parallel model with a fixed π and three different values of θ, where the solid line is corresponding to $\theta = 1/3$; the dotted line is corresponding to $\theta = 0.5$; and the dashed line is corresponding to $\theta = 2/3$. (i) $\pi = 0.05$; (ii) $\pi = 0.20$; (iii) $\pi = 0.50$; (iv) $\pi = 0.95$.

Define
$$
Y_{\text{v}} = \begin{cases} -1, & \text{if a tick is put in the circle,} \\ 0, & \text{if a tick is put in the triangle,} \\ 1, & \text{if a tick is put in the upper square.} \end{cases}
$$

Let $\text{DPP}_\bigcirc(\pi, \theta, p)$ (or $\text{DPP}_\triangle(\pi, \theta, p)$) represent the conditional probability of a respondent belonging to the sensitive class $\{Y = 1\}$ given that a tick is put in the circle (or triangle) in Table 10.1. Clearly, we have

$$
\text{DPP}_\bigcirc(\pi, \theta, p) \;=\; \Pr(Y = 1 | Y_{\text{v}} = -1) = 0 \quad \text{and}
$$
$$
\text{DPP}_\triangle(\pi, \theta, p) \;=\; \Pr(Y = 1 | Y_{\text{v}} = 0) = 0.
$$

Similarly, let $\text{DPP}_\square(\pi, \theta, p)$ denote the conditional probability of a respondent belonging to the sensitive class when a tick is put in the upper square in Table 10.1. We have

$$
\text{DPP}_\square(\pi, \theta, p) = \Pr(Y = 1 | Y_{\text{v}} = 1) = \frac{\pi p}{\pi p + \theta(1 - p)}. \tag{10.6}
$$

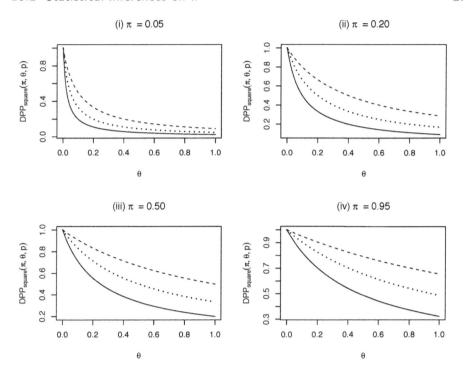

Figure 10.3 Plots of $\mathrm{DPP}_{\square}(\pi, \theta, p)$ defined by (10.6) against θ for the variant of the parallel model with a fixed π and three different values of p, where the solid line is corresponding to $p = 1/3$; the dotted line is corresponding to $p = 0.5$; and the dashed line is corresponding to $p = 2/3$. (i) $\pi = 0.05$; (ii) $\pi = 0.20$; (iii) $\pi = 0.50$; (iv) $\pi = 0.95$.

In particular, when $p = 1$, we have $\mathrm{DPP}_{\square}(\pi, \theta, 1) = 1$, which equals to the DPP for the design of direct questioning. For any fixed π and θ, $\mathrm{DPP}_{\square}(\pi, \theta, p)$ is a monotonically increasing function of p. Figure 10.2 shows three curves (corresponding to $\theta = 1/3$, 0.5 and $2/3$) of $\mathrm{DPP}_{\square}(\pi, \theta, p)$ against p with a fixed π, where $\pi = 0.05$, 0.20, 0.50 and 0.95, respectively.

In addition, for any fixed π and p, $\mathrm{DPP}_{\square}(\pi, \theta, p)$ is a monotonically decreasing function of θ. Figure 10.3 shows three curves (corresponding to $p = 1/3$, 0.5 and $2/3$) of $\mathrm{DPP}_{\square}(\pi, \theta, p)$ against θ with a fixed π, where $\pi = 0.05$, 0.20, 0.50 and 0.95, respectively.

10.2 Statistical Inferences on π

In this section, we first provide an unbiased estimator for the variance of $\hat{\pi}_{\mathrm{v}}$ in Theorem 10.1 below. Second, we construct three asymptotic confidence intervals (i.e., Wald, Wilson, and likelihood ratio confidence intervals) of π by using this unbiased estimator. Third, the exact or Clopper–Pearson confidence interval of π is also derived. Finally, a modified MLE of π is presented and the corresponding asymptotic property is investigated.

10.2.1 An unbiased estimator of the variance of $\hat{\pi}_V$

Theorem 10.1 Let $\overline{\mathrm{Var}}(\hat{\pi}_V) = \hat{\lambda}_2(1 - \hat{\lambda}_2)/\{(n-1)p^2\}$. We have

$$\overline{\mathrm{Var}}(\hat{\pi}_V) = \frac{\hat{\pi}_V(1 - \hat{\pi}_V)}{n-1} + \frac{(1 - \hat{\pi}_V)(1 - p)}{(n-1)p} \tag{10.7}$$

and it is an unbiased estimator of $\mathrm{Var}(\hat{\pi}_V) = \lambda_2(1 - \lambda_2)/(np^2)$. ¶

Proof. From (10.3), we know that $\hat{\lambda}_2 = p(1 - \hat{\pi}_V)$, where $\hat{\pi}_V$ is given by (10.2). Hence,

$$
\begin{aligned}
\overline{\mathrm{Var}}(\hat{\pi}_V) &= \frac{\hat{\lambda}_2(1 - \hat{\lambda}_2)}{(n-1)p^2} \\
&= \frac{p(1 - \hat{\pi}_V)(1 - p + p\hat{\pi}_V)}{(n-1)p^2} \\
&= \frac{\hat{\pi}_V(1 - \hat{\pi}_V)}{n-1} + \frac{(1 - \hat{\pi}_V)(1 - p)}{(n-1)p},
\end{aligned}
$$

which implies (10.7). Next, we prove the second part. Since

$$n_2 \sim \mathrm{Binomial}\,(n; \lambda_2),$$

we have

$$E(\hat{\lambda}_2) = E\left(\frac{n_2}{n}\right) = \lambda_2$$

and

$$\mathrm{Var}(\hat{\lambda}_2) = \frac{\mathrm{Var}(n_2)}{n^2} = \frac{\lambda_2(1 - \lambda_2)}{n},$$

so that

$$
\begin{aligned}
E\{\hat{\lambda}_2(1 - \hat{\lambda}_2)\} &= E(\hat{\lambda}_2) - \{E(\hat{\lambda}_2)\}^2 - \mathrm{Var}(\hat{\lambda}_2) \\
&= \frac{(n-1)\lambda_2(1 - \lambda_2)}{n}.
\end{aligned}
$$

Thus, we have

$$E\left\{\overline{\mathrm{Var}}(\hat{\pi}_V)\right\} = \frac{E\{\hat{\lambda}_2(1 - \hat{\lambda}_2)\}}{(n-1)p^2} = \frac{\lambda_2(1 - \lambda_2)}{np^2},$$

i.e., $\overline{\mathrm{Var}}(\hat{\pi}_V)$ is an unbiased estimator of $\mathrm{Var}(\hat{\pi}_V)$. □

10.2.2 Three asymptotic confidence intervals of π for large sample sizes

(a) Wald confidence interval

Let z_α denote the upper α-th quantile of the standard normal distribution. From the Central Limit Theorem, as $n \to \infty$, the $(1 - \alpha)100\%$ Wald confidence interval of π based on the unbiased estimate $\overline{\mathrm{Var}}(\hat{\pi}_{\mathrm{V}})$ is given by

$$[\hat{\pi}_{\mathrm{V,WL}}, \ \hat{\pi}_{\mathrm{V,WU}}] = \left[\hat{\pi}_{\mathrm{V}} - z_{\alpha/2}\sqrt{\mathrm{Var}(\hat{\pi}_{\mathrm{V}})}, \ \hat{\pi}_{\mathrm{V}} + z_{\alpha/2}\sqrt{\mathrm{Var}(\hat{\pi}_{\mathrm{V}})}\right]. \quad (10.8)$$

(b) Wilson (score) confidence interval

One drawback for the Wald confidence interval (10.8) is that the lower bound may be less than zero when the true value of π is close to zero while the upper bound may be beyond one when the true value of π is near to one. For these situations, we can construct the $(1 - \alpha)100\%$ Wilson confidence interval of π based on

$$
\begin{aligned}
1 - \alpha \quad &= \quad \mathrm{Pr}\left\{\left|\frac{\hat{\pi}_{\mathrm{V}} - \pi}{\sqrt{\mathrm{Var}(\hat{\pi}_{\mathrm{V}})}}\right| \leqslant z_{\alpha/2}\right\} \\[2mm]
&= \quad \mathrm{Pr}\left\{(\hat{\pi}_{\mathrm{V}} - \pi)^2 \leqslant z_{\alpha/2}^2 \mathrm{Var}(\hat{\pi}_{\mathrm{V}})\right\} \\[2mm]
&\overset{(10.4)}{=} \quad \mathrm{Pr}\left[(\hat{\pi}_{\mathrm{V}} - \pi)^2 \leqslant \frac{z_{\alpha/2}^2}{n}\left\{\pi(1 - \pi) + \frac{(1 - p)(1 - \pi)}{p}\right\}\right] \\[2mm]
&= \quad \mathrm{Pr}\left\{\hat{\pi}_{\mathrm{V}}^2 - 2\hat{\pi}_{\mathrm{V}}\pi + \pi^2 \leqslant \frac{z_{\alpha/2}^2(-\pi^2 + \rho_1\pi + \rho_2)}{n}\right\} \\[2mm]
&= \quad \mathrm{Pr}\left\{(1 + z_*)\pi^2 - (2\hat{\pi}_{\mathrm{V}} + z_*\rho_1)\pi + \hat{\pi}_{\mathrm{V}}^2 - z_*\rho_2 \leqslant 0\right\}, \quad (10.9)
\end{aligned}
$$

where $z_* \hat{=} z_{\alpha/2}^2/n$, $\rho_1 \hat{=} 1 - \rho_2$ and

$$\rho_2 \hat{=} \frac{1 - p}{p}. \quad (10.10)$$

Solving the quadratic inequality inside the probability in (10.9), we obtain the Wilson (score) confidence interval $[\hat{\pi}_{\mathrm{V,WSL}}, \ \hat{\pi}_{\mathrm{V,WSU}}]$ of π as follows:

$$\frac{2\hat{\pi}_{\mathrm{V}} + z_*\rho_1 \pm \sqrt{(2\hat{\pi}_{\mathrm{V}} + z_*\rho_1)^2 - 4(1 + z_*)(\hat{\pi}_{\mathrm{V}}^2 - z_*\rho_2)}}{2(1 + z_*)}, \quad (10.11)$$

which is, in general, within $[0, 1]$. The Wilson confidence interval has been shown to have better performance than the Wald confidence interval and the exact (Clopper–Pearson) confidence interval. See Agresti & Coull (1998), Newcombe (1998), and Brown, Cai & DasGupta (2001) for more detail.

(c) Likelihood ratio confidence interval

When the true value of π is small, the likelihood ratio confidence interval could provide better performance than other alternatives. To construct the likelihood ratio confidence interval of π, we test the following hypotheses

$$H_0: \pi = \pi_0 \quad \text{against} \quad H_1: \pi \neq \pi_0.$$

Let $\hat{\theta}_{\mathrm{R}}$ denote the restricted MLE of θ under H_0. Then

$$\hat{\theta}_{\mathrm{R}} = \frac{n_3(1 - p) - n_1\pi_0 p}{(n_1 + n_3)(1 - p)}.$$

When $n \to \infty$, it is well known that

$$\Lambda_1(\pi_0) = -2\{\ell_{\mathrm{v}}(\pi_0, \hat{\theta}_{\mathrm{R}} | Y_{\mathrm{obs}}) - \ell_{\mathrm{v}}(\hat{\pi}_{\mathrm{v}}, \hat{\theta} | Y_{\mathrm{obs}})\} \stackrel{.}{\sim} \chi^2(1),$$

where $\hat{\pi}_{\mathrm{v}}$ and $\hat{\theta}$ denote the unrestricted MLEs of π and θ specified by (10.2), respectively. Since

$$\Lambda_1(\pi_0) = -2\Big[n_1 \log(1 - \hat{\theta}_{\mathrm{R}}) + n_2 \log(1 - \pi_0) + n_3 \log\{\hat{\theta}_{\mathrm{R}}(1 - p) + \pi_0 p\}$$
$$- n_1 \log(1 - \hat{\theta}) - n_2 \log(1 - \hat{\pi}_{\mathrm{v}}) - n_3 \log\{\hat{\theta}(1 - p) + \hat{\pi}_{\mathrm{v}} p\}\Big],$$

it is easy to verify that $\Lambda_1(\pi_0)$ is an increasing function of π_0 when $\pi_0 \in [0, 1 - n_2/(np)]$ and a decreasing function of π_0 when $\pi_0 \in [1 - n_2/(np), 1]$. Therefore, for a given significance level α, the $(1 - \alpha)100\%$ likelihood ratio confidence interval for π is given by

$$[\hat{\pi}_{\mathrm{V,LRL}}, \ \hat{\pi}_{\mathrm{V,LRU}}], \tag{10.12}$$

where $\hat{\pi}_{\mathrm{V,LRL}}$ and $\hat{\pi}_{\mathrm{V,LRU}}$ are two roots of π_0 to the following equation

$$\Lambda_1(\pi_0) = \chi^2(\alpha, 1),$$

and $\chi^2(\alpha, 1)$ denotes the upper α-th quantile of chi-squared distribution with one degree of freedom.

The asymptotic confidence intervals (10.8), (10.11) and (10.12) are appropriate for cases with large sample sizes. When n is small to moderate, we could use the bootstrap confidence intervals (10.28) and/or (10.29).

10.2.3 The exact (Clopper–Pearson) confidence interval

When the sample size is small to moderate, Clopper & Pearson (1934) proposed a method to calculate the exact confidence limits for the binomial proportion by inverting the equal-tailed test based on the binomial distribution. In this subsection, we employ this method to compute the confidence interval of $\pi = 1 - \lambda_2/p$, see (10.3). Note that

$$n_2 \sim \text{Binomial}\,(n; \lambda_2),$$

the $(1-\alpha)100\%$ exact (Clopper–Pearson) confidence interval $[\hat{\lambda}_{2,\,\text{EL}}, \hat{\lambda}_{2,\,\text{EU}}]$ of λ_2 satisfies the following equations:

$$\hat{\lambda}_{2,\,\text{EL}} \;=\; 0, \qquad \text{when } n_2 = 0,$$

$$\sum_{x=n_2}^{n} \binom{n}{x} \hat{\lambda}_{2,\,\text{EL}}^{x} (1 - \hat{\lambda}_{2,\,\text{EL}})^{n-x} \;=\; \frac{\alpha}{2}, \qquad 1 \leqslant n_2 \leqslant n-1, \qquad (10.13)$$

$$\sum_{x=0}^{n_2} \binom{n}{x} \hat{\lambda}_{2,\,\text{EU}}^{x} (1 - \hat{\lambda}_{2,\,\text{EU}})^{n-x} \;=\; \frac{\alpha}{2}, \qquad 1 \leqslant n_2 \leqslant n-1, \qquad (10.14)$$

$$\hat{\lambda}_{2,\,\text{EU}} \;=\; 1, \qquad \text{when } n_2 = n.$$

By solving (10.13) and (10.14), we obtain

$$\hat{\lambda}_{2,\,\text{EL}} \;=\; \left\{ 1 + \frac{n - n_2 + 1}{n_2 F(1 - \alpha/2;\; 2n_2,\; 2(n - n_2 + 1))} \right\}^{-1} \quad \text{and}$$

$$\hat{\lambda}_{2,\,\text{EU}} \;=\; \left\{ 1 + \frac{n - n_2}{(n_2 + 1) F(\alpha/2;\; 2(n_2 + 1),\; 2(n - n_2))} \right\}^{-1},$$

where $F(\alpha;\; k_1,\; k_2)$ denotes the upper α-th quantile of the F distribution $F(k_1,\; k_2)$. Thus, the $(1-\alpha)100\%$ exact confidence interval of π is given by

$$\hat{\pi}_{\text{V, EL}} = 1 - \frac{\hat{\lambda}_{2,\,\text{EU}}}{p} \quad \text{and} \quad \hat{\pi}_{\text{V, EU}} = 1 - \frac{\hat{\lambda}_{2,\,\text{EL}}}{p}. \qquad (10.15)$$

Since this corresponds to a discrete distribution, the confidence coefficient (or coverage probability) of the exact confidence interval is not exactly $1-\alpha$ but is at least $1-\alpha$. Thus, this exact confidence interval is conservative.

10.2.4 A modified MLE of π and its asymptotic property

(a) An example for MLE of π beyond the unit interval

The MLE of π specified by (10.2) may be beyond the unit interval $[0, 1]$. For example, let

$$(n_1, n_2, n_3)^\top = (15, 20, 35)^\top \quad \text{and} \quad p = \frac{1}{4}.$$

From (10.2), we obtain

$$\hat{\pi}_V = -0.1429 < 0 \quad \text{and} \quad \hat{\theta} = 0.7143.$$

For these cases, we can apply an EM algorithm to calculate the MLEs of π and θ. In Section 10.5.2, we will introduce an EM algorithm to find the posterior modes for both π and θ by using two independent beta prior distributions. Specifically, when two independent uniform distributions on $[0, 1]$ are adopted as the priors, the posterior modes of π and θ are identical to their MLEs. In (10.36)–(10.38) let

$$a_1 = b_1 = a_2 = b_2 = 1,$$

and let $\pi^{(0)} = \theta^{(0)} = 0.5$ be the initial values of π and θ, the EM algorithm converges to

$$\hat{\pi}_V = 2.22 \times 10^{-17} \approx 0 \quad \text{and} \quad \hat{\theta} = 0.70 < 0.7143$$

in 197 iterations.

(b) A modified MLE of π

From (10.2), it can be seen that $0 \leqslant \hat{\pi}_V \leqslant 1$ if and only if $0 \leq n_2 \leq np$. Therefore, a modified MLE of π is

$$\hat{\pi}_{VM} = \max(0, \hat{\pi}_V) = \begin{cases} 0, & \text{if } n_2 > np, \\ \hat{\pi}_V, & \text{if } n_2 \leqslant np. \end{cases} \tag{10.16}$$

The following result shows that the $\hat{\pi}_{VM}$ and $\hat{\pi}_V$ are asymptotically equivalent.

Theorem 10.2 If $0 < \pi < 1$, then $\sqrt{n}\,(\hat{\pi}_{VM} - \pi)$ and $\sqrt{n}\,(\hat{\pi}_V - \pi)$ have the same asymptotic distribution as $n \to \infty$.   ¶

Proof. It suffices to show that $\sqrt{n}\,(\hat{\pi}_{VM} - \pi) - \sqrt{n}\,(\hat{\pi}_V - \pi)$ converges to zero in probability as $n \to \infty$, i.e.,

$$\Pr\{|\sqrt{n}\,(\hat{\pi}_{VM} - \hat{\pi}_V)| > 0\} \to 0, \quad \text{as } n \to \infty. \tag{10.17}$$

When $n_2 \leqslant np$, from (10.16), we have $\hat{\pi}_{VM} = \hat{\pi}_V$. Hence, (10.17) follows immediately.

Now, we consider the case of $n_2 > np$, i.e.,

$$\hat{\lambda}_2 > p. \tag{10.18}$$

Note that $\hat{\lambda}_2$ is the MLE of $\lambda_2 = (1 - \pi)p$. It is easy to show that

$$\Pr(|\hat{\lambda}_2 - \lambda_2| > \varepsilon) \to 0$$

for any given $\varepsilon > 0$ as $n \to \infty$. Thus, we need only to prove

$$\Pr\{|\sqrt{n}\,(\hat{\pi}_{VM} - \hat{\pi}_V)| > 0\} \leqslant \Pr(|\hat{\lambda}_2 - \lambda_2| > \varepsilon) \tag{10.19}$$

for any $\varepsilon < \pi p = p - \lambda_2$. Since $\hat{\pi}_{VM} = 0$, we have

$$|\sqrt{n}\,(\hat{\pi}_{VM} - \hat{\pi}_V)| > 0$$

$$\Rightarrow \quad |\sqrt{n}\,\{0 - (1 - \hat{\lambda}_2/p)\}| > 0$$

$$\Rightarrow \quad |\hat{\lambda}_2 - p| > 0$$

$$\Rightarrow \quad 0 < |\hat{\lambda}_2 - p| \overset{(10.18)}{=} \hat{\lambda}_2 - p = (\hat{\lambda}_2 - \lambda_2) - (p - \lambda_2)$$

$$\Rightarrow \quad |\hat{\lambda}_2 - \lambda_2| \geqslant \hat{\lambda}_2 - \lambda_2 > p - \lambda_2 > \varepsilon.$$

Consequently, (10.19) follows immediately. □

10.3 Statistical Inferences on θ

10.3.1 Three asymptotic confidence intervals of θ for large sample sizes

(a) Wald confidence interval

From (10.2), the variance of $\hat{\theta}$ is

$$\mathrm{Var}(\hat{\theta}) = \frac{\mathrm{Var}(n_1)}{n^2(1-p)^2} = \frac{\lambda_1(1-\lambda_1)}{n(1-p)^2}. \tag{10.20}$$

Similar to Theorem 10.1, it is easy to verify that

$$\overline{\mathrm{Var}}(\hat{\theta}) = \frac{\hat{\lambda}_1(1-\hat{\lambda}_1)}{(n-1)(1-p)^2}$$

is an unbiased estimator of $\text{Var}(\hat{\theta})$. Based on this unbiased estimator, the $(1-\alpha)100\%$ Wald confidence interval of θ is

$$[\hat{\theta}_{\text{WL}}, \hat{\theta}_{\text{WU}}] = \left[\hat{\theta} - z_{\alpha/2}\sqrt{\widehat{\text{Var}}(\hat{\theta})}, \ \hat{\theta} + z_{\alpha/2}\sqrt{\widehat{\text{Var}}(\hat{\theta})}\right]. \tag{10.21}$$

(b) Wilson (score) confidence interval

The $(1-\alpha)100\%$ Wilson (score) CI of θ can be constructed based on

$$
\begin{aligned}
1 - \alpha \quad &= \quad \Pr\left\{\left|\frac{\hat{\theta} - \theta}{\sqrt{\text{Var}(\hat{\theta})}}\right| \leqslant z_{\alpha/2}\right\} \\
&\stackrel{(10.20)}{=} \Pr\left[(\hat{\theta} - \theta)^2 \leqslant \frac{z_{\alpha/2}^2(1-\theta)(1-p)\{1-(1-\theta)(1-p)\}}{n(1-p)^2}\right] \\
&= \quad \Pr\left\{(1+z_*)\theta^2 - (2\hat{\theta} + 2z_* - z_*\rho_3)\theta + \hat{\theta}^2 + z_* - z_*\rho_3 \leqslant 0\right\},
\end{aligned}
$$

where $z_* \stackrel{\wedge}{=} z_{\alpha/2}^2/n$ and $\rho_3 \stackrel{\wedge}{=} 1/(1-p)$. Solving the quadratic inequality inside the above probability, we obtain the Wilson (score) confidence interval $[\hat{\theta}_{\text{WSL}}, \hat{\theta}_{\text{WSU}}]$ of π as follows:

$$\frac{2\hat{\theta} + 2z_* - z_*\rho_3 \pm \sqrt{(2\hat{\theta} + 2z_* - z_*\rho_3)^2 - 4(1+z_*)(\hat{\theta}^2 + z_* - z_*\rho_3)}}{2(1+z_*)}.$$
$$\tag{10.22}$$

which is, in general, within $[0, 1]$.

(c) Likelihood ratio confidence interval

To construct the likelihood ratio confidence interval of θ, we test the following hypotheses

$$H_0\colon \theta = \theta_0 \quad \text{against} \quad H_1\colon \theta \neq \theta_0.$$

Let $\hat{\pi}_{\text{R}}$ denote the restricted MLE of π under H_0. We have

$$\hat{\pi}_{\text{R}} = \frac{n_3 p - n_2\theta_0(1-p)}{(n_2 + n_3)p}.$$

When $n \to \infty$, it is well known that

$$\Lambda_2(\theta_0) = -2\{\ell_{\text{v}}(\hat{\pi}_{\text{R}}, \theta_0 | Y_{\text{obs}}) - \ell_{\text{v}}(\hat{\pi}_{\text{v}}, \hat{\theta} | Y_{\text{obs}})\} \sim \chi^2(1),$$

where $\hat{\pi}_V$ and $\hat{\theta}$ denote the unrestricted MLEs of π and θ specified by (10.2). Since

$$
\begin{aligned}
\Lambda_2(\theta_0) = -2 \Big[& n_1 \log(1 - \theta_0) + n_2 \log(1 - \hat{\pi}_R) + n_3 \log\{\theta_0(1 - p) + \hat{\pi}_R p\} \\
& - n_1 \log(1 - \hat{\theta}) - n_2 \log(1 - \hat{\pi}_V) - n_3 \log\{\hat{\theta}(1 - p) + \hat{\pi}_V p\} \Big],
\end{aligned}
$$

it is easy to verify that $\Lambda_2(\theta_0)$ is an increasing function of θ_0 when

$$
\theta_0 \in \left[0, 1 - \frac{n_1}{n(1 - p)} \right]
$$

and a decreasing function of θ_0 when

$$
\theta_0 \in \left[1 - \frac{n_1}{n(1 - p)}, 1 \right].
$$

Therefore, for a given significance level α, the $(1 - \alpha)100\%$ likelihood ratio confidence interval for θ is given by

$$
[\hat{\theta}_{LRL}, \hat{\theta}_{LRU}], \tag{10.23}
$$

where $\hat{\theta}_{LRL}$ and $\hat{\theta}_{LRU}$ are two roots of θ_0 to the following equation

$$
\Lambda_2(\theta_0) = \chi^2(\alpha, 1).
$$

10.3.2 The exact (Clopper–Pearson) confidence interval

Similar to Section 10.2.3, the $(1 - \alpha)100\%$ exact CI of θ is given by

$$
\hat{\theta}_{EL} = 1 - \frac{\hat{\lambda}_{1,EU}}{1 - p} \quad \text{and} \quad \hat{\theta}_{EU} = 1 - \frac{\hat{\lambda}_{1,EL}}{1 - p}, \tag{10.24}
$$

where

$$
\hat{\lambda}_{1,EL} = \left\{ 1 + \frac{n - n_1 + 1}{n_1 F(1 - \alpha/2; 2n_1, 2(n - n_1 + 1))} \right\}^{-1} \quad \text{and}
$$

$$
\hat{\lambda}_{1,EU} = \left\{ 1 + \frac{n - n_1}{(n_1 + 1)F(\alpha/2; 2(n_1 + 1), 2(n - n_1))} \right\}^{-1}.
$$

10.3.3 Testing Hypotheses

Sometimes, we may have a certain knowledge on the unknown parameter $\theta = \Pr(U = 1)$ before our investigation. For example, we may define $U = 1$ if the respondent's birthday is in the second half of a month and $U = 0$ otherwise. Usually, we assume that $\theta \approx 0.5$. To test whether or not this assumption is valid, in this subsection, we focus on testing the following hypotheses:

$$H_0: \theta = \theta_0 \quad \text{against} \quad H_1: \theta \neq \theta_0. \tag{10.25}$$

(a) Hypothesis test for large sample sizes

Let n_1 represent the number of respondents who put a tick in the circle in Table 10.1 and X be the corresponding random variable. We have

$$X \sim \text{Binomial}(n, \lambda_1).$$

Since $\lambda_1 = (1 - \theta)(1 - p)$, the null and alternative hypotheses in (10.25) are reduced to

$$H_0^*: \lambda_1 = \lambda_{10} \quad \text{against} \quad H_1^*: \lambda_1 \neq \lambda_{10},$$

where $\lambda_{10} = (1 - \theta_0)(1 - p)$.

For large sample sizes, we can use the normal distribution to approximate the binomial distribution. The test statistic and the corresponding z value are given by

$$Z = \frac{X - n\lambda_{10}}{\sqrt{n\lambda_{10}(1 - \lambda_{10})}} \quad \text{and} \quad z = \frac{n_1 - n\lambda_{10}}{\sqrt{n\lambda_{10}(1 - \lambda_{10})}}.$$

Under H_0^*, we have $Z \sim N(0, 1)$. Hence, the corresponding p-value is

$$
\begin{aligned}
p_{v1} &= 2 \Pr(Z > |z|) \\
&= \Pr(Z^2 > z^2) \\
&= \Pr\{\chi^2(1) > z^2\}, \tag{10.26}
\end{aligned}
$$

where $\chi^2(\nu)$ denotes the chi-squared distribution with ν degrees of freedom. When $p_{v1} \geq \alpha$, we cannot reject the null hypothesis H_0^* (equivalently, H_0) at the α level of significance.

(b) Hypothesis test for small to moderate sample sizes

When the sample size is not too large, we need to compute the exact p-value for testing H_0 against H_1. Note that

$$X | H_0^* \sim \text{Binomial}(n, \lambda_{10}),$$

we define

$$\beta_x \hat{=} \Pr(X = x | H_0^*) = \binom{n}{x} \lambda_{10}^x (1 - \lambda_{10})^{n-x}, \quad x = 0, 1, \ldots, n.$$

Thus, the exact two-sided p-value is calculated by

$$p_{v2} = \sum_{x=0}^{n} \beta_x I_{(\beta_x \leq \beta_{n_1})}, \tag{10.27}$$

where $I_{(\cdot)}$ denotes the indicate function.

10.4 Bootstrap Confidence Intervals

In Section 10.2.2, we provided two asymptotic confidence intervals (10.8) and (10.11) of π, which are available only for large sample sizes. Next, the Wilson confidence interval (10.11) is still possible to be beyond the unit interval $[0, 1]$. Third, although the exact confidence interval (10.15) is available for small to moderate sample sizes, its performance can be shown (Agresti & Coull, 1998) to be even inferior to that of the Wilson confidence interval specified in (10.11). For these cases, we could employ the bootstrap method to find bootstrap confidence intervals of π for cases with small to moderate sample sizes. Finally, in Section 10.2.4 (a), we mentioned that if the MLE of π calculated by (10.2) is less than zero, then the EM algorithm (10.36)–(10.38) with $a_1 = b_1 = a_2 = b_2 = 1$ can be used to compute the MLEs of π and θ. For these situations, the bootstrap method is also a useful tool to find confidence intervals for an arbitrary function of π and θ, say, $\vartheta = h(\pi, \theta)$.

Let $\hat{\vartheta} = h(\hat{\pi}_V, \hat{\theta})$ denote the MLE of ϑ, where $\hat{\pi}_V$ and $\hat{\theta}$ respectively represent the MLEs of π and θ calculated by means of either (10.2) or the EM algorithm (10.36)–(10.38) with $a_1 = b_1 = a_2 = b_2 = 1$. Based on the obtained MLEs $\hat{\pi}_V$ and $\hat{\theta}$, we can generate

$$(n_1^*, n_2^*, n_3^*)^\top \sim \text{Multinomial}\,(n;\ (1 - \hat{\theta})(1 - p), (1 - \hat{\pi}_V)p, \hat{\theta}(1 - p) + \hat{\pi}_V p).$$

Having obtained $Y_{\text{obs}}^* = \{n;\ n_1^*, n_2^*, n_3^*\}$, we can calculate a bootstrap replication $\hat{\pi}_V^*$ and $\hat{\theta}^*$ and calculate $\hat{\vartheta}^* = h(\hat{\pi}_V^*, \hat{\theta}^*)$. Independently repeating this process G times, we obtain G bootstrap replications $\{\hat{\vartheta}_g^*\}_{g=1}^G$. Consequently, the standard error, $\text{se}(\hat{\vartheta})$, of $\hat{\vartheta}$ can be estimated by the sample standard deviation of the G replications, i.e.,

$$\widehat{\text{se}}(\hat{\vartheta}) = \left\{ \frac{1}{G-1} \sum_{g=1}^{G} \left(\hat{\vartheta}_g^* - \frac{\hat{\vartheta}_1^* + \cdots + \hat{\vartheta}_G^*}{G} \right)^2 \right\}^{\frac{1}{2}}.$$

If $\{\hat\vartheta_g^*\}_{g=1}^G$ is approximately normally distributed, a $(1-\alpha)100\%$ bootstrap confidence interval for ϑ is

$$\left[\hat\vartheta - z_{\alpha/2} \times \widehat{se}(\hat\vartheta),\ \hat\vartheta + z_{\alpha/2} \times \widehat{se}(\hat\vartheta)\right]. \tag{10.28}$$

Alternatively, if $\{\hat\vartheta_g^*\}_{g=1}^G$ is non-normally distributed, a $(1-\alpha)100\%$ bootstrap confidence interval of ϑ can be obtained as

$$[\hat\vartheta_{\mathrm{L}},\ \hat\vartheta_{\mathrm{U}}], \tag{10.29}$$

where $\hat\vartheta_{\mathrm{L}}$ and $\hat\vartheta_{\mathrm{U}}$ are the $100(\alpha/2)$ and $100(1-\alpha/2)$ percentiles of $\{\hat\vartheta_g^*\}_{g=1}^G$, respectively.

10.5 Bayesian Inferences

In this section, we first derive posterior moments of π and θ when a certain prior information is available. Second, we utilize the EM algorithm to calculate the posterior modes of π and θ when their posterior distributions are highly skewed. Finally, we generate i.i.d. posterior samples of π and θ via the exact IBF sampling.

10.5.1 Posterior moments with explicit expressions

By ignoring the normalizing constant and the known factor $(1-p)^{n_1}p^{n_2+n_3}$, we write the kernel of (10.1) as

$$l_{\mathrm{V}}(\pi,\theta|Y_{\mathrm{obs}}) = (1-\theta)^{n_1}(1-\pi)^{n_2}(\theta\rho_2 + \pi)^{n_3}, \tag{10.30}$$

where $0 < \pi < 1$, $0 < \theta < 1$ and ρ_2 is defined in (10.10). If two independent beta distributions $\mathrm{Beta}(a_1, b_1)$ and $\mathrm{Beta}(a_2, b_2)$ are adopted as the prior distributions of π and θ, respectively, then the joint posterior distribution of π and θ is

$$f(\pi,\theta|Y_{\mathrm{obs}}) = \frac{\pi^{a_1-1}(1-\pi)^{b_1-1}\theta^{a_2-1}(1-\theta)^{b_2-1} \times l_{\mathrm{V}}(\pi,\theta|Y_{\mathrm{obs}})}{c_{\mathrm{V}}(a_1,b_1,a_2,b_2;\ n_1,n_2,n_3)}, \tag{10.31}$$

where the normalizing constant is given by

$$c_{\mathrm{V}}(a_1,b_1,a_2,b_2;\ n_1,n_2,n_3)$$

$$= \int_0^1\int_0^1 \pi^{a_1-1}(1-\pi)^{b_1-1}\theta^{a_2-1}(1-\theta)^{b_2-1} \times l_{\mathrm{V}}(\theta,\pi|Y_{\mathrm{obs}})\ \mathrm{d}\pi\,\mathrm{d}\theta$$

$$= \sum_{i=0}^{n_3} \binom{n_3}{i} \rho_2^i \int_0^1 \pi^{a_1+n_3-i-1}(1-\pi)^{b_1+n_2-1}\ \mathrm{d}\pi$$

$$\times \int_0^1 \theta^{a_2+i-1}(1-\theta)^{b_2+n_1-1}\, d\theta$$

$$= \sum_{i=0}^{n_3} \binom{n_3}{i} \rho_2^i B(a_1+n_3-i,\ b_1+n_2)B(a_2+i,\ b_2+n_1). \quad (10.32)$$

Therefore, the r-th posterior moments of π and θ are given by

$$E(\pi^r|Y_{\mathrm{obs}}) = \frac{c_{\mathrm{V}}(a_1+r,b_1,a_2,b_2;\ n_1,n_2,n_3)}{c_{\mathrm{V}}(a_1,b_1,a_2,b_2;\ n_1,n_2,n_3)} \quad \text{and}$$

$$E(\theta^r|Y_{\mathrm{obs}}) = \frac{c_{\mathrm{V}}(a_1,b_1,a_2+r,b_2;\ n_1,n_2,n_3)}{c_{\mathrm{V}}(a_1,b_1,a_2,b_2;\ n_1,n_2,n_3)}, \quad (10.33)$$

respectively.

10.5.2 Calculation of the posterior modes via the EM algorithm

The EM algorithm is a useful tool for computing MLEs in the presence of missing or latent data. Let Z denote the number of respondents belonging to the sensitive subclass $\{Y = 1,\ W = 1\}$ in Table 10.1. Since Z is unobservable, it is natural to treat Z as a latent variable. In addition, let z denote the realization of Z. Thus, the likelihood function of π and θ for the complete data $\{Y_{\mathrm{obs}}, z\}$ is

$$L_{\mathrm{V}}(\pi,\theta|Y_{\mathrm{obs}},z) = \binom{n}{n_1,n_2,n_3-z,z}\{(1-\theta)(1-p)\}^{n_1}$$

$$\times \{(1-\pi)p\}^{n_2}\{\theta(1-p)\}^{n_3-z}(\pi p)^z,$$

$$\propto \pi^z(1-\pi)^{n_2}\theta^{n_3-z}(1-\theta)^{n_1}.$$

Again, the product of two independent beta densities $\mathrm{Beta}\,(\pi|a_1,b_1)$ and $\mathrm{Beta}\,(\theta|a_2,b_2)$ is adopted as the joint prior density of π and θ. Hence, the complete-data posterior distribution and the conditional predictive distribution are

$$f(\pi,\theta|Y_{\mathrm{obs}},z) = \mathrm{Beta}\,(\pi|a_1+z,b_1+n_2)$$

$$\times \mathrm{Beta}\,(\theta|a_2+n_3-z,b_2+n_1) \quad \text{and} \quad (10.34)$$

$$f(z|Y_{\mathrm{obs}},\pi,\theta) = \mathrm{Binomial}\left(z\,\middle|\,n_3,\ \frac{\pi p}{\theta(1-p)+\pi p}\right), \quad (10.35)$$

respectively. The M-step of the EM algorithm is to calculate the complete-data posterior modes of π and θ as

$$\tilde{\pi}_{\mathrm{V}} = \frac{a_1+z-1}{a_1+b_1+n_2+z-2} \quad (10.36)$$

and

$$\tilde{\theta} = \frac{a_2 + n_3 - z - 1}{a_2 + b_2 + n_1 + n_3 - z - 2}, \tag{10.37}$$

and the E-step is to replace z in (10.36) and (10.37) by the conditional expectation

$$E(Z|Y_{\text{obs}}, \pi, \theta) = \frac{n_3 \pi p}{\theta(1 - p) + \pi p}. \tag{10.38}$$

10.5.3 Generation of i.i.d. posterior samples via the exact IBF sampling

According to the exact IBF sampling presented in Appendix B, to generate i.i.d. posterior samples of π and θ we simply need to identify the conditional support $\mathcal{S}_{(Z|Y_{\text{obs}}, \pi, \theta)}$ and calculate the weights $\{\omega_k\}_{k=1}^K$. From (10.35), we have

$$\mathcal{S}_{(Z|Y_{\text{obs}})} = \mathcal{S}_{(Z|Y_{\text{obs}}, \pi, \theta)} = \{z_1, \dots, z_K\} = \{0, 1, \dots, n_3\},$$

where $K = n_3 + 1$. Setting $\pi_0 = \theta_0 = 0.5$, from (B.2) and (B.3), we obtain

$$q_k(\pi_0, \theta_0) = \frac{f(Z = z_k|Y_{\text{obs}}, \pi_0, \theta_0)}{f(\pi_0, \theta_0|Y_{\text{obs}}, z_k)}, \tag{10.39}$$

and $\omega_k = q_k(0.5, 0.5) / \sum_{k'=1}^K q_{k'}(0.5, 0.5)$ for $k = 1, \dots, K$.

10.6 Comparison with the Crosswise Model

In this section, we will compare the variant of the parallel model with the crosswise model. The criteria of the difference of variances and the ratio of variances are considered. Theoretical and numerical results are provided.

10.6.1 The difference of variances

Let $\hat{\pi}_{\text{C}}$ denote the MLE of $\pi = \Pr(Y = 1)$ under the crosswise model with $p = \Pr(W = 1) \neq 0.5$. From (2.9) and (10.4), we have

$$\begin{aligned}
\text{Var}(\hat{\pi}_{\text{C}}) - \text{Var}(\hat{\pi}_{\text{V}}) &= \frac{p(1 - p)}{n(2p - 1)^2} - \frac{(1 - p)(1 - \pi)}{np} \\
&= \frac{1 - p}{np(2p - 1)^2} \times h_{\text{CV}}(p|\pi), \quad p \neq 0.5, \tag{10.40}
\end{aligned}$$

where

$$h_{\text{CV}}(p|\pi) \triangleq (4\pi - 3)p^2 + 4(1 - \pi)p + \pi - 1$$

is a quadratic function of p for any fixed π $(\pi \neq 3/4)$. The discriminant of the $h_{\mathrm{CV}}(p|\pi)$ is given by

$$
\begin{aligned}
D(h_{\mathrm{CV}}) &= 16(1-\pi)^2 - 4(4\pi - 3)(\pi - 1) \\
&= 4(1-\pi) > 0.
\end{aligned}
$$

We then have the following results.

Theorem 10.3 Let $\pi \in (0,1)$ and $p \in (0,1)$.

(1) When $\pi = 3/4$, the variant of the parallel model is always more efficient than the crosswise model for any $p > 1/4$.

(2) When $\pi > 3/4$, the variant of the parallel model is more efficient than the crosswise model for any $p \in (p_\pi, 1)$, where

$$
p_\pi = \frac{-2(1-\pi) + \sqrt{1-\pi}}{4\pi - 3} \tag{10.41}
$$

is a monotonically decreasing function of $\pi \in (3/4, 1)$ and $0 < p_\pi < 0.25$.

(3) When $\pi < 3/4$, the variant of the parallel model is always more efficient than the crosswise model for any $p \in (p_\pi, 1)$, where p_π defined by (10.41) is a monotonically decreasing function of $\pi \in (0, 3/4)$ and $0.25 < p_\pi < 1/3$. ¶

Proof. (1) When $\pi = 3/4$, we have $h_{\mathrm{CV}}(p|\pi) = p - 1/4$. Hence, $h_{\mathrm{CV}}(p|\pi) > 0$ if and only if $p > 1/4$. From (10.40), we obtain

$$
\mathrm{Var}(\hat{\pi}_{\mathrm{C}}) > \mathrm{Var}(\hat{\pi}_{\mathrm{V}})
$$

for $p > 1/4$.

(2) When $3/4 < \pi < 1$, it can be shown that the equation $h_{\mathrm{CV}}(p|\pi) = 0$ has two roots

$$
p_{\mathrm{L}} = \frac{-2(1-\pi) - \sqrt{1-\pi}}{4\pi - 3}
$$

and p_π, which is defined by (10.41). It is clear that $p_{\mathrm{L}} < 0$. Since

$$
\frac{\mathrm{d}p_\pi}{\mathrm{d}\pi} = \frac{-(2\sqrt{1-\pi} - 1)^2}{2\sqrt{1-\pi}(4\pi - 3)^2} < 0, \tag{10.42}
$$

p_π is a monotonically decreasing function of π. The infimum of p_π equals to $\lim_{\pi \to 1} p_\pi = 0$ and the supremum of p_π is equal to

$$
\lim_{\pi \to 0.75} p_\pi = \lim_{\pi \to 0.75} \frac{2 - \dfrac{1}{2\sqrt{1-\pi}}}{4} = \frac{1}{4},
$$

so that $0 < p_\pi < 0.25$. Thus, $h_{\text{CV}}(p|\pi) > 0$ if and only if $p_\pi < p < 1$.

(3) When $0 < \pi < 3/4$, it can see that the equation $h_{\text{CV}}(p|\pi) = 0$ has two roots p_π defined by (10.41) and

$$
\begin{aligned}
p_{\text{U}}(\pi) &= \frac{-2(1-\pi) - \sqrt{1-\pi}}{4\pi - 3} \\
&= \frac{2(1-\pi) + \sqrt{1-\pi}}{3 - 4\pi}.
\end{aligned}
$$

Note that

$$
\frac{dp_{\text{U}}(\pi)}{d\pi} = \frac{(2\sqrt{1-\pi} + 1)^2}{2\sqrt{1-\pi}\,(4\pi - 3)^2} > 0.
$$

Hence, $p_{\text{U}}(\pi)$ is a monotonically increasing function of $\pi \in (0, 3/4)$ so that $p_{\text{U}}(\pi) > p_{\text{U}}(0) = 1$. From (10.42), we know that p_π is also a monotonically decreasing function of $\pi \in (0, 3/4)$. The infimum of p_π is

$$
\lim_{\pi \to 0.75} p_\pi = \frac{1}{4}
$$

and the supremum of p_π is

$$
\lim_{\pi \to 0} p_\pi = \frac{1}{3}.
$$

In other words, we have

$$
0.25 < p_\pi < \frac{1}{3}.
$$

Thus, $h_{\text{CV}}(p|\pi) > 0$ if and only if $p_\pi < p < 1$.   □

From Theorem 10.3, we have immediately the following result.

Corollary 10.1 The variant of the parallel model is always more efficient than the crosswise model for any $\pi \in (0, 1)$ and $p > 1/3$.   ¶

10.6.2   Relative efficiency of the crosswise model to the variant of the parallel model

The RE of the crosswise model to the variant of the parallel model is

$$
\begin{aligned}
\text{RE}_{\text{C} \to \text{V}}(\pi, p) &= \frac{\text{Var}(\hat{\pi}_{\text{C}})}{\text{Var}(\hat{\pi}_{\text{V}})} \\
&= \frac{\pi(1-\pi) + p(1-p)/(2p-1)^2}{\pi(1-\pi) + (1-\pi)(1-p)/p},
\end{aligned}
$$

Table 10.3 Relative efficiency $\mathrm{RE}_{\mathrm{C}\to\mathrm{V}}(\pi, p)$ for various combinations of π and p

π	p					
	1/3	0.40	0.45	0.55	0.60	2/3
0.05	1.0513	4.1070	20.5174	30.0659	8.8825	3.9187
0.10	1.1058	4.2292	20.8739	30.0594	8.8261	3.8704
0.20	1.2273	4.5294	21.8936	30.5815	8.8846	3.8571
0.30	1.3727	4.9286	23.4244	31.8885	9.1773	3.9464
0.40	1.5556	5.4737	25.6747	34.1903	9.7500	4.1481
0.50	1.8000	6.2500	29.0320	37.9310	10.714	4.5000
0.60	2.1538	7.4286	34.2851	44.0529	12.316	5.0909
0.70	2.7284	9.4091	43.2832	54.8024	15.146	6.1389
0.80	3.8571	13.391	61.5907	76.9691	21.000	8.3077
0.90	7.2069	25.375	117.047	144.571	38.872	14.929
0.95	13.881	49.367	228.315	280.486	74.814	28.241

which is independent of the sample size n.

Table 10.3 reports some values of $\mathrm{RE}_{\mathrm{C}\to\mathrm{V}}(\pi, p)$ for various combinations of π and p. For example, when $\pi = 0.95$ and $p = 0.55$, we have $\mathrm{RE}_{\mathrm{C}\to\mathrm{V}}(0.95, 0.55) = 280.486$, which implies that the efficiency of the variant of the parallel model greatly outweighs that of the crosswise model. When $\pi = 0.80$ and $p = 0.60$, we have $\mathrm{RE}_{\mathrm{C}\to\mathrm{V}}(0.80, 0.60) = 21.000$, implying that the efficiency of the variant of the parallel model is 21 times that of the crosswise model.

10.7 Comparison with the Triangular Model

10.7.1 The difference of variances

Let $\hat{\pi}_{\mathrm{T}}$ denote the MLE of $\pi = \Pr(Y = 1)$ under the triangular model with $p = \Pr(W = 1)$. From (3.3) and (10.4), we have

$$
\begin{aligned}
\mathrm{Var}(\hat{\pi}_{\mathrm{T}}) - \mathrm{Var}(\hat{\pi}_{\mathrm{V}}) &= \frac{(1-\pi)p}{n(1-p)} - \frac{(1-\pi)(1-p)}{np} \\
&= \frac{(1-\pi)(2p-1)}{np(1-p)}, \quad (10.43)
\end{aligned}
$$

where $p \in (0, 1)$.

Table 10.4   Relative efficiency $\mathrm{RE}_{\mathrm{T}\to\mathrm{V}}(\pi, p)$ for various combinations of π and p

π	p						
	1/3	0.40	0.45	0.5	0.55	0.60	2/3
0.05	0.2683	0.4624	0.6824	1	1.4654	2.1628	3.7273
0.10	0.2857	0.4792	0.6944	1	1.4400	2.0870	3.5000
0.20	0.3182	0.5098	0.7159	1	1.3968	1.9615	3.1429
0.30	0.3478	0.5370	0.7346	1	1.3613	1.8621	2.8750
0.40	0.3750	0.5614	0.7509	1	1.3317	1.7813	2.6667
0.50	0.4000	0.5833	0.7654	1	1.3065	1.7143	2.5000
0.60	0.4231	0.6032	0.7783	1	1.2849	1.6579	2.3636
0.70	0.4444	0.6212	0.7898	1	1.2661	1.6098	2.2500
0.80	0.4643	0.6377	0.8002	1	1.2497	1.5682	2.1538
0.90	0.4828	0.6528	0.8096	1	1.2352	1.5319	2.0714
0.95	0.4915	0.6599	0.8140	1	1.2285	1.5155	2.0345

Theorem 10.4   For any $\pi \in (0, 1)$ and $p \in (0.5, 1)$, the variant of the parallel model is always more efficient than the triangular model, i.e.,

$$\mathrm{Var}(\hat{\pi}_{\mathrm{T}}) > \mathrm{Var}(\hat{\pi}_{\mathrm{V}}).$$ ¶

10.7.2   Relative efficiency of the triangular model to the variant of the parallel model

The RE of the triangular model to the variant of the parallel model is

$$\mathrm{RE}_{\mathrm{T}\to\mathrm{V}}(\pi, p) = \frac{\mathrm{Var}(\hat{\pi}_{\mathrm{T}})}{\mathrm{Var}(\hat{\pi}_{\mathrm{V}})} = \frac{\pi + p/(1-p)}{\pi + (1-p)/p},$$

which is independent of the sample size n.

Table 10.4 reports some values of $\mathrm{RE}_{\mathrm{T}\to\mathrm{V}}(\pi, p)$ for various combinations of π and p. We observe from Table 10.4 that, for any $\pi \in (0, 1)$, $\mathrm{RE}_{\mathrm{T}\to\mathrm{V}}(\pi, p) > 1$ if $p > 0.5$, while $\mathrm{RE}_{\mathrm{T}\to\mathrm{V}}(\pi, p) < 1$ if $p < 0.5$. In other words, when $p > 0.5$, the efficiency of the variant of the parallel model is superior to that of the triangular model and when $p < 0.5$ the efficiency of the variant of the parallel model is inferior to that of the triangular model. In particular, when $p = 0.5$, the efficiency of the two models is equivalent.

10.8   The Noncompliance Behavior

The noncompliance behavior encountered in randomized response practice is that some respondents are not willing to follow the design instructions even though interviewers provide them with secret answer sheets, sealed envelopes, and sincere promises of confidentiality (Mangat, 1994). However, in our opinion, a possible/partial reason for such noncompliance behaviors may be caused by the use of randomizing devices which are, in general, controlled by interviewers. One aim for developing non-randomized response techniques is trying to alleviate the noncompliance behavior. For example, for the crosswise model (Chapter 2 of this book) and the parallel model (Chapter 7 of this book) with two sensitive categories (i.e., both $\{Y = 0\}$ and $\{Y = 1\}$ are sensitive), we in general believe that for those respondents not refusing, they are willing to follow the instruction since their privacy is well protected. However, for the triangular model (Chapter 3 of this book), a tick put in the triangle indicates that the respondent may belong to the sensitive class. Therefore, the noncompliance may occur in the triangular model. Tang & Wu (2013) developed two design techniques (i.e., the dual non-randomized response triangular model and the alternating non-randomized response triangular model) which incorporate noncompliance into the triangular model.

And actually, the noncompliance behavior can also occur in the variant of the parallel model. We note that only the sub-category $\{Y = 1, W = 1\}$ (i.e., the lower square in Table 10.1) contains sensitive information. Respondents belonging to this sub-category and having insufficient confidence on such a survey may put a tick in the triangle in Table 10.1, resulting in the noncompliance. Taking the noncompliance into consideration, we denote the probability of the respondents who have the sensitive characteristic and belong to $\{W = 1\}$ following the design instruction in Table 10.1 by ω. Because the new parameter ω is added, the respondents need be randomly assigned into one of two groups. For the first group, we utilize the variant of parallel model with two non-sensitive binary variates W and U and the sensitive binary variate Y. However, for the second group, we employ the parallel model (Tian, 2012) with the same W, U and Y to estimate the sensitive proportion $\pi = \Pr(Y = 1)$.

Suppose that in the first group, we observe n_{11}, n_{12} and n_{13} ($n_1 = n_{11} + n_{12} + n_{13}$) respondents put ticks in the circle, triangle and upper square, respectively. Thus, the cell probabilities for the three categories are given by

$$\lambda_1^* = \Pr\{U = 0, W = 0\} = (1 - \theta)(1 - p),$$
$$\lambda_2^* = \Pr\{Y = 0, W = 1\} = (1 - \pi\omega)p, \quad \text{and}$$

$$\lambda_3^* = \Pr\{U = 1, W = 0\} + \Pr\{Y = 1, W = 1\} = \theta(1-p) + \pi\omega p.$$

From the first equation, the MLE of θ is

$$\hat{\theta} = 1 - \frac{n_{11}}{(1-p)n_1}. \qquad (10.44)$$

From the second/third equation, it is clear that only $\pi\omega$ is estimable. The corresponding estimate is $1 - n_{12}/(pn_1)$. This is why we need the second group. Assume that in the second group, we observe n_{21} and n_{22} ($n_2 = n_{21} + n_{22}$) individuals put ticks in the upper circle and upper square, respectively. Thus, the MLEs of π and ω are given by

$$\hat{\pi}_{\mathrm{P}} = \frac{n_{22}/n_2 - \hat{\theta}(1-p)}{p} \quad \text{and} \quad \hat{\omega} = \frac{1}{\hat{\pi}_{\mathrm{P}}}\left(1 - \frac{n_{12}}{pn_1}\right), \qquad (10.45)$$

respectively. If at least one of the values of $\hat{\theta}$, $\hat{\pi}_{\mathrm{P}}$ and $\hat{\omega}$ are beyond the unit interval $[0, 1]$, we employ the EM algorithm to calculate the MLEs.

10.9 An Illustrative Example of Sexual Practices

As a sensitive topic, talking about individual sexual practices is still embarrassing even in countries with open minds. Consequently, it is very difficult to estimate the average numbers of sexual partners or the cell probabilities of having x ($x \leqslant 1$ or $x \geqslant 2$) sexual partners in a targeted population based on survey data from direct questionnaires. However, gathering information from this kind of sensitive topic plays a crucial role in assisting researchers to investigate the relationship between sexual behaviors and some sexual transmission diseases.

Consider a subset of the sexual practice data from the study of Monto (2001), in which participants were men arrested for trying to hire prostitutes in three cities (San Francisco, Las Vegas, and Portland, Oregon) of the United States. From participants' background characteristics shown in Table 1 of Monto (2001), we can see that 343 individuals graduated at most from some high schools and 927 individuals received at least some college training. Also, there are 593 respondents having no more than one sexual partner and 668 respondents having no fewer than two sexual partners.

To demonstrate the proposed design for the variant of the parallel model presented in Table 10.1, we define $Y = 1$ if the respondent has at least two sexual partners and $Y = 0$ otherwise. To estimate the unknown proportion $\pi = \Pr(Y = 1)$, we employ two non-sensitive binary variables U and W, where $U = 1$ if the respondent received at least some college training and $U = 0$ otherwise; and $W = 1$ if the respondent's birthday is from September to December and $W = 0$ otherwise. Thus, it is reasonable to assume that $p = \Pr(W = 1) \approx 1/3$.

Table 10.5 Survey data from Monto (2001)

Level of education	The number of sexual partners		Total
	$Y = 0\ (\leqslant 1)$	$Y = 1\ (> 2)$	
$U = 0$	160 (m_1)	180 (m_2)	340
$U = 1$	433 (m_3)	488 (m_4)	921
Total	593	668	1261

Table 10.6 Six 95% confidence intervals of π

Type of CIs	Confidence interval	Width
Wald CI (10.8)	[0.4715823, 0.5915091]	0.1199268
Wilson CI (10.11)	[0.4684999, 0.5883603]	0.1198604
Likelihood ratio CI (10.12)	[0.4695520, 0.5893770]	0.1198250
Exact CI (10.15)	[0.4680426, 0.5902233]	0.1221806
Bootstrap CI† (10.28)	[0.4716088, 0.5914962]	0.1198874
Bootstrap CI‡ (10.29)	[0.4700315, 0.5906940]	0.1206625

Table 10.7 Six 95% confidence intervals of θ

Type of CIs	Confidence interval	Width
Wald CI (10.21)	[0.6973280, 0.7608739]	0.06354587
Wilson CI (10.22)	[0.6959085, 0.7593993]	0.06349080
Likelihood ratio CI (10.23)	[0.6963906, 0.7598780]	0.06348745
Exact CI (10.24)	[0.6956505, 0.7603133]	0.06466284
Bootstrap CI† (10.28)	[0.6972551, 0.7609398]	0.06368466
Bootstrap CI‡ (10.29)	[0.6971609, 0.7610410]	0.06388013

First, we need to verify the independence between the level of education and the number of sexual partners. Table 10.5 displays the survey data of Monto (2001). The MLE of the odds ratio is given by

$$\hat{\psi} = \frac{m_1 m_4}{m_2 m_3} = 1.001796.$$

We would like to test the null hypothesis H_0: $\psi = 1$ against the alternative hypothesis H_1: $\psi \neq 1$. The corresponding p-value is

$$p\text{-value} = 2 \times \Pr\left(Z < \frac{-|L|}{\widehat{se}} \right) = 0.9887377,$$

Table 10.8 Posterior estimates of parameters for the data of sexual practices

Parameter	Posterior mode	Posterior mean	Posterior std	95% Bayesian credible interval
π	0.5315	0.5302	0.0303	[0.4688, 0.5881]
θ	0.7291	0.7285	0.0163	[0.6959, 0.7596]

where Z denotes the standard normal random variable,

$$L = \log\left(\frac{m_1 m_4}{m_2 m_3}\right)$$

and

$$\widehat{se} = \sqrt{\frac{1}{m_1} + \frac{1}{m_2} + \frac{1}{m_3} + \frac{1}{m_4}}\,.$$

Since the p-value $= 0.9887377 \gg 0.05$, we have reason to believe that there is no association between the level of education and the number of sexual partners.

As a result, the observed data can be constructed as

$$(n_1, n_2, n_3)^{\top} = (343 \times (1-p),\ 593 \times p,\ 927 \times (1-p) + 668 \times p)^{\top}$$

$$\approx (229, 198, 841)^{\top},$$

where $n = n_1 + n_2 + n_3 = 1268$.

According to (10.2), the MLEs of π and θ are given by

$$\hat{\pi}_{\mathrm{V}} = 0.5315 \quad \text{and} \quad \hat{\theta} = 0.7291.$$

The confidence intervals of π based on (10.8), (10.11), (10.12), (10.15), (10.28) and (10.29) are shown in Table 10.6. Similarly, the confidence intervals of θ based on (10.21), (10.22), (10.23), (10.24), (10.28) and (10.29) are shown in Table 10.7.

Suppose that we want to test the null hypothesis

$$H_0\colon \theta = \theta_0 = 0.73 \quad \text{against} \quad H_1\colon \theta \neq 0.73.$$

Let $\alpha = 0.05$, from (10.26) and (10.27), we have

$$p_{\mathrm{v1}} = 0.9557 \quad \text{and} \quad p_{\mathrm{v2}} = 0.9418.$$

Since both p-values are larger than 0.05, we fail to reject H_0. If we set $\theta_0 = 0.69$, then

$$p_{\mathrm{v1}} = 0.0219 \quad \text{and} \quad p_{\mathrm{v2}} = 0.0220.$$

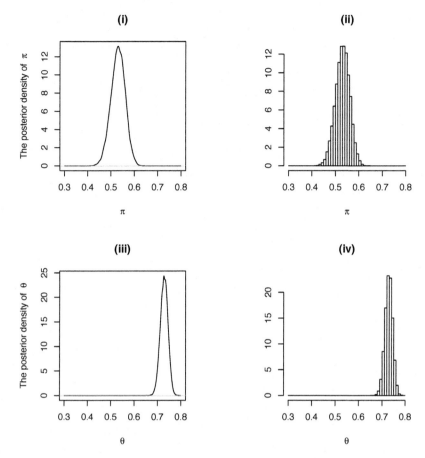

Figure 10.4 Posterior densities of π and θ via a kernel density smoother based on $L = 20{,}000$ i.i.d. posterior samples generated by the IBF sampling with two independent uniform distributions on $(0, 1)$ as the prior distributions of π and θ. (i) The posterior density of π; (ii) the histogram of π; (iii) the posterior density of θ; (iv) the histogram of θ.

As a result, H_0 is rejected at the level of $\alpha = 0.025$.

We adopt two independent uniform distributions (i.e., $a_1 = b_1 = a_2 = b_2 = 1$) as the prior distributions of π and θ, respectively. Using $\pi^{(0)} = \theta^{(0)} = 0.5$ as the initial values, the EM algorithm specified by (10.36)–(10.38) converges to the posterior modes

$$\tilde{\pi}_{\mathrm{V}} = 0.5315 \quad \text{and} \quad \tilde{\theta} = 0.7291$$

in 19 iterations.

Based on (10.39), we employ the IBF sampler to generate $L = 20{,}000$ i.i.d. posterior samples of π and θ. The posterior means, the posterior standard deviations and 95% Bayesian credible intervals of π and θ are given in the third, fourth and fifth columns of Table 10.8. Figure 10.4 shows the posterior densities of π and θ and their histograms.

10.10   Case Studies on Cheating Behavior in Examinations

10.10.1   Design and analysis under the assumption of complete compliance

Cheating in examinations in universities and colleges around the world definitely results in unfairness and it has been considered as a sensitive issue from which we can hardly obtain reliable answer via direct asking. To investigate the proportion of undergraduates who have ever cheated in examinations, we used the variant of parallel model to conduct a survey in March 2013 among 150 undergraduates at the University of Hong Kong (HKU) in Hong Kong, P. R. China. The questions in the questionnaire include:

(1)   If your birthday is in the first half of the year and you are not a Hong Kong permanent resident, please circle 1;

(2)   If your birthday is in the second half of the year and you had never cheated in examinations at HKU, please circle 2;

(3)   If your birthday is in the first half of the year and you are a Hong Kong permanent resident OR if your birthday is in the second half of the year and you had ever cheated in examinations at HKU, please circle 3.

At the end of the study, 115 students (52 female and 63 male) returned the completed questionnaire, where 1 student was from the Faculty of Arts, 22 were from the Faculty of Business and Economics, 2 were from the Faculty of Engineering, 89 are from the Faculty of Science and 1 student did not report his/her faculty. Among these students, 99 were Year 1 students, 2 were Year 2 student, 13 were Year 3 students and 1 was Year 4 student. It was observed that 22 circles on 1, 54 circles on 2 and 39 circles on 3. Let $\pi = \Pr(Y = 1)$ denote the unknown proportion of undergraduates with cheating behavior in examinations at HKU and $\theta = \Pr(U = 1)$ denote the unknown proportion of undergraduates being Hong Kong permanent residents. The observed data can be represented by

$$Y_{\text{obs}} = \{n; \, n_1, \, n_2, \, n_3\} = \{115; 22, 54, 39\}.$$

According to (10.2), the MLEs of π and θ are given by

$$\hat{\pi}_{\text{V}} = 0.0609 \quad \text{and} \quad \hat{\theta} = 0.6174.$$

Table 10.9 Six 95% confidence intervals of π

Type of CIs	95% CI	Width
Wald CI (10.8)	[−0.1223575, 0.2440966]	0.3664541
Wilson CI (10.11)	[−0.1205648, 0.2383688]	0.3589336
Likelihood ratio CI (10.12)	[−0.1213907, 0.2404458]	0.3618365
Exact CI (10.15)	[−0.1297411, 0.2482693]	0.3780104
Bootstrap CI[†] (10.28)	[−0.0676406, 0.2186684]	0.2863091
Bootstrap CI[‡] (10.29)	[6.832142×10^{-18}, 0.2347826]	0.2347826

Table 10.10 Six 95% confidence intervals of θ

Type of CIs	95% CI	Width
Wald CI (10.21)	[0.4729868, 0.7617958]	0.2888089
Wilson CI (10.22)	[0.4546011, 0.7402681]	0.2856670
Likelihood ratio CI (10.23)	[0.4608817, 0.7467020]	0.2858203
Exact CI (10.24)	[0.4496279, 0.7521093]	0.3024814
Bootstrap CI[†] (10.28)	[0.4725176, 0.7512286]	0.2787111
Bootstrap CI[‡] (10.29)	[0.4608696, 0.7407407]	0.2798712

Table 10.11 Posterior estimates of parameters for the data of cheating behavior in examinations at HKU

Parameter	Posterior mode	Posterior mean	Posterior std	95% Bayesian credible interval
π	0.0609	0.1040	0.0678	[0.0061, 0.2256]
θ	0.6174	0.5977	0.0704	[0.4503, 0.7261]

Six 95% confidence intervals of π based on (10.8), (10.11), (10.12), (10.15), (10.28) and (10.29) are shown in Table 10.9. Similarly, six 95% confidence intervals of θ based on (10.21), (10.22), (10.23), (10.24), (10.28) and (10.29) are shown in Table 10.10.

Suppose that we want to test the null hypothesis

$$H_0: \theta = \theta_0 = 0.35 \quad \text{against} \quad H_1: \theta \neq 0.35.$$

Let $\alpha = 0.05$, from (10.26) and (10.27), we have

$$p_{v1} = 0.0022 \quad \text{and} \quad p_{v2} = 0.0019.$$

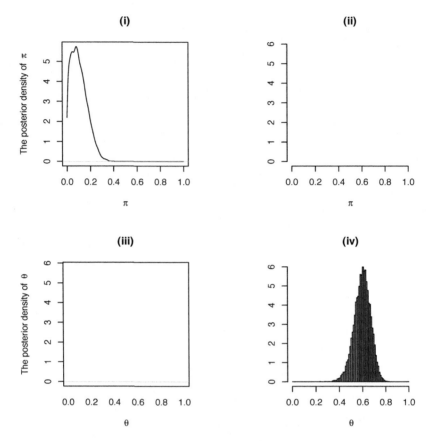

Figure 10.5 Posterior densities of π and θ via a kernel density smoother based on $L = 20{,}000$ i.i.d. posterior samples generated by the IBF sampling with two independent uniform distributions on $(0, 1)$ as the prior distributions of π and θ. (i) The posterior density of π; (ii) the histogram of π; (iii) the posterior density of θ; (iv) the histogram of θ.

As a result, H_0 should be rejected at the level of $\alpha = 0.01$. If we set $\theta_0 = 0.60$, then

$$p_{v1} = 0.8157 \quad \text{and} \quad p_{v2} = 0.9073.$$

Since both p-values are larger than 0.05, we cannot reject the H_0.

For the Bayesian analysis, we adopt two independent uniform distributions (i.e., $a_1 = b_1 = a_2 = b_2 = 1$) as the prior distributions of π and θ, respectively. Using $\pi^{(0)} = \theta^{(0)} = 0.5$ as the initial values, the EM algorithm specified by (10.36)–(10.38) converged to the posterior modes

$$\tilde{\pi}_{\mathrm{v}} = 0.0609 \quad \text{and} \quad \tilde{\theta} = 0.6174$$

in 133 iterations.

Based on (10.39), we employ the IBF sampling to generate $L = 20{,}000$ i.i.d. posterior samples of π and θ. The posterior means, the posterior standard deviations, and 95% Bayesian credible intervals of π and θ are given in the third, fourth, and fifth columns of Table 10.11. Figure 10.5 shows the posterior densities of π and θ and their histograms.

10.10.2 Design and analysis under the consideration of noncompliance

To account for the noncompliance behavior in the questionnaire, we conducted the second survey by using the parallel model in March 2013 among 100 undergraduates at HKU. The questions in the questionnaire include:

(1) If your birthday is in the first half of the year and you are not a Hong Kong permanent resident OR if your birthday is in the second half of the year and you had NEVER cheated in examinations at HKU, please circle 'no';

(2) If your birthday is in the first half of the year and you are a Hong Kong permanent resident OR if your birthday is in the second half of the year and you had EVER cheated in examinations at HKU, please circle 'yes'.

At the end of the data collection, 77 students (27 female and 50 male) returned the completed questionnaire, where 2 were from the Faculty of Law, 7 were from the Faculty of Business and Economics, 67 are from the Faculty of Science and 1 was from an un-specified faculty. Among these students, 43 were Year 1 students, 10 were Year 2 students, 20 were Year 3 students, and 4 were Year 4 students. It was observed that 40 circles on 'no' and 37 circles on 'yes'. Let $\pi = \Pr(Y = 1)$ denote the unknown proportion of undergraduates with cheating behavior in examinations at HKU, $\theta = \Pr(U = 1)$ denote the unknown proportion of undergraduates being Hong Kong permanent resident, and ω denote the unknown proportion of undergraduates (who have cheating behavior in examinations at HKU and their birthdays are in the second half of the year) following the design instruction in Table 10.1. The observed data can be denoted by

$$Y_{\text{obs}} = \{n_2; \; n_{21}, \; n_{22}\} = \{77; 40, 37\}.$$

According to (10.44) and (10.45), we have

$$\hat{\theta} = 0.6174, \quad \hat{\pi}_{\text{p}} = 0.3436, \quad \text{and} \quad \hat{\omega} = 0.1771.$$

The MLE of π obtained from the combined data of two groups is significantly higher than that obtained only from the first sample. Since $\hat{\omega} = 0.1771$, we can see that about

$$\hat{\pi}(1 - \hat{\omega})p = 0.3436 * (1 - 0.1771) * 0.5 = 14.14\%$$

students did not follow the instruction of the design for the variant of parallel model in our surveys.

10.11   Discussion

This chapter presents a new development for the parallel model originally proposed by Tian (2012) in sample surveys with sensitive questions. The basic idea is to use two additional non-sensitive binary variates in conjunction with the sensitive binary response variable to create a scenario under which confidentiality of the respondent is preserved and partial information on the sensitive response variable is also obtained. The proposed model assumes that the population mean (proportion) of one of the non-sensitive variates is known but the other one is unknown. The last fact is new and provides certain flexibility in choosing the non-sensitive binary variate. Point and variance estimates of the population proportion of the sensitive response are derived, and several asymptotic confidence intervals are provided. Theoretical and numerical comparisons show that the proposed variant of the parallel model outperforms the existing non-randomized response crosswise and triangular models for most of the possible parameter ranges as shown in Corollary 10.1, Theorem 10.4, Tables 10.3, and 10.4. A possible explanation for these conclusions is that the variant of the parallel design can gather exact information (instead of mixing information) for two cells (i.e., the circle and the triangle in Table 10.1) because of the introduction of an additional non-sensitive binary variate U when comparing with the crosswise and triangular models. Finally, we provide a simple way to handle the possible noncompliance behavior in the proposed model.

CHAPTER 11

The Combination Questionnaire Model

In Chapter 6, we introduced the hidden sensitivity model for evaluating the association of two sensitive binary variates. However, in practice, we sometimes need to assess the association between one sensitive binary variate and one non-sensitive binary variate. In this chapter, we introduce a combination questionnaire model, which consists of a main questionnaire and a supplemental questionnaire without using any randomizing device (Yu, Lu & Tian, 2013). The corresponding statistical inference approaches are also presented.

11.1 The Survey Design

Assume that X is a sensitive binary random variable, Y is a non-sensitive binary random variable, and they are correlated. Let $\{X = 1\}$ denote the sensitive class (e.g., $X = 1$ if a respondent is a drug user) and $\{X = 0\}$ denote the non-sensitive class (e.g., $X = 0$ if a respondent is not a drug user). Furthermore, let both $\{Y = 1\}$ (e.g., $Y = 1$ if a respondent receives at least some college training) and $\{Y = 0\}$ (e.g., $Y = 0$ if a respondent graduates at most from some high school) be non-sensitive classes. Define $\boldsymbol{\theta} = (\theta_1, \ldots, \theta_4)^{\top}$, where

$$\left\{ \begin{array}{rcl} \theta_1 & = & \Pr(X = 0,\ Y = 0), \\ \theta_2 & = & \Pr(X = 1,\ Y = 0), \\ \theta_3 & = & \Pr(X = 1,\ Y = 1), \\ \theta_4 & = & \Pr(X = 0,\ Y = 1), \end{array} \right.$$

then $\boldsymbol{\theta} \in \mathbb{T}_4$. The objective is to make inferences on $\boldsymbol{\theta}$,

$$\theta_x \ \hat{=} \ \Pr(X = 1) = \theta_2 + \theta_3,$$

$$\theta_y \ \hat{=} \ \Pr(Y = 1) = \theta_4 + \theta_3$$

and the odds ratio

$$\psi = \frac{\theta_1 \theta_3}{\theta_2 \theta_4}.$$

Table 11.1 The main questionnaire for the combination questionnaire model

Category	$W = 1$	$W = 2$	$W = 3$
I: $\{X = 0,\ Y = 0\}$	Block 1: _____	Block 2: _____	Block 3: _____
II: $\{X = 1,\ Y = 0\}$	Category II: Please put a tick in Block 2		
III: $\{X = 1,\ Y = 1\}$	Category III: Please put a tick in Block 3		
IV: $\{X = 0,\ Y = 1\}$	Block 4: _____		

Note: Only $\{X = 1\}$ is a sensitive class, while $\{X = 0\}$, $\{Y = 0\}$ and $\{Y = 1\}$ are non-sensitive classes.

Table 11.2 The supplemental questionnaire for the combination questionnaire model

Category	
$\{Y = 0\}$	Please put a tick in Block 5: _____ if you belong to $\{Y = 0\}$
$\{Y = 1\}$	Please put a tick in Block 6: _____ if you belong to $\{Y = 1\}$

Note: Both the $\{Y = 0\}$ and $\{Y = 1\}$ are non-sensitive classes.

The survey scheme consists of a main questionnaire and a supplemental questionnaire. To design the main questionnaire which is to be assigned to group 1 with n respondents (n is specified by the investigators), we first introduce a non-sensitive question (say, Q_{W}) with three possible answers. Assume that W is a non-sensitive variate with trichotomous outcomes associated with the Q_{W} and W is independent of both X and Y. Define

$$p_i = \Pr(W = i) \quad i = 1,\ 2,\ 3.$$

For example, let $W = 1\,(2,\ 3)$ if a respondent was born in January–April (May–August, September–December) and thus we could assume that $p_i \approx 1/3$. The main questionnaire is shown in Table 11.1, under which each respondent is asked to answer the non-sensitive question Q_{W}.

On the one hand, since Category I (i.e., $\{X = 0,\ Y = 0\}$) and Category IV(i.e., $\{X = 0,\ Y = 1\}$) are non-sensitive to each respondent, it is reasonable to assume that a respondent is willing to provide his/her truthful answer by putting a tick in Block i ($i = 1, \ldots, 4$) according to his/her true status. On the other hand, Category II (i.e., $\{X = 1,\ Y = 0\}$) and Category III (i.e., $\{X = 1,\ Y = 1\}$) are usually sensitive to respondents. In this case,

Table 11.3 Cell probabilities, observed and unobservable counts for the main questionnaire

Category	$W = 1$	$W = 2$	$W = 3$	Total
I: $\{X = 0,\ Y = 0\}$	$p_1\theta_1$	$p_2\theta_1$	$p_3\theta_1$	$\theta_1\ (Z_1)$
II: $\{X = 1,\ Y = 0\}$				$\theta_2\ (Z_2)$
III: $\{X = 1,\ Y = 1\}$				$\theta_3\ (Z_3)$
IV: $\{X = 0,\ Y = 1\}$		$\theta_4\ (n_4)$		$\theta_4\ (n_4)$
Total	$p_1\theta_1\ (n_1)$	$p_2\theta_1 + \theta_2\ (n_2)$	$p_3\theta_1 + \theta_3\ (n_3)$	$1\ (n)$

Note: $n = \sum_{i=1}^{4} n_i$, $Z_1 = n - (Z_2 + Z_3) - n_4$, where (Z_2, Z_3) are unobservable.

Table 11.4 Cell probabilities and observed counts for the supplemental questionnaire

Category		Total
$\{Y = 0\}$	Block 5: _____	$\theta_1 + \theta_2\ (m_0)$
$\{Y = 1\}$	Block 6: _____	$\theta_3 + \theta_4\ (m_1)$
Total		$1\ (m)$

Note: $m = m_0 + m_1$.

if a respondent belongs to Category II (III), he/she is designed to put a tick in Block 2 (3).

The supplemental questionnaire is designed as shown in Table 11.2, under which m respondents (m is also specified by the investigators) in group 2 are asked to put a tick in Block 5 or Block 6 depending on their true status; that is, $\{Y = 0\}$ or $\{Y = 1\}$. Since both the $\{Y = 0\}$ and $\{Y = 1\}$ are non-sensitive, the supplemental questionnaire is in fact a design of direct questioning. Thus, we call this design the combination questionnaire model.

Table 11.3 shows the cell probabilities $\{\theta_i\}_{i=1}^{4}$, the observed frequencies $\{n_i\}_{i=1}^{4}$ and the unobservable frequencies $\{Z_i\}_{i=1}^{3}$ for the main questionnaire. The observed frequency n_2 is the sum of the frequency of respondents belonging to Block 2 and the frequency of those belonging to Category II. The observed frequency n_3 is the sum of the frequency of respondents belonging to Block 3 and the frequency of those belonging to Category III. Note that $n = \sum_{i=1}^{4} n_i = \sum_{i=1}^{3} Z_i + n_4$, we have $Z_1 = n - (Z_2 + Z_3) - n_4$. Thus, only Z_2 and Z_3 are unobservable.

Table 11.4 shows the cell probabilities and observed counts for the supplemental questionnaire.

11.2 Likelihood-based Inferences

In this section, MLEs of the $\boldsymbol{\theta}$ and the odds ratio ψ are derived by using the EM algorithm. In addition, asymptotic confidence intervals and the bootstrap confidence intervals of an arbitrary function of $\boldsymbol{\theta}$ are also provided. Finally, a likelihood ratio test is presented for testing association between the two binary random variables.

11.2.1 MLEs via the EM algorithm

A total of $N = n+m$ respondents are classified into two groups by a randomization approach such that n respondents answer the questions in the main questionnaire and m respondents answer the questions in the supplemental questionnaire. Let $Y_{\mathrm{obs,\,M}} = \{n; \, n_1, \ldots, n_4\}$ denote the observed counts collected in the main questionnaire (see Table 11.3), where $\sum_{i=1}^4 n_i = n$. The likelihood function of $\boldsymbol{\theta}$ based on $Y_{\mathrm{obs,\,M}}$ is

$$L(\boldsymbol{\theta}|Y_{\mathrm{obs,\,M}}) = \binom{n}{n_1, n_2, n_3, n_4} (p_1\theta_1)^{n_1} \left\{ \prod_{i=2}^3 (p_i\theta_1 + \theta_i)^{n_i} \right\} \theta_4^{n_4}.$$

Let $Y_{\mathrm{obs,\,S}} = \{m; \, m_0, m_1\}$ denote the observed counts gathered in the supplemental questionnaire (see Table 11.4), where $m_0 + m_1 = m$. The likelihood function of $\boldsymbol{\theta}$ based on $Y_{\mathrm{obs,\,S}}$ is

$$L(\boldsymbol{\theta}|Y_{\mathrm{obs,\,S}}) = \binom{m}{m_0} (\theta_1 + \theta_2)^{m_0} (\theta_3 + \theta_4)^{m_1}.$$

Let $Y_{\mathrm{obs}} = \{Y_{\mathrm{obs,\,M}}, \, Y_{\mathrm{obs,\,S}}\}$. Since $Y_{\mathrm{obs,\,M}}$ and $Y_{\mathrm{obs,\,S}}$ are independent, the observed-data likelihood function of $\boldsymbol{\theta} \in \mathbb{T}_4$ is

$$L_{\mathrm{CQ}}(\boldsymbol{\theta}|Y_{\mathrm{obs}}) \propto \theta_1^{n_1} \left\{ \prod_{i=2}^3 (p_i\theta_1 + \theta_i)^{n_i} \right\} \theta_4^{n_4} \times (\theta_1 + \theta_2)^{m_0} (\theta_3 + \theta_4)^{m_1}, \quad (11.1)$$

where the subscript 'CQ' denotes the 'combination questionnaire' model.

By treating the observed counts n_2 and n_3 as incomplete data, we use the EM algorithm to find the MLE $\hat{\boldsymbol{\theta}}$ of $\boldsymbol{\theta}$. The counts $\{Z_i\}_{i=1}^3$ in Table 11.3 can be viewed as missing data. Briefly, Z_1, Z_2, Z_3 and n_4 represent the counts of the respondents belonging to Categories I, II, III and IV, respectively. Thus, we denote the latent data by $Y_{\mathrm{mis}} = \{z_2, z_3\}$ and the complete data by $Y_{\mathrm{com}} = \{Y_{\mathrm{obs}}, Y_{\mathrm{mis}}\}$. Note that all $\{p_i\}$ are known. Consequently, the complete-data likelihood function for $\boldsymbol{\theta}$ is

$$L_{\mathrm{CQ}}(\boldsymbol{\theta}|Y_{\mathrm{com}}) \propto \left(\prod_{i=1}^3 \theta_i^{z_i} \right) \theta_4^{n_4} \times (\theta_1 + \theta_2)^{m_0} (\theta_3 + \theta_4)^{m_1}, \quad (11.2)$$

where $z_1 = n - (z_2 + z_3) - n_4$.

By treating $\{\theta_i\}_{i=1}^4$ as random variables, we note that the complete-data likelihood function (11.2) has the density form of a grouped Dirichlet distribution (Tian, Ng & Geng, 2003). Ng et al. (2008) derived the mode of a grouped Dirichlet density with explicit expressions (see Section 11.5.1). Hence, from (11.21) and (11.22), the complete-data MLEs for $\boldsymbol{\theta}$ are given by

$$\begin{cases} \theta_1 &= 1 - \theta_2 - \theta_3 - \theta_4, \\[2mm] \theta_2 &= \dfrac{z_2}{N}\left(1 + \dfrac{m_0}{n - z_3 - n_4}\right), \\[2mm] \theta_3 &= \dfrac{z_3}{N}\left(1 + \dfrac{m_1}{z_3 + n_4}\right), \\[2mm] \theta_4 &= \dfrac{n_4}{N}\left(1 + \dfrac{m_1}{z_3 + n_4}\right). \end{cases} \quad (11.3)$$

Given Y_{obs} and $\boldsymbol{\theta}$, Z_i follows the binomial distribution with parameters n_i and $\theta_i/(p_i\theta_1 + \theta_i)$; that is,

$$Z_i|(Y_{\mathrm{obs}}, \boldsymbol{\theta}) \sim \mathrm{Binomial}\left(n_i, \frac{\theta_i}{p_i\theta_1 + \theta_i}\right), \quad i = 2, 3.$$

Therefore, the E-step of the EM algorithm computes the following conditional expectations

$$E(Z_i|Y_{\mathrm{obs}}, \boldsymbol{\theta}) = \frac{n_i\theta_i}{p_i\theta_1 + \theta_i}, \quad i = 2, 3, \quad (11.4)$$

and the M-step updates (11.3) by replacing z_2 and z_3 with above conditional expectations.

11.2.2 Asymptotic confidence intervals

Let $\boldsymbol{\theta}_{-4} = (\theta_1, \theta_2, \theta_3)^{\mathsf{T}}$. The asymptotic variance-covariance matrix of the MLE $\hat{\boldsymbol{\theta}}_{-4}$ is then given by $\mathbf{I}_{\mathrm{obs}}^{-1}(\hat{\boldsymbol{\theta}}_{-4})$, where

$$\mathbf{I}_{\mathrm{obs}}(\boldsymbol{\theta}_{-4}) = -\frac{\partial^2 \ell_{\mathrm{CQ}}(\boldsymbol{\theta}|Y_{\mathrm{obs}})}{\partial\boldsymbol{\theta}_{-4}\partial\boldsymbol{\theta}_{-4}^{\mathsf{T}}}$$

denotes the observed information matrix and $\ell_{\mathrm{CQ}}(\boldsymbol{\theta}|Y_{\mathrm{obs}}) = \log L_{\mathrm{CQ}}(\boldsymbol{\theta}|Y_{\mathrm{obs}})$ is the observed-data log-likelihood function. From (11.1), we have

$$\begin{aligned} \ell_{\mathrm{CQ}}(\boldsymbol{\theta}|Y_{\mathrm{obs}}) &= n_1 \log\theta_1 + n_2 \log(p_2\theta_1 + \theta_2) + n_3 \log(p_3\theta_1 + \theta_3) \\ &\quad + n_4 \log(1 - \theta_1 - \theta_2 - \theta_3) + m_0 \log(\theta_1 + \theta_2) \\ &\quad + m_1 \log(1 - \theta_1 - \theta_2). \end{aligned}$$

It is easy to show that

$$\frac{\partial \ell_{\mathrm{CQ}}(\boldsymbol{\theta}|Y_{\mathrm{obs}})}{\partial \theta_1} = \frac{n_1}{\theta_1} - \frac{n_4}{\theta_4} + \sum_{i=2}^{3} \frac{n_i p_i}{p_i \theta_1 + \theta_i} + \frac{m_0}{\theta_1 + \theta_2} - \frac{m_1}{1 - \theta_1 - \theta_2},$$

$$\frac{\partial \ell_{\mathrm{CQ}}(\boldsymbol{\theta}|Y_{\mathrm{obs}})}{\partial \theta_2} = -\frac{n_4}{\theta_4} + \frac{n_2}{p_2 \theta_1 + \theta_2} + \frac{m_0}{\theta_1 + \theta_2} - \frac{m_1}{1 - \theta_1 - \theta_2},$$

$$\frac{\partial \ell_{\mathrm{CQ}}(\boldsymbol{\theta}|Y_{\mathrm{obs}})}{\partial \theta_3} = -\frac{n_4}{\theta_4} + \frac{n_3}{p_3 \theta_1 + \theta_3},$$

and

$$-\frac{\partial^2 \ell_{\mathrm{CQ}}(\boldsymbol{\theta}|Y_{\mathrm{obs}})}{\partial \theta_1^2} = \frac{n_1}{\theta_1^2} + \frac{n_4}{\theta_4^2} + \sum_{i=2}^{3} \frac{n_i p_i^2}{(p_i \theta_1 + \theta_i)^2} + \phi,$$

$$-\frac{\partial^2 \ell_{\mathrm{CQ}}(\boldsymbol{\theta}|Y_{\mathrm{obs}})}{\partial \theta_2^2} = \frac{n_4}{\theta_4^2} + \frac{n_2}{(p_2 \theta_1 + \theta_2)^2} + \phi,$$

$$-\frac{\partial^2 \ell_{\mathrm{CQ}}(\boldsymbol{\theta}|Y_{\mathrm{obs}})}{\partial \theta_3^2} = \frac{n_4}{\theta_4^2} + \frac{n_3}{(p_3 \theta_1 + \theta_3)^2},$$

$$-\frac{\partial^2 \ell_{\mathrm{CQ}}(\boldsymbol{\theta}|Y_{\mathrm{obs}})}{\partial \theta_1 \partial \theta_2} = \frac{n_4}{\theta_4^2} + \frac{n_2 p_2}{(p_2 \theta_1 + \theta_2)^2} + \phi,$$

$$-\frac{\partial^2 \ell_{\mathrm{CQ}}(\boldsymbol{\theta}|Y_{\mathrm{obs}})}{\partial \theta_1 \partial \theta_3} = \frac{n_4}{\theta_4^2} + \frac{n_3 p_3}{(p_3 \theta_1 + \theta_3)^2},$$

$$-\frac{\partial^2 \ell_{\mathrm{CQ}}(\boldsymbol{\theta}|Y_{\mathrm{obs}})}{\partial \theta_2 \partial \theta_3} = \frac{n_4}{\theta_4^2},$$

where

$$\phi = \frac{m_0}{(\theta_1 + \theta_2)^2} + \frac{m_1}{(1 - \theta_1 - \theta_2)^2}.$$

Hence, the observed information matrix can be expressed as

$$\mathbf{I}_{\mathrm{obs}}(\boldsymbol{\theta}_{-4}) = \mathrm{diag}\left(\frac{n_1}{\theta_1^2}, 0, 0\right) + \frac{n_4}{\theta_4^2} \times \mathbf{1}_3 \mathbf{1}_3^{\top} + \mathbf{A}, \qquad (11.5)$$

where

$$\mathbf{A} = \begin{pmatrix} \displaystyle\sum_{i=2}^{3} \frac{n_i p_i^2}{(p_i \theta_1 + \theta_i)^2} + \phi, & \dfrac{n_2 p_2}{(p_2 \theta_1 + \theta_2)^2} + \phi, & \dfrac{n_3 p_3}{(p_3 \theta_1 + \theta_3)^2} \\[4mm] \dfrac{n_2 p_2}{(p_2 \theta_1 + \theta_2)^2} + \phi, & \dfrac{n_2}{(p_2 \theta_1 + \theta_2)^2} + \phi, & 0 \\[4mm] \dfrac{n_3 p_3}{(p_3 \theta_1 + \theta_3)^2}, & 0, & \dfrac{n_3}{(p_3 \theta_1 + \theta_3)^2} \end{pmatrix}. \qquad (11.6)$$

Let $\mathrm{se}(\hat{\theta}_i)$ denote the standard error of $\hat{\theta}_i$ for $i = 1, 2, 3$. Note that $\mathrm{se}(\hat{\theta}_i)$ can be estimated by the square root of the i-th diagonal element of $\mathbf{I}_{\mathrm{obs}}^{-1}(\hat{\boldsymbol{\theta}}_{-4})$. We denote the estimated value of $\mathrm{se}(\hat{\theta}_i)$ by $\widehat{\mathrm{se}}(\hat{\theta}_i)$. Thus, a 95% normal-based asymptotic confidence interval for θ_i can be constructed as

$$[\hat{\theta}_i - 1.96 \times \widehat{\mathrm{se}}(\hat{\theta}_i), \ \hat{\theta}_i + 1.96 \times \widehat{\mathrm{se}}(\hat{\theta}_i)], \quad i = 1, 2, 3. \qquad (11.7)$$

Let $\vartheta = h(\boldsymbol{\theta}_{-4})$ be an arbitrary differentiable function of $\boldsymbol{\theta}_{-4}$. For example, $\theta_4 = 1 - \sum_{i=1}^{3} \theta_i$ and the odds ratio $\psi = \theta_1 \theta_3 / (\theta_2 \theta_4)$. The delta method (e.g., Tanner, 1996, p. 34) can be used to approximate the standard error of $\hat{\vartheta} = h(\hat{\boldsymbol{\theta}}_{-4})$ and a 95% normal-based asymptotic confidence interval for ϑ is given by

$$[\hat{\vartheta} - 1.96 \times \widehat{\mathrm{se}}(\hat{\vartheta}), \ \hat{\vartheta} + 1.96 \times \widehat{\mathrm{se}}(\hat{\vartheta})], \qquad (11.8)$$

where

$$\widehat{\mathrm{se}}(\hat{\vartheta}) = \left\{ \left(\frac{\partial \vartheta}{\partial \boldsymbol{\theta}_{-4}} \right)^{\top} \mathbf{I}_{\mathrm{obs}}^{-1}(\boldsymbol{\theta}_{-4}) \left(\frac{\partial \vartheta}{\partial \boldsymbol{\theta}_{-4}} \right) \Big|_{\boldsymbol{\theta}_{-4} = \hat{\boldsymbol{\theta}}_{-4}} \right\}^{\frac{1}{2}}. \qquad (11.9)$$

11.2.3 Bootstrap confidence intervals

When the normal-based asymptotic confidence interval like (11.7) is beyond the low bound zero or the upper bound one, the bootstrap approach can be used to construct the bootstrap confidence interval of $\vartheta = h(\boldsymbol{\theta}_{-4})$. Based on the obtained MLE $\hat{\boldsymbol{\theta}}$, we independently generate

$$(n_1^*, \ldots, n_4^*)^{\top} \sim \mathrm{Multinomial}\,(n;\ p_1\hat{\theta}_1,\ p_2\hat{\theta}_1 + \hat{\theta}_2,\ p_3\hat{\theta}_1 + \hat{\theta}_3,\ \hat{\theta}_4) \qquad (11.10)$$

and

$$m_0^* \sim \mathrm{Binomial}\,(m,\ \hat{\theta}_1 + \hat{\theta}_2). \qquad (11.11)$$

Having obtained $Y_{\mathrm{obs,\,M}}^* = \{n;\ n_1^*, \ldots, n_4^*\}$ and $Y_{\mathrm{obs,\,S}}^* = \{m;\ m_0^*, m_1^*\}$, where $m_1^* = m - m_0^*$, we can calculate the bootstrap replication $\hat{\vartheta}^* = h(\boldsymbol{\theta}_{-4}^*)$ based on $Y_{\mathrm{obs}}^* = \{Y_{\mathrm{obs,\,M}}^*,\ Y_{\mathrm{obs,\,S}}^*\}$ via the EM algorithm specified by (11.3) and (11.4). Independently repeating this process G times, we obtain G bootstrap replications $\{\hat{\vartheta}_g^*\}_{g=1}^{G}$. Consequently, a $(1 - \alpha)100\%$ bootstrap confidence interval for ϑ is given by

$$[\hat{\vartheta}_{\mathrm{L}},\ \hat{\vartheta}_{\mathrm{U}}], \qquad (11.12)$$

where $\hat{\vartheta}_{\mathrm{L}}$ and $\hat{\vartheta}_{\mathrm{U}}$ are the $100(\alpha/2)$ and $100(1 - \alpha/2)$ percentiles of $\{\hat{\vartheta}_g^*\}_{g=1}^{G}$, respectively.

11.2.4   The likelihood ratio test for testing association

The likelihood ratio statistic can be used to test whether the two binary random variables X and Y are correlated. The corresponding null and alternative hypotheses are

$$H_0: \psi = 1 \quad \text{against} \quad H_1: \psi \neq 1.$$

The likelihood ratio statistic is defined by

$$\Lambda = -2\{\ell_{\mathrm{CQ}}(\hat{\boldsymbol{\theta}}_{\mathrm{R}}|Y_{\mathrm{obs}}) - \ell_{\mathrm{CQ}}(\hat{\boldsymbol{\theta}}|Y_{\mathrm{obs}})\}, \tag{11.13}$$

where $\hat{\boldsymbol{\theta}}_{\mathrm{R}}$ denotes the restricted MLE of $\boldsymbol{\theta}$ under H_0, $\hat{\boldsymbol{\theta}}$ denotes the MLE of $\boldsymbol{\theta}$, which can be obtained by the EM algorithm specified by (11.3) and (11.4), and $\ell_{\mathrm{CQ}}(\boldsymbol{\theta}|Y_{\mathrm{obs}}) = \log L_{\mathrm{CQ}}(\boldsymbol{\theta}|Y_{\mathrm{obs}})$.

To find the restricted MLE $\hat{\boldsymbol{\theta}}_{\mathrm{R}}$, we also employ the EM algorithm. Under H_0: $\theta_1\theta_3 = \theta_2\theta_4$, we have

$$\left\{ \begin{array}{rcl} \theta_1 & = & (1 - \theta_x)(1 - \theta_y), \\ \theta_2 & = & \theta_x(1 - \theta_y), \\ \theta_3 & = & \theta_x\theta_y, \\ \theta_4 & = & (1 - \theta_x)\theta_y. \end{array} \right. \tag{11.14}$$

In other words, under H_0 we only have two free parameters θ_x and θ_y. Having obtained the restricted MLEs $\hat{\theta}_{x,\mathrm{R}}$ and $\hat{\theta}_{y,\mathrm{R}}$, we can compute the restricted MLE $\hat{\boldsymbol{\theta}}_{\mathrm{R}} = (\hat{\theta}_{1,\mathrm{R}}, \ldots, \hat{\theta}_{4,\mathrm{R}})^{\top}$ from (11.14) by

$$\left\{ \begin{array}{rcl} \hat{\theta}_{1,\mathrm{R}} & = & (1 - \hat{\theta}_{x,\mathrm{R}})(1 - \hat{\theta}_{y,\mathrm{R}}), \\ \hat{\theta}_{2,\mathrm{R}} & = & \hat{\theta}_{x,\mathrm{R}}(1 - \hat{\theta}_{y,\mathrm{R}}), \\ \hat{\theta}_{3,\mathrm{R}} & = & \hat{\theta}_{x,\mathrm{R}}\hat{\theta}_{y,\mathrm{R}}, \\ \hat{\theta}_{4,\mathrm{R}} & = & (1 - \hat{\theta}_{x,\mathrm{R}})\hat{\theta}_{y,\mathrm{R}}. \end{array} \right. \tag{11.15}$$

In what follows, we consider the computation of the restricted MLEs $\hat{\theta}_{x,\mathrm{R}}$ and $\hat{\theta}_{y,\mathrm{R}}$. Now, the complete-data likelihood function (11.2) becomes

$$\begin{aligned} L_{\mathrm{CQ}}(\theta_x, \theta_y|Y_{\mathrm{com}}, H_0) & \propto & \{(1 - \theta_x)(1 - \theta_y)\}^{z_1}\{\theta_x(1 - \theta_y)\}^{z_2} \\ & & \times\ (\theta_x\theta_y)^{z_3}\{(1 - \theta_x)\theta_y\}^{n_4}(1 - \theta_y)^{m_0}\theta_y^{m_1} \\ & = & \theta_x^{z_2+z_3}(1 - \theta_x)^{z_1+n_4}\theta_y^{z_3+n_4+m_1}(1 - \theta_y)^{z_1+z_2+m_0} \end{aligned}$$

so that the restricted MLEs of θ_x and θ_y based on the complete-data are given by

$$\theta_x = \frac{z_2 + z_3}{n} \quad \text{and} \quad \theta_y = \frac{z_3 + n_4 + m_1}{N}, \tag{11.16}$$

respectively. Thus, the M-step of the EM algorithm calculates (11.16) and the E-step computes the conditional expectations given in (11.4), where $\{\theta_i\}$ are defined in (11.14). Finally, under H_0, Λ asymptotically follows chi-squared distribution with one degree of freedom.

11.3   Bayesian Inferences

To derive the posterior mode of $\boldsymbol{\theta}$, we employ the EM algorithm again. The latent data $Y_{\text{mis}} = \{z_2, z_3\}$ are the same as those in Section 11.2.1. Based on the complete-data likelihood function (11.2), if the Dirichlet distribution Dirichlet$(a_1, \ldots, a_4)$ is adopted as the prior distribution of $\boldsymbol{\theta}$, then the complete-data posterior distribution is

$$f(\boldsymbol{\theta}|Y_{\text{obs}}, Y_{\text{mis}}) = \text{GD}_{4,2,2}(\boldsymbol{\theta}_{-4}|\boldsymbol{a}, \boldsymbol{b}), \tag{11.17}$$

where $\boldsymbol{a} = (a_1 + z_1, a_2 + z_2, a_3 + z_3, a_4 + n_4)^{\mathsf{T}}$, $\boldsymbol{b} = (m_0, m_1)^{\mathsf{T}}$, and $z_1 = n - (z_2 + z_3) - n_4$. The conditional predictive distribution is

$$f(Y_{\text{mis}}|Y_{\text{obs}}, \boldsymbol{\theta}) = \prod_{i=2}^{3} \text{Binomial}\left(z_i \middle| n_i, \frac{\theta_i}{p_i\theta_1 + \theta_i}\right). \tag{11.18}$$

Therefore, the M-step of the EM algorithm is to calculate the complete-data posterior mode:

$$\begin{cases} \theta_1 &= 1 - \theta_2 - \theta_3 - \theta_4, \\[2mm] \theta_2 &= \dfrac{a_2 - 1 + z_2}{a_+ - 4 + N}\left(1 + \dfrac{m_0}{a_1 + a_2 - 2 + n - z_3 - n_4}\right), \\[2mm] \theta_3 &= \dfrac{a_3 - 1 + z_3}{a_+ - 4 + N}\left(1 + \dfrac{m_1}{a_3 + a_4 - 2 + z_3 + n_4}\right), \\[2mm] \theta_4 &= \dfrac{a_4 - 1 + n_4}{a_+ - 4 + N}\left(1 + \dfrac{m_1}{a_3 + a_4 - 2 + z_3 + n_4}\right). \end{cases} \tag{11.19}$$

where $a_+ = \sum_{i=1}^{4} a_i$, and the E-step is to replace $\{z_i\}_{i=2}^{3}$ by the conditional expectations given by (11.4).

In addition, based on (11.17) and (11.18), the data augmentation algorithm of Tanner and Wong (1987) can be used to generate posterior samples of $\boldsymbol{\theta}$. A sampling method from (11.17) is given in Section 11.5.2.

11.4   Analyzing Cervical Cancer Data in Atlanta

Williamson & Haber (1994) reported a study which examined the relationship between disease status of cervical cancer and the number of sex partners and other risk factors. Cases were 20–70-year-old women of Fulton or Dekalb counties in Atlanta, Georgia. They were diagnosed and were ascertained to have invasive cervical cancer. Controls were randomly chosen from the same counties and the same age ranges. Table 11.5 gives the cross classification of number of sex partners ('few, 0–3' or 'many, $\geqslant 4$', denoted by $X = 0$ or $X = 1$) and disease status (control or case, denoted by $Y = 0$ or $Y = 1$). Generally, a sizable proportion (13.5% in this example) of the responses would be missing because of the sensitive question about the number of sex partners in a telephone interview. The objective is to examine if an association exists between the number of sex partners and disease status of cervical cancer.

Table 11.5 Cervical cancer data from Williamson & Haber (1994)

Number of	Disease status of cervical cancer	
sex partners	$Y = 0$ (control)	$Y = 1$ (case)
$X = 0$ (few, 0–3)	165 $(r_1,\ \theta_1)$	103 $(r_4,\ \theta_4)$
$X = 1$ (many, $\geqslant 4$)	164 $(r_2,\ \theta_2)$	221 $(r_3,\ \theta_3)$
Missing	43 $(r_{12},\ \theta_1 + \theta_2)$	59 $(r_{34},\ \theta_3 + \theta_4)$

Note: The observed counts and the corresponding cell probabilities are in parentheses. X is a sensitive binary variate and Y is a non-sensitive binary variate.

To illustrate the proposed design and approaches, let $W = 1\,(2, 3)$ if a respondent was born in January–April (May–August, September–December). It is then reasonable to assume that $p_k = \Pr(W = k) \approx 1/3$ for $k = 1, 2, 3$, and W is independent of the sensitive binary variate X and the the non-sensitive binary variate Y. For the ideal situation (i.e., no sampling errors), the observed counts from the main questionnaire as shown in Tables 11.1 and 11.3 would be

$$
\begin{aligned}
n_1 &= \frac{r_1}{3} = 55, \\[6pt]
n_2 &= 55 + r_2 = 219, \\[6pt]
n_3 &= 55 + r_3 = 276, \quad \text{and} \\[6pt]
n_4 &= r_4 = 103;
\end{aligned}
$$

that is,

$$
Y_{\text{obs, M}} = \{n;\ n_1, \ldots, n_4\} = \{653;\ 55, 219, 276, 103\}.
$$

On the other hand, we can view the missing data in Table 11.5 as the observed counts from the supplemental questionnaire as shown in Table 11.2. From Table 11.4, we have $m_0 = r_{12} = 43$ and $m_1 = r_{34} = 59$; that is,

$$Y_{\text{obs, S}} = \{m; \, m_0, \, m_1\} = \{102; \, 43, \, 59\}.$$

Therefore, we obtain the observed data $Y_{\text{obs}} = \{Y_{\text{obs, M}}, \, Y_{\text{obs, S}}\}$.

11.4.1 Likelihood-based inferences

Using $\boldsymbol{\theta}^{(0)} = \mathbf{1}_4/4$ as the initial values, the EM algorithm in (11.3) and (11.4) converged in 29 iterations. The resultant MLEs of $\hat{\boldsymbol{\theta}}$ and ψ are given in the second column of Table 11.6. From (11.5) and (11.6), the asymptotic variance-covariance matrix of the MLEs $\hat{\boldsymbol{\theta}}_{-4}$ is

$$\mathbf{I}_{\text{obs}}^{-1}(\hat{\boldsymbol{\theta}}_{-4}) = \begin{pmatrix} 0.00086302 & -0.000442447 & -0.000384452 \\ -0.00044245 & 0.000511253 & -0.000010026 \\ -0.00038445 & -0.000010026 & 0.000505241 \end{pmatrix}.$$

The estimated standard errors of $\hat{\theta}_i$ $(i = 1, 2, 3)$ are square roots of the main diagonal elements of the above matrix. From (11.9), the estimated standard errors of $\hat{\theta}_4 = 1 - \hat{\theta}_1 - \hat{\theta}_2 - \hat{\theta}_3$ and $\hat{\psi} = \hat{\theta}_1\hat{\theta}_3/(\hat{\theta}_2\hat{\theta}_4)$ are given by

$$\widehat{\text{se}}(\hat{\theta}_4) = \left\{ \mathbf{1}_3^{\top} \mathbf{I}_{\text{obs}}^{-1}(\hat{\boldsymbol{\theta}}_{-4}) \mathbf{1}_3 \right\}^{\frac{1}{2}}$$

and

$$\widehat{\text{se}}(\hat{\psi}) = \left\{ \boldsymbol{\alpha}^{\top} \mathbf{I}_{\text{obs}}^{-1}(\hat{\boldsymbol{\theta}}_{-4}) \boldsymbol{\alpha} \right\}^{\frac{1}{2}},$$

respectively, where

$$\boldsymbol{\alpha} = \left(\frac{\hat{\theta}_3(1 - \hat{\theta}_2 - \hat{\theta}_3)}{\hat{\theta}_2\hat{\theta}_4^2}, \; \frac{\hat{\theta}_1\hat{\theta}_3(\hat{\theta}_2 - \hat{\theta}_4)}{(\hat{\theta}_2\hat{\theta}_4)^2}, \; \frac{\hat{\theta}_1(1 - \hat{\theta}_1 - \hat{\theta}_2)}{\hat{\theta}_2\hat{\theta}_4^2} \right)^{\top}.$$

These estimated standard errors are listed in the third column of Table 11.6. From (11.7) and (11.8), we can obtain the 95% asymptotic confidence intervals of $\boldsymbol{\theta}$ and ψ, which are showed in the fourth column of Table 11.6.

Based on (11.10) and (11.11), we generate $G = 10,000$ bootstrap samples. The corresponding 95% bootstrap confidence intervals of $\boldsymbol{\theta}$ and ψ are displayed in the last column of Table 11.6.

Table 11.6 MLEs and 95% confidence intervals of $\boldsymbol{\theta}$ and ψ

Parameter	MLE	std	95% Asymptotic CI	95% Bootstrap CI
θ_1	0.23793	0.02937	[0.18036, 0.29551]	[0.18006, 0.29879]
θ_2	0.24916	0.02261	[0.20484, 0.29348]	[0.20375, 0.29359]
θ_3	0.35196	0.02247	[0.30790, 0.39601]	[0.30731, 0.39666]
θ_4	0.16095	0.01434	[0.13284, 0.18905]	[0.13400, 0.18908]
ψ	2.08830	0.43971	[1.22646, 2.95011]	[1.35361, 3.18593]

To test the null hypothesis H_0: $\psi = 1$ against H_1: $\psi \neq 1$, we need to obtain the restricted MLE $\hat{\boldsymbol{\theta}}_{\mathrm{R}}$. Using

$$(\theta_x^{(0)},\ \theta_y^{(0)})^\top = (0.5,\ 0.5)^\top$$

as the initial values, the EM algorithm in (11.16) converged to

$$\hat{\theta}_{x,\mathrm{R}} = 0.64807 \quad \text{and} \quad \hat{\theta}_{y,\mathrm{R}} = 0.52948$$

in 19 iterations. From (11.15), the restricted MLEs of $\boldsymbol{\theta}$ are obtained as

$$\hat{\boldsymbol{\theta}}_{\mathrm{R}} = (0.16559,\ 0.30493,\ 0.34314,\ 0.18634)^\top.$$

The log-likelihood ratio statistic Λ is equal to 12.469 and the p-value is 0.0004137. Since this p-value is far less than 0.05, the H_0 is rejected at the 0.05 level of significance. Thus, we can conclude that there is an association between sex partners and cervical cancer status based on the current data. This conclusion is identical to that from the two 95% confidence intervals of the odds ratio as shown in Table 11.6, where both confidence intervals exclude the value 1.

Table 11.7 Posterior modes and estimates of parameters for the cervical cancer data

Parameter	Posterior mode	Bayesian mean	Bayesian std	95% Bayesian credible interval
θ_1	0.23793	0.24025	0.02934	[0.18574, 0.30050]
θ_2	0.24916	0.24789	0.02251	[0.20365, 0.29192]
θ_3	0.35196	0.35029	0.02246	[0.30621, 0.39432]
θ_4	0.16095	0.16155	0.01431	[0.13448, 0.19042]
ψ	2.08830	2.14538	0.45838	[1.39399, 3.18678]

11.4.2   Bayesian inferences

When Dirichlet $(1, 1, 1, 1)$ (i.e., the uniform distribution on $\mathbb{T}_4$) is adopted as the prior distribution of $\boldsymbol{\theta}$, the posterior modes of $\boldsymbol{\theta}$ are equal to the corresponding MLEs. Using $\boldsymbol{\theta}^{(0)} = \mathbf{1}_4/4$ as the initial values, we employ the data augmentation algorithm to generate 1,000,000 posterior samples of $\boldsymbol{\theta}$ and discard the first half of the samples. The Bayesian estimates of $\boldsymbol{\theta}$ and ψ are given in Table 11.7. Since the lower bound of the Bayesian credible interval of the ψ is larger than 1, we believe that there is an association between the number of sexual partners and cervical cancer status.

The posterior densities of the $\{\theta_i\}_{i=1}^4$ and ψ estimated by a kernel density smoother are plotted in Figures 11.1 and 11.2.

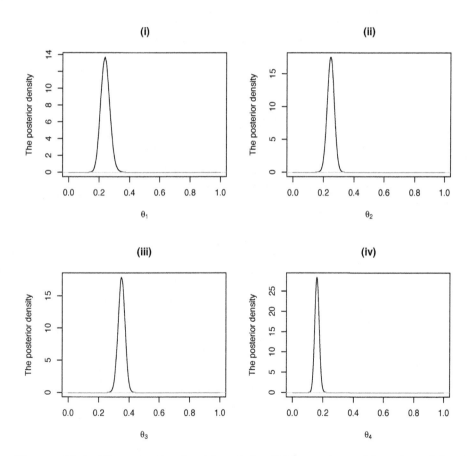

Figure 11.1 The posterior densities of the $\{\theta_i\}_{i=1}^4$ estimated by a kernel density smoother based on the second 500,000 posterior samples of $\boldsymbol{\theta}$ generated by the data augmentation algorithm when the prior distribution is Dirichlet $(1, 1, 1, 1)$. (i) The posterior density of θ_1; (ii) The posterior density of θ_2; (iii) The posterior density of θ_3; (iv) The posterior density of θ_4.

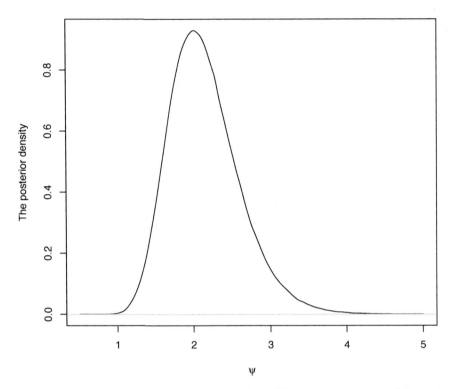

Figure 11.2 The posterior density of the odds ratio ψ estimated by a kernel density smoother based on the second 500,000 posterior samples of $\boldsymbol{\theta}$ generated by the data augmentation algorithm when the prior distribution is Dirichlet $(1, 1, 1, 1)$.

11.5   Group Dirichlet Distribution

11.5.1   The mode of a group Dirichlet density

Let

$$\mathbb{V}_{n-1} = \left\{ (x_1, \ldots, x_{n-1})^\top : \; x_i \geqslant 0, \; i = 1, \ldots, n-1, \; \sum_{i=1}^{n-1} x_i \leqslant 1 \right\}.$$

denote the $(n-1)$-dimensional open simplex in $\mathbb{R}^{n-1}$. A random vector $\mathbf{x} = (X_1, \ldots, X_n)^\top \in \mathbb{T}_n$ is said to follow a group Dirichlet distribution with two partitions, if the density of $\mathbf{x}_{-n} = (X_1, \ldots, X_{n-1})^\top \in \mathbb{V}_{n-1}$ is

$$\mathrm{GD}_{n,2,s}(\boldsymbol{x}_{-n}|\boldsymbol{a}, \boldsymbol{b}) = c_{\mathrm{GD}}^{-1} \left(\prod_{i=1}^{n} x_i^{a_i-1} \right) \left(\sum_{i=1}^{s} x_i \right)^{b_1} \left(\sum_{i=s+1}^{n} x_i \right)^{b_2}, \quad (11.20)$$

where $\boldsymbol{x}_{-n} = (x_1, \ldots, x_{n-1})^\top$, $\boldsymbol{a} = (a_1, \ldots, a_n)^\top$ is a positive parameter vector, $\boldsymbol{b} = (b_1, b_2)^\top$ is a non-negative parameter vector, s is a known positive integer less than n, and the normalizing constant is given by

$$c_{\mathrm{GD}} = B(a_1, \ldots, a_s)\, B(a_{s+1}, \ldots, a_n)\, B(\textstyle\sum_{i=1}^{s} a_i + b_1,\ \sum_{i=s+1}^{n} a_i + b_2).$$

We write $\mathbf{x} \sim \mathrm{GD}_{n,2,s}(\boldsymbol{a}, \boldsymbol{b})$ on $\mathbb{T}_n$ or $\mathbf{x}_{-n} \sim \mathrm{GD}_{n,2,s}(\boldsymbol{a}, \boldsymbol{b})$ on $\mathbb{V}_{n-1}$ to distinguish the two equivalent representations.

If $a_i \geqslant 1$, then the mode of the grouped Dirichlet density (11.20) is given by (Ng $et\ al.$, 2008; Ng, Tian & Tang, 2011)

$$\hat{x}_i \;=\; \frac{a_i - 1}{N}\left\{ 1 + \frac{b_1}{\sum_{j=1}^{s}(a_j - 1)} \right\}, \qquad 1 \leqslant i \leqslant s, \tag{11.21}$$

$$\hat{x}_i \;=\; \frac{a_i - 1}{N}\left\{ 1 + \frac{b_2}{\sum_{j=s+1}^{n}(a_j - 1)} \right\}, \qquad s+1 \leqslant i \leqslant n. \tag{11.22}$$

where $N = \sum_{i=1}^{n} a_i - n + b_1 + b_2$.

11.5.2   Sampling from a group Dirichlet distribution

The following procedure can be used to generate i.i.d. samples from a group Dirichlet distribution (Ng $et\ al.$, 2008). Let

(1) $\mathbf{y}^{(1)} \sim \mathrm{Dirichlet}\,(a_1, \ldots, a_s)$ on $\mathbb{T}_s$;

(2) $\mathbf{y}^{(2)} \sim \mathrm{Dirichlet}\,(a_{s+1}, \ldots, a_n)$ on $\mathbb{T}_{n-s}$;

(3) $R \sim \mathrm{Beta}\,(\sum_{i=1}^{s} a_i + b_1,\ \sum_{i=s+1}^{n} a_i + b_2)$; and

(4) $\mathbf{y}^{(1)}$, $\mathbf{y}^{(2)}$ and R are mutually independent.

Define

$$\mathbf{x}^{(1)} = R \times \mathbf{y}^{(1)} \quad \text{and} \quad \mathbf{x}^{(2)} = (1 - R) \times \mathbf{y}^{(2)}. \tag{11.23}$$

Then,

$$\mathbf{x} = \begin{pmatrix} \mathbf{x}^{(1)} \\ \mathbf{x}^{(2)} \end{pmatrix} \sim \mathrm{GD}_{n,2,s}(\boldsymbol{a}, \boldsymbol{b}) \tag{11.24}$$

on $\mathbb{T}_n$, where $\boldsymbol{a} = (a_1, \ldots, a_n)^{\top}$ and $\boldsymbol{b} = (b_1, b_2)^{\top}$.

The EM and DA Algorithms

A.1 The EM Algorithm

The *expectation–maximization* (EM) algorithm is an iterative deterministic method for finding *maximum likelihood estimates* (MLEs) or posterior modes, and is remarkably simple both conceptually and computationally in in incomplete data problems (Dempster, Laird & Rubin, 1977). The basic principle behind the EM is that instead of performing a complicated optimization, one augments the observed data with latent data to perform a series of simple optimizations.

Let $\boldsymbol{\theta}$ denote a parameter vector of interest and $\ell(\boldsymbol{\theta}|Y_{\text{obs}})$ denote the observed-data log-likelihood function. Usually, directly solving the MLE $\hat{\boldsymbol{\theta}}$ is extremely difficult. We augment the observed data Y_{obs} with missing data Y_{mis} (or a latent variable Z, or a latent vector $\mathbf{z}$) so that both the complete-data log-likelihood $\ell(\boldsymbol{\theta}|Y_{\text{obs}}, \mathbf{z})$ and the conditional predictive distribution $f(\mathbf{z}|Y_{\text{obs}}, \boldsymbol{\theta})$ are available, where $\mathbf{z}$ denotes the realization of $\mathbf{z}$. Each iteration of the EM algorithm consists of an *expectation step* (E-step) and a *maximization step*.

Specifically, let $\boldsymbol{\theta}^{(t)}$ be the current best guess at the MLE $\hat{\boldsymbol{\theta}}$. The E-step is to compute the Q function defined by

$$Q(\boldsymbol{\theta}|\boldsymbol{\theta}^{(t)}) = E\left\{\ell(\boldsymbol{\theta}|Y_{\text{obs}}, \mathbf{z})\middle|Y_{\text{obs}}, \boldsymbol{\theta}^{(t)}\right\}$$

$$= \int \ell(\boldsymbol{\theta}|Y_{\text{obs}}, \mathbf{z}) \times f(\mathbf{z}|Y_{\text{obs}}, \boldsymbol{\theta}^{(t)})\, \mathrm{d}\mathbf{z}, \qquad (A.1)$$

and the M-step is to maximize Q with respect to $\boldsymbol{\theta}$ to obtain

$$\boldsymbol{\theta}^{(t+1)} = \arg\max_{\Theta} Q(\boldsymbol{\theta}|\boldsymbol{\theta}^{(t)}). \qquad (A.2)$$

The two-step process is repeated until convergence occurs.

Example A.1 (Genetic linkage model). In this study (Rao, 1973; Lange, 2002), AB/ab animals are crossed to measure the recombination fraction (r) between loci with alleles A and a at the first locus and alleles B and b at the second locus. Then the offspring of an $AB/ab \times AB/ab$ mating fall into the four categories AB, Ab, aB and ab with cell probabilities

$$\frac{\theta + 2}{4}, \frac{1 - \theta}{4}, \frac{1 - \theta}{4}, \frac{\theta}{4}, \quad 0 \leqslant \theta \leqslant 1,$$

where $\theta = (1 - r)^2$. Observed frequencies $(Y_1, \ldots, Y_4)$ follow a multinomial distribution with the above cell probabilities, i.e.,

$$(Y_1, \ldots, Y_4)^\top \sim \text{Multinomial}\left(n; \frac{\theta + 2}{4}, \frac{1 - \theta}{4}, \frac{1 - \theta}{4}, \frac{\theta}{4}\right),$$

where $n = \sum_{i=1}^{4} y_i$ and $\{y_i\}_{i=1}^{4}$ denote the corresponding observed values of $\{Y_i\}_{i=1}^{4}$.

Augmenting the observed data $Y_{\text{obs}} = \{y_1, \ldots, y_4\}$ with a latent variable Z by splitting $y_1 = Z + (y_1 - Z)$, we have

$$f(z|Y_{\text{obs}}, \theta) = \text{Binomial}\left(z|y_1, \theta/(\theta + 2)\right)$$

$$= \binom{y_1}{z} \left(\frac{\theta}{\theta + 2}\right)^z \left(\frac{2}{\theta + 2}\right)^{y_1 - z}, \quad z = 0, 1, \ldots, y_1. \quad \text{(A.3)}$$

The likelihood of the complete-data $\{Y_{\text{obs}}, z\} = \{z, y_1 - z, y_2, y_3, y_4\}$ is

$$L(\theta|Y_{\text{obs}}, z) = \binom{n}{z, y_1 - z, y_2, y_3, y_4} \left(\frac{\theta}{4}\right)^z \left(\frac{2}{4}\right)^{y_1 - z} \left(\frac{1 - \theta}{4}\right)^{y_2 + y_3} \left(\frac{\theta}{4}\right)^{y_4}$$

so that the complete-data log-likelihood function is given by

$$\ell(\theta|Y_{\text{obs}}, z) = c(z) + (z + y_4) \log \theta + (y_2 + y_3) \log(1 - \theta), \quad \text{(A.4)}$$

where $c(z)$ is a function of z, not depending on θ. From (A.1), we obtain

$$
\begin{aligned}
Q(\theta|\theta^{(t)}) &= \int \ell(\theta|Y_{\text{obs}}, z) \times f(z|Y_{\text{obs}}, \theta^{(t)}) \, dz \\
&= \sum_{z=0}^{y_1} \{c(z) + (z + y_4) \log \theta + (y_2 + y_3) \log(1 - \theta)\} \\
&\qquad \times \binom{y_1}{z} \left(\frac{\theta^{(t)}}{\theta^{(t)} + 2}\right)^z \left(\frac{2}{\theta^{(t)} + 2}\right)^{y_1 - z} \\
&= E\left\{c(Z)\middle|Y_{\text{obs}}, \theta^{(t)}\right\} + \left\{E(Z|Y_{\text{obs}}, \theta^{(t)}) + y_4\right\} \log \theta \\
&\qquad + (y_2 + y_3) \log(1 - \theta).
\end{aligned}
$$

From (A.2), we have

$$\theta^{(t+1)} = \frac{E(Z|Y_{\text{obs}}, \theta^{(t)}) + y_4}{E(Z|Y_{\text{obs}}, \theta^{(t)}) + y_2 + y_3 + y_4}$$

$$= \frac{y_1 \theta^{(t)} / (\theta^{(t)} + 2) + y_4}{y_1 \theta^{(t)} / (\theta^{(t)} + 2) + y_2 + y_3 + y_4}. \tag{A.5}$$

Let $y_1 = 125$, $y_2 = 18$, $y_3 = 20$ and $y_4 = 34$. If $\theta^{(0)} = 0.5$, using (A.5), we obtain

$$\theta^{(1)} = 0.608247,$$

$$\theta^{(2)} = 0.624321,$$

$$\theta^{(3)} = 0.626489,$$

$$\theta^{(4)} = 0.626777,$$

$$\theta^{(5)} = 0.626816,$$

$$\theta^{(6)} = 0.626821,$$

$$\theta^{(7)} = 0.626821.$$

A summary of the EM algorithm:

Step 1: Find the complete-data log-likelihood function $\ell(\theta|Y_{\text{obs}}, z)$ given by (A.4);

Step 2: Find the conditional predictive distribution $f(z|Y_{\text{obs}}, \theta)$ given by (A.3);

E-step: From (A.3), compute the conditional expectation

$$E(Z|Y_{\text{obs}}, \theta) = \frac{y_1 \theta}{\theta + 2};$$

M-step: From (A.4), find the complete-data MLE

$$\hat{\theta} = \frac{z + y_4}{z + y_2 + y_3 + y_4}$$

and update $\hat{\theta}$ by replacing z with $E(Z|Y_{\text{obs}}, \theta)$. ‖

Example A.2 (Two-parameter multinomial model). Gelfand & Smith (1990) extend the genetic-linkage model in Example A.1 to a two-parameter multinomial model. Let $\mathbf{y} = (Y_1, \ldots, Y_5)^\top$ and

$$\mathbf{y} \sim \text{Multinomial}(n; \, a_1\theta_1 + b_1, a_2\theta_1 + b_2, a_3\theta_2 + b_3, a_4\theta_2 + b_4, c\theta_3)$$

where $a_i, b_i \geqslant 0$ are known,

$$0 \leqslant c = 1 - \sum_{i=1}^{4} b_i = a_1 + a_2 = a_3 + a_4 \leqslant 1$$

and $\boldsymbol{\theta} = (\theta_1, \theta_2, \theta_3)^\top \in \mathbb{T}_3$. To split the first four cells, we introduce a latent vector $\mathbf{z} = (Z_1, \ldots, Z_4)^\top$ so that the augmented sampling distribution is

$$\begin{pmatrix} \mathbf{y} \\ \mathbf{z} \end{pmatrix} \sim \text{Multinomial}(n; \, a_1\theta_1, b_1, a_2\theta_1, b_2, a_3\theta_2, b_3, a_4\theta_2, b_4, c\theta_3).$$

Let $Y_{\text{obs}} = \{y_1, \ldots, y_5\}$ denote the observed data, where $\{y_i\}_{i=1}^5$ denote the corresponding observed values of $\{Y_i\}_{i=1}^5$. Equivalently, the complete-data likelihood function is

$$L(\boldsymbol{\theta}|Y_{\text{obs}}, \mathbf{z}) \propto \theta_1^{z_1+z_2} \theta_2^{z_3+z_4} \theta_3^{y_5}. \tag{A.6}$$

In addition, the conditional predictive density is

$$f(\mathbf{z}|Y_{\text{obs}}, \boldsymbol{\theta}) = \prod_{i=1}^{4} \text{Binomial}(z_i|y_i, p_i),$$

where

$$p_i = \frac{a_i\theta_1 I_{(1 \leqslant i \leqslant 2)}}{a_i\theta_1 + b_i} + \frac{a_i\theta_2 I_{(3 \leqslant i \leqslant 4)}}{a_i\theta_2 + b_i}. \tag{A.7}$$

Now suppose we are interested in finding the MLEs of $\boldsymbol{\theta}$. From (A.6), the complete-data MLEs are given by

$$\hat{\theta}_1 = \frac{z_1 + z_2}{\Delta}, \quad \hat{\theta}_2 = \frac{z_3 + z_4}{\Delta}, \quad \hat{\theta}_3 = \frac{y_5}{\Delta}, \tag{A.8}$$

where $\Delta = \sum_{i=1}^4 z_i + y_5$. Thus, the E-step of the EM algorithm computes

$$E(Z_i|Y_{\text{obs}}, \boldsymbol{\theta}) = y_i p_i$$

with p_i being defined by (A.7), and the M-step updates (A.8) by replacing $\{z_i\}_{i=1}^4$ with $E(Z_i|Y_{\text{obs}}, \boldsymbol{\theta})$.

For illustrative purpose, we consider the same dataset as that of Gelfand & Smith (1990):

$$(y_1, \ldots, y_5)^\top = (14, 1, 1, 1, 5)^\top, \quad a_1 = \cdots = a_4 = 0.25,$$

$$b_1 = \frac{1}{8}, \qquad b_2 = b_3 = 0, \qquad b_4 = \frac{3}{8}.$$

Using $\theta^{(0)} = \mathbf{1}_3/3$ as initial values, we obtain

$$\theta^{(1)} = (0.516358, \ 0.0924609, \ 0.391181)^\top,$$

$$\theta^{(2)} = (0.572495, \ 0.0746653, \ 0.352840)^\top,$$

$$\theta^{(3)} = (0.583528, \ 0.0721333, \ 0.344339)^\top,$$

$$\theta^{(4)} = (0.585487, \ 0.0717070, \ 0.342806)^\top,$$

$$\theta^{(5)} = (0.585828, \ 0.0716332, \ 0.342538)^\top,$$

$$\theta^{(6)} = (0.585888, \ 0.0716204, \ 0.342492)^\top,$$

$$\theta^{(7)} = (0.585898, \ 0.0716182, \ 0.342484)^\top,$$

$$\theta^{(8)} = (0.585900, \ 0.0716178, \ 0.342482)^\top,$$

$$\theta^{(9)} = (0.585900, \ 0.0716178, \ 0.342482)^\top.$$

The EM algorithm converged in 8 iterations. The resultant MLE is

$$\hat{\boldsymbol{\theta}} = (0.585900, \ 0.0716178, \ 0.342482)^\top. \qquad\qquad \|$$

A.2 The DA Algorithm

Based on the observed posterior distribution $p(\boldsymbol{\theta}|Y_{\text{obs}})$, suppose that we want to make Bayesian inference on the parameter vector $\boldsymbol{\theta}$ of interest. The *data augmentation* (DA) algorithm is designed for simulating from $p(\boldsymbol{\theta}|Y_{\text{obs}})$. The principle of the DA is that instead of a complicated simulation, the observed data is augmented with latent data so that a series of simple simulations can be performed.

As a stochastic version of the EM algorithm, the DA algorithm was originally proposed by Tanner & Wong (1987). The basic idea is to introduce a latent vector $\mathbf{z}$ with realization z so that the complete-data posterior distribution

$$p(\boldsymbol{\theta}|Y_{\text{obs}}, z)$$

and the conditional predictive distribution

$$f(z|Y_{\mathrm{obs}}, \boldsymbol{\theta})$$

are available, where by available it means either the samples can be easily generated from both, or each can be easily evaluated at any given point. The DA algorithm can be summarized as follows.

THE DA ALGORITHM:

I-step: Draw $z^{(t+1)} \sim f(z|Y_{\mathrm{obs}}, \boldsymbol{\theta}^{(t)})$.

P-step: Draw $\boldsymbol{\theta}^{(t+1)} \sim p(\boldsymbol{\theta}|Y_{\mathrm{obs}}, z^{(t+1)})$.

APPENDIX B

The Exact IBF Sampling

Let Y_{obs} denote the observed data, Z the missing or latent data and π the parameter of interest. We consider the data augmentation structure as in Tanner & Wong (1987) with available complete-data posterior distribution $p(\pi|Y_{\mathrm{obs}}, z)$ and conditional predictive distribution $f(z|Y_{\mathrm{obs}}, \pi)$, where 'available' means that the sampling from the two conditional distributions and the evaluation of the two conditional densities can routinely be implemented and z denotes the realization of Z. The goal is to obtain i.i.d. samples from the observed posterior distribution $p(\pi|Y_{\mathrm{obs}})$.

The fundamental conditional sampling principle states that: If we could obtain independent samples $\{z^{(\ell)}\}_{\ell=1}^{L}$ from $f(z|Y_{\mathrm{obs}})$ and generate $\pi^{(\ell)} \sim p(\pi|Y_{\mathrm{obs}}, z^{(\ell)})$ for $\ell = 1, \ldots, L$, then $\{\pi^{(\ell)}\}_{1}^{L}$ are i.i.d. samples from the observed posterior distribution $p(\pi|Y_{\mathrm{obs}})$. In other words, the key issue is to generate independent samples from $f(z|Y_{\mathrm{obs}})$.

Let $\mathcal{S}_{(\pi|Y_{\mathrm{obs}})}$ and $\mathcal{S}_{(Z|Y_{\mathrm{obs}})}$ denote the conditional supports of $\pi|Y_{\mathrm{obs}}$ and $Z|Y_{\mathrm{obs}}$, respectively. The sampling-wise *inverse Bayes formula* (IBF) states that (Ng, 1997; Tan, Tian & Ng, 2003)

$$f(z|Y_{\mathrm{obs}}) \propto \frac{f(z|Y_{\mathrm{obs}}, \pi_0)}{p(\pi_0|Y_{\mathrm{obs}}, z)}, \tag{B.1}$$

for any arbitrary $\pi_0 \in \mathcal{S}_{(\pi|Y_{\mathrm{obs}})}$ and all $z \in \mathcal{S}_{(Z|Y_{\mathrm{obs}})}$.

When Z is a discrete random variable/vector taking finite values on the domain, we denote the conditional support of $Z|(Y_{\mathrm{obs}}, \pi)$ by

$$\mathcal{S}_{(Z|Y_{\mathrm{obs}}, \pi)} = \{z_1, \ldots, z_K\}.$$

Since $f(z|Y_{\mathrm{obs}}, \pi)$ is available, we can first directly identify $\{z_k\}_{k=1}^{K}$ from the model specification and all $\{z_k\}_{k=1}^{K}$ become known. Noting that $\{z_k\}_{k=1}^{K}$ generally do not depend on π, we have

$$\mathcal{S}_{(Z|Y_{\mathrm{obs}})} = \mathcal{S}_{(Z|Y_{\mathrm{obs}}, \pi)} = \{z_1, \ldots, z_K\}.$$

Due to the discreteness of Z, the notation $f(z_k|Y_{\mathrm{obs}})$ will be used to denote the probability mass function, i.e., $f(z_k|Y_{\mathrm{obs}}) = \Pr(Z = z_k|Y_{\mathrm{obs}})$. Thus, it suffices to find $\omega_k = f(z_k|Y_{\mathrm{obs}})$ for $k = 1, \ldots, K$. For any $\pi_0 \in \mathcal{S}_{(\pi|Y_{\mathrm{obs}})}$, let

$$q_k(\pi_0) = \frac{\Pr(Z = z_k|Y_{\mathrm{obs}}, \pi_0)}{p(\pi_0|Y_{\mathrm{obs}}, z_k)}, \quad k = 1, \ldots, K. \tag{B.2}$$

From the sampling-wise IBF (B.1), we immediately obtain

$$\omega_k = \frac{q_k(\pi_0)}{\sum_{k'=1}^{K} q_{k'}(\pi_0)}, \quad k = 1, \ldots, K. \tag{B.3}$$

and $\{\omega_k\}_1^K$ are independent of π_0. Thus, it is easy to sample from $f(z|Y_{\text{obs}})$ since it is a discrete distribution with probability ω_k on z_k for $k = 1, \ldots, K$. We summarize the algorithm as follows (Tian, Tan & Ng, 2007; Tan, Tian & Ng, 2010).

THE EXACT IBF SAMPLING:

(1) Identify $\mathcal{S}_{(Z|Y_{\text{obs}})} = \mathcal{S}_{(Z|Y_{\text{obs}}, \pi)} = \{z_1, \ldots, z_K\}$ from $f(z|Y_{\text{obs}}, \pi)$ and calculate $\{\omega_k\}_{k=1}^K$ according to (B.3) and (B.2).

(2) Generate i.i.d. samples $\{z^{(\ell)}\}_{\ell=1}^L$ of Z from the probability mass function $f(z|Y_{\text{obs}})$ with probabilities $\{\omega_k\}_{k=1}^K$ on $\{z_k\}_{k=1}^K$.

(3) Generate $\pi^{(\ell)} \sim f(\pi|Y_{\text{obs}}, z^{(\ell)})$ for $\ell = 1, \ldots, L$, then $\{\pi^{(\ell)}\}_{\ell=1}^L$ are i.i.d. samples from the observed posterior distribution $p(\pi|Y_{\text{obs}})$.

Some Statistical Distributions

C.1 Discrete Distributions

C.1.1 General finite distribution

Notation: $X \sim \text{Finite}_n(\boldsymbol{x}, \boldsymbol{p})$, where $\boldsymbol{x} = (x_1, \ldots, x_n)^\top$, $\{x_i\}_{i=1}^n$ are real numbers, $\boldsymbol{p} = (p_1, \ldots, p_n)^\top \in \mathbb{T}_n$, which is defined by (5.1).

Density: $\Pr(X = x_i) = p_i, \quad i = 1, \ldots, n.$

Moments: $E(X) = \sum_{i=1}^n x_i p_i$ and $\text{Var}(X) = \sum_{i=1}^n x_i^2 p_i - (\sum_{i=1}^n x_i p_i)^2.$

Note: (1) The *uniform discrete* distribution is a special case of the general finite distribution with $p_i = 1/n$ for all i.

(2) As an extension of a binary random variable, a *categorical random variable* can take on one of n values $\{1, \ldots, n\}$, where n is a fixed positive integer.

(3) As a generalization of the Bernoulli distribution, the *categorical* distribution is a special member of the family of general finite distributions with $x_i = i$ for $i = 1, \ldots, n$. We write $X \sim \text{Finite}_n(\{i\}, \{p_i\})$.

C.1.2 Hypergeometric distribution

Notation: $X \sim \text{Hgeometric}(m, n, k)$, where m, n, k are positive integers.

Density: $\text{Hgeometric}(x | m, n, k) = \binom{m}{x}\binom{n}{k-x} \Big/ \binom{m+n}{k}$, where $x = \max(0, k - n), \ldots, \min(m, k).$

Moments: $E(X) = km/N'$ and $\text{Var}(X) = kmn(N' - k)/\{N'^2(N' - 1)\}$, where $N' \hat{=} m + n.$

C.1.3 Poisson distribution

Notation: $X \sim \text{Poisson}(\lambda)$, where $\lambda > 0.$

Density: $\text{Poisson}(x|\lambda) = \lambda^x \, \mathrm{e}^{-\lambda}/x!, \; x = 0, 1, 2, \ldots.$

Moments: $E(X) = \lambda$ and $\text{Var}(X) = \lambda.$

Properties: (1) If $\{X_i\}_{i=1}^n \overset{\text{ind}}{\sim} \text{Poisson}(\lambda_i)$, then

$$\sum_{i=1}^n X_i \sim \text{Poisson}\left(\sum_{i=1}^n \lambda_i\right)$$

and

$$(X_1, \ldots, X_n) \left| \left(\sum_{i=1}^n X_i = m\right) \right. \sim \text{Multinomial}_n(m, \boldsymbol{p}),$$

where $\boldsymbol{p} = (\lambda_1, \ldots, \lambda_n)^\top / \sum_{i=1}^n \lambda_i.$

(2) The Poisson and gamma distributions have the relationship:

$$\sum_{x=k}^\infty \text{Poisson}(x|\lambda) = \int_0^\lambda \text{Gamma}(y|k, 1) \, \mathrm{d}y.$$

C.1.4 Binomial distribution

Notation: $X \sim \text{Binomial}(n, p)$, where n is a positive integer and $p \in (0, 1)$.

Density: $\text{Binomial}(x|n, p) = \binom{n}{x} p^x (1 - p)^{n-x}, \; x = 0, 1, \ldots, n.$

Moments: $E(X) = np$ and $\text{Var}(X) = np(1 - p).$

Properties: (1) If $\{X_i\}_{i=1}^d \overset{\text{ind}}{\sim} \text{Binomial}(n_i, p)$, then

$$\sum_{i=1}^d X_i \sim \text{Binomial}\left(\sum_{i=1}^d n_i, p\right).$$

(2) The binomial and beta distributions have the relationship:

$$\sum_{x=0}^k \text{Binomial}(x|n, p) = \int_0^{1-p} \text{Beta}(x|n - k, k + 1) \, \mathrm{d}x,$$

where $0 \leqslant k \leqslant n.$

Note: When $n = 1$, the binomial distribution is called *Bernoulli* distribution, denoted by $X \sim \text{Bernoulli}(p)$.

C.1.5 Multinomial distribution

Notation: $\mathbf{x} = (X_1, \ldots, X_n)^\top \sim \text{Multinomial}(N; p_1, \ldots, p_n)$ or
$\mathbf{x} = (X_1, \ldots, X_n)^\top \sim \text{Multinomial}_n(N, \boldsymbol{p})$, where
N is a positive integer and $\boldsymbol{p} = (p_1, \ldots, p_n)^\top \in \mathbb{T}_n$.

Density: $\text{Multinomial}_n(\boldsymbol{x} | N, \boldsymbol{p}) = \begin{pmatrix} N \\ x_1, \ldots, x_n \end{pmatrix} \prod_{i=1}^{n} p_i^{x_i}$, where
$\boldsymbol{x} = (x_1, \ldots, x_n)^\top$, $x_i \geqslant 0$, $\sum_{i=1}^{n} x_i = N$.

Moments: $E(X_i) = Np_i$, $\text{Var}(X_i) = Np_i(1 - p_i)$ and $\text{Cov}(X_i, X_j) = -Np_i p_j$.

Properties: (1) If X is a categorical random variable and $X \sim \text{Finite}_n(\{i\}, \{p_i\})$, then there is a unique n-dimensional random vector $\mathbf{x} = (X_1, \ldots, X_n)^\top$ such that $\mathbf{x} \sim \text{Multinomial}_n(1, \boldsymbol{p})$, where $X_i = I_{(X=i)}$ (i.e., $X_i = 1$ if X is in Category i and $X_i = 0$ otherwise) and $\sum_{i=1}^{n} X_i = 1$.

(2) Let $\mathbf{x}_j \overset{\text{ind}}{\sim} \text{Multinomial}_n(N_j, \boldsymbol{p})$ for $j = 1, \ldots, m$. Then

$$\mathbf{x}_1 + \cdots + \mathbf{x}_m \sim \text{Multinomial}_n(N_1 + \cdots + N_m, \boldsymbol{p}).$$

Note: The binomial distribution is a special case of the multinomial with $n = 2$.

C.2 Continuous Distributions

C.2.1 Uniform distribution

Notation: $X \sim U(a, b)$, where $a < b$.

Density: $U(x | a, b) = 1/(b - a)$, $a < x < b$.

Moments: $E(X) = (a + b)/2$ and $\text{Var}(X) = (b - a)^2/12$.

Properties: If $Y \sim U(0, 1)$, then $X = a + (b - a)Y \sim U(a, b)$.

C.2.2 Beta distribution

Notation: $X \sim \text{Beta}(a, b)$, where $a > 0$ and $b > 0$.

Density: $\text{Beta}(x|a, b) = x^{a-1}(1 - x)^{b-1}/B(a, b)$, $0 < x < 1$.

Moments: $E(X) = a/(a + b)$, $E(X^2) = a(a + 1)/\{(a + b)(a + b + 1)\}$ and $\text{Var}(X) = ab/\{(a + b)^2(a + b + 1)\}$.

Properties: If $Y_1 \sim \text{Gamma}(a, 1)$, $Y_2 \sim \text{Gamma}(b, 1)$, and $Y_1 \perp\!\!\!\perp Y_2$, then $Y_1/(Y_1 + Y_2) \sim \text{Beta}(a, b)$.

Note: When $a = b = 1$, $\text{Beta}(1, 1) = U(0, 1)$.

C.2.3 Exponential distribution

Notation: $X \sim \text{Exponential}(\beta)$, where $\beta > 0$ is the rate parameter.

Density: $\text{Exponential}(x|\beta) = \beta\,e^{-\beta x}$, $x > 0$.

Moments: $E(X) = 1/\beta$ and $\text{Var}(X) = 1/\beta^2$.

Properties: (1) If $U \sim U(0, 1)$, then $-\log(U)/\beta \sim \text{Exponential}(\beta)$.

(2) If $\{X_i\}_{i=1}^{n} \overset{iid}{\sim} \text{Exponential}(\beta)$, then

$$\sum_{i=1}^{n} X_i \sim \text{Gamma}(n, \beta).$$

C.2.4 Gamma distribution

Notation: $X \sim \text{Gamma}(\alpha, \beta)$, where $\alpha > 0$ is the shape parameter and $\beta > 0$ is the rate parameter.

Density: $\text{Gamma}(x|\alpha, \beta) = \dfrac{\beta^\alpha}{\Gamma(\alpha)} x^{\alpha-1}\,e^{-\beta x}$, $x > 0$.

Moments: $E(X) = \alpha/\beta$ and $\text{Var}(X) = \alpha/\beta^2$.

Properties: (1) If $X \sim \text{Gamma}(\alpha, \beta)$, then for any $c > 0$, we have

$$cX \sim \text{Gamma}(\alpha, \beta/c).$$

(2) If $\{X_i\}_{i=1}^{n} \overset{ind}{\sim} \text{Gamma}(\alpha_i, \beta)$, then

$$\sum_{i=1}^{n} X_i \sim \text{Gamma}\left(\sum_{i=1}^{n} \alpha_i, \beta\right).$$

(3) $\Gamma(\alpha + 1) = \alpha\Gamma(\alpha)$, $\Gamma(1) = 1$ and $\Gamma(1/2) = \sqrt{\pi}$.

Note: Gamma$(1, \beta)$ = Exponential(β).

C.2.5 Chi-squared distribution

Notation: $X \sim \chi^2(\nu) \equiv$ Gamma$(\nu/2, 1/2)$, where $\nu > 0$ is the degree of freedom.

Density: $\chi^2(x|\nu) = \dfrac{2^{-\nu/2}}{\Gamma(\nu/2)} x^{\nu/2-1} e^{-x/2}$, $x > 0$.

Moments: $E(X) = \nu$ and Var$(X) = 2\nu$.

Properties: (1) If $Y \sim N(0, 1)$, then $X = Y^2 \sim \chi^2(1)$.

     (2) If $\{X_i\}_{i=1}^n \overset{\text{ind}}{\sim} \chi^2(\nu_i)$, then $\sum_{i=1}^n X_i \sim \chi^2(\sum_{i=1}^n \nu_i)$.

C.2.6 t- or Student's t-distribution

Notation: $X \sim t(\nu)$, where ν is the degree of freedom.

Density: $t(x|\nu) = \dfrac{\Gamma(\frac{\nu+1}{2})}{\sqrt{\pi\nu}\,\Gamma(\frac{\nu}{2})}\left(1 + \dfrac{x^2}{\nu}\right)^{-\frac{\nu+1}{2}}$, $-\infty < x < \infty$.

Moments: $E(X) = 0$ (if $\nu > 1$) and Var$(X) = \frac{\nu}{\nu-2}$ (if $\nu > 2$).

Properties: Let $Z \sim N(0, 1)$, $Y \sim \chi^2(\nu)$ and $Z \perp\!\!\!\perp Y$, then

$$\frac{Z}{\sqrt{Y/\nu}} \sim t(\nu).$$

Note: When $\nu = 1$, $t(\nu) = t(1)$ is called *standard Cauchy* distribution, whose mean and variance do not exist.

C.2.7 F- or Fisher's F-distribution

Notation: $X \sim F(\nu_1, \nu_2)$, where ν_1 and ν_2 are two degrees of freedom.

Density: $F(x|\nu_1, \nu_2) = \dfrac{(\nu_1/\nu_2)^{\nu_1/2}}{B(\frac{\nu_1}{2}, \frac{\nu_2}{2})} x^{\frac{\nu_1}{2}-1}\left(1 + \dfrac{\nu_1 x}{\nu_2}\right)^{-\frac{\nu_1+\nu_2}{2}}$, $x > 0$.

Moments: $E(X) = \dfrac{\nu_2}{\nu_2 - 2}$ (if $\nu_2 > 2$) and

    Var$(X) = \dfrac{2\nu_2^2(\nu_1 + \nu_2 - 2)}{\nu_1(\nu_2 - 4)(\nu_2 - 2)^2}$ (if $\nu_2 > 4$).

Properties: Let $X_i \sim \chi^2(\nu_i)$, $i = 1, 2$, and $X_1 \perp\!\!\!\perp X_2$, then

$$\frac{X_1/\nu_1}{X_2/\nu_2} \sim F(\nu_1, \nu_2).$$

C.2.8 Normal or Gaussian distribution

Notation: $X \sim N(\mu, \sigma^2)$, where $-\infty < \mu < \infty$ and $\sigma^2 > 0$.

Density: $N(x|\mu, \sigma^2) = \dfrac{1}{\sqrt{2\pi}\sigma} \exp\left\{ -\dfrac{(x-\mu)^2}{2\sigma^2} \right\}$, $-\infty < x < \infty$.

Moments: $E(X) = \mu$ and $\mathrm{Var}(X) = \sigma^2$.

Properties: (1) If $\{X_i\}_{i=1}^n \overset{\text{ind}}{\sim} N(\mu_i, \sigma_i^2)$, then

$$\sum_{i=1}^n a_i X_i \sim N\left(\sum_{i=1}^n a_i \mu_i, \ \sum_{i=1}^n a_i^2 \sigma_i^2 \right).$$

(2) If $X_1|X_2 \sim N(X_2, \sigma_1^2)$ and $X_2 \sim N(\mu_2, \sigma_2^2)$, then

$$X_1 \sim N(\mu_2, \sigma_1^2 + \sigma_2^2).$$

C.2.9 Multivariate normal or Gaussian distribution

Notation: $\mathbf{x} = (X_1, \ldots, X_n)^\top \sim N_n(\boldsymbol{\mu}, \boldsymbol{\Sigma})$, where $\boldsymbol{\mu} \in \mathbb{R}^n$ and $\boldsymbol{\Sigma} > 0$.

Density: $N_n(\boldsymbol{x}|\boldsymbol{\mu}, \boldsymbol{\Sigma}) = \dfrac{1}{(\sqrt{2\pi})^n |\boldsymbol{\Sigma}|^{1/2}} \exp\left\{ -\dfrac{1}{2}(\boldsymbol{x} - \boldsymbol{\mu})^\top \boldsymbol{\Sigma}^{-1} (\boldsymbol{x} - \boldsymbol{\mu}) \right\}$,
where $\boldsymbol{x} = (x_1, \ldots, x_n)^\top \in \mathbb{R}^n$.

Moments: $E(\mathbf{x}) = \boldsymbol{\mu}$ and $\mathrm{Var}(\mathbf{x}) = \boldsymbol{\Sigma}$.

List of Figures

List of Tables

References

Abernathy, J.R., Greenberg, B.G. and Horvitz, D.G. (1970). Estimates of induced abortion in urban North Carolina. *Demography* **7**(1), 19–29.

Abul-Ela, A.L.A., Greenberg, B.G. and Horvitz, D.G. (1967). A multi-proportions randomized response model. *Journal of the American Statistical Association* **62**, 990–1008.

Agresti, A. and Coull, B.A. (1998). Approximate is better than "exact" for interval estimation of binomial proportions. *The American Statistician* **52**(2), 119–126.

Ahart, A.M. and Sackett, P.R. (2004). A new method of examining relationships between individual difference measures and sensitive behavior criteria: Evaluating the unmatched count technique. *Organizational Research Methods* **7**(1), 101–114.

Albert, J.H. and Gupta, A.K. (1983). Estimation in contingency tables using prior information. *Journal of the Royal Statistical Society, B* **45**, 60–69.

Albert, J.H. and Gupta, A.K. (1985). Bayesian methods for binomial data with applications to a non-response problem. *Journal of the American Statistical Association* **80**, 167–174.

Anderson, D.A., Simmons, A.M., Milnes, S.M. and Earleywine, M. (2007). Effect of response format on endorsement of eating disordered attitudes and behaviors. *International Journal of Eating Disorders* **40**(1), 90–93.

Arnold, S.F. (1993). Gibbs sampling. In *Handbook of Statistics: Computational Statistics* (C.R. Rao, ed.), Vol. 9, 599–625. Elsevier Science Publishers B.V., New York.

Atkinson, A.C. and Donev, A.N. (1992). *Optimum Experimental Designs*. Clarendon Press, Oxford.

Avetisyan, M. and Fox, J.P. (2012). The Dirichlet–multinomial model for multivariate randomized response data and small samples. *Psicológica* **33**, 362–390.

Bar-Lev, S.K., Bobovitch, E. and Boukai, B. (2003). A common conjugate prior structure for several randomized response models. *Test* **12**(1), 101–113.

Bar-Lev, S.K., Bobovitch, E. and Boukai, B. (2004). A note on randomized response models for quantitative data. *Metrika* **60**(3), 255–260.

Barton, A.H. (1958). Asking the embarrassing question. *Public Opinion Quarterly* **22**(1), 67–68.

Berman, J., McCombs, H. and Boruch, R. (1977). Notes on the contamination method: Two small experiments in assuring confidentiality of Responses. *Sociological Methods & Research* **6**(1), 45–62.

Bhargava, M. and Singh, R. (2002). On the efficiency comparison of certain randomized response strategies. *Metrika* **55**, 191–197.

Böckenholt, U. and van der Heijden, P.G.M. (2007). Item randomized-response models for measuring noncompliance: Risk-return perceptions, social influences, and self-protective responses. *Psychometrika* **72**(2), 245–262.

Boruch, R.F. (1971). Assuring confidentiality of responses in social research: A note on strategies. *The American Sociologist* **6**(4), 308–311.

Bourke, P.D. (1974). Multi-proportions randomized response using the unrelated question technique. Report No. 74 of the Errors in Surveys Research Project. Institute of Statistics, University of Stockholm (Mimeo).

Bourke, P.D. (1982). Randomized response multivariate designs for categorical data. *Communications in Statistics—Theory and Methods* **11**, 2889–2901.

Bourke, P.D. and Dalenius, T. (1973). Multi-proportions randomized response using a single sample. Report No. 68 of the Errors in Surveys Research Project. Institute of Statistics, University of Stockholm (Mimeo).

Bourke, P.D. and Moran, M.A. (1988). Estimating proportions from randomized response data using the EM algorithm. *Journal of the American Statistical Association* **83**, 964–968.

Bradburn, N.M., Sudman, S. and Associates (1979). *Improving Interview Method and Questionnaire Design: Response Effects to Threatening Questions in Survey Interviews*. Jossey-Bass, A Publishing Unit of John Wiley & Sons, San Francisco, California.

Brown, L.D., Cai, T.T. and DasGupta, A. (2001). Interval estimation for a binomial proportion. *Statistical Science* **16**(2), 101–133.

Casella, G. and Berger, R.L. (2002). *Statistical Inference* (2nd Ed.). Duxbury, Boston.

Chang, H.J. and Liang, D.H. (1996). A two-stage unrelated randomized response procedure. *Australian Journal of Statistics* **38**, 43–51.

Chaudhuri, A. (2011). *Randomized Response and Indirect Questioning Techniques in Surveys*. Chapman & Hall/CRC, Boca Raton.

Chaudhuri, A. and Christofides, T.C. (2007). Item count technique in estimating the proportion of people with a sensitive feature. *Journal of Statistical Planning and Inference* **137**, 589–593.

Chaudhuri, A. and Dihidar, K. (2009). Estimating means of stigmatizing qualitative and quantitative variables from discretionary responses randomized or direct. *Sankhyā, B* **71**(1), 123–136.

Chaudhuri, A. and Mukerjee, R. (1988). *Randomized Response: Theory and Technique*. Marcel Dekker, New York.

Chaudhuri, A. and Stenger, H. (1992). *Survey Sampling: Theory and Methods*. Marcel Dekker, New York.

Chow, S.C., Shao, J. and Wang, H.S. (2003). *Sample Size Calculations in Clinical Research*. Chapman & Hall/CRC, Boca Raton.

Christofides, T.C. (2005). Randomized response technique for two sensitive characteristics at the same time. *Metrika* **62**(1), 53–63.

Clark, S.J. and Desharnais, R.A. (1998). Honest answers to embarrassing questions: Detecting cheating in the randomized response model. *Psychological Methods* **3**(2), 160–168.

Clopper, C.J. and Pearson, E.S. (1934). The use of confidence or fiducial limits illustrated in the case of binomial. *Biometrika* **26**, 404–413.

Cochran, W.G. (1977). *Sampling Techniques* (3rd Ed.). Wiley, New York.

Coutts, E. and Jann, B. (2011). Sensitive questions in online surveys: Experimental results for the randomized response technique and the unmatched count technique (UCT). *Sociological Methods & Research* **40**(1), 169–193.

Dalenius, T. (1980). Privacy and telephone surveys. *Proceedings of the Business and Economics Statistics Section*, American Statistical Association, 83–85.

Dalton, D.R., Wimbush, J.C. and Daily, C.M. (1994). Using the unmatched count technique (UCT) to estimate base rates for sensitive behavior. *Personnel Psychology* **47**(4), 817–829.

Dalton, D.R., Daily, C.M. and Wimbush, J.C. (1997). Collecting "sensitive" data in business ethics research: A case for the unmatched count technique (UCT). *Journal of Business Ethics* **16**, 1049–1057.

Daniel, W.W. (1993). *Collecting Sensitive Data by Randomized Response: An Annotated Bibliography* (2nd Ed.). Research Monograph No. 107. Georgia State University Business Press, Atlanta.

Dempster, A.P., Laird, N.M. and Rubin, D.B. (1977). Maximum likelihood from Incomplete Data via the EM algorithm (with discussions). *Journal of the Royal Statistical Society, B* **39**(1), 1–38.

Diana, G. and Perri, P.F. (2011). A class of estimators for quantitative sensitive data. *Statistical Papers* **52**(3), 633–650.

DiPietro, M. (2004). Bayesian randomized response as a class project. *The American Statistician* **58**(4), 303–309.

Dowling, T.A. and Shachtman, R.H. (1975). On the relative efficiency of randomized response models. *Journal of the American Statistical Association* **70**, 84–87.

Droitcour, J., Caspar, R.A., Hubbard, M.L., Parsley, T.L., Visscher, W. and Ezzati, T.M. (1991). The item count technique as a method of indirect questioning: A review of its development and a case study application. In *Measurement Errors in Surveys* (P.P. Biemer, R.M. Groves, L.E. Lyberg, N.A. Mathiowetz and S. Sudman, eds.), 185–210. Wiley, New York.

Efron, B. and Tibshirani, R.J. (1993). *An Introduction to the Bootstrap.* Chapman & Hall/CRC, Boca Raton.

Eichhorn, B.H. and Hayre, L.S. (1983). Scrambled randomized response methods for obtaing sensitive quantitative data. *Journal of Statistical Planning and Inference* **7**, 307–316.

Erdfelder, E., Auer, T.S., Hilbig, B.E., Aßfalg, A., Moshagen, M. and Nadarevic, L. (2009). Multinomial processing tree models: A review of the literature. *Zeitschrift für Psychologie/Journal of Psychology* **217**, 108–124.

Eriksson, S.A. (1973). A new model for randomizing response. *International Statistical Review* **41**, 101–113.

Esponda, F. and Guerreroc, V.M. (2009). Surveys with negative questions for sensitive items. *Statistics & Probability Letters* **79**(24), 2456–2461.

Esponda, F., Forrest, S. and Helman, P. (2009). Negative representations of information. *International Journal of Information Security* **8**(5), 331–345.

Fidler, D.S. and Kleinknecht, R.E. (1977). Randomized response versus direct questioning: Two data-collection methods for sensitive information. *Psychological Bulletin* **84**(5), 1045–1049.

Folsom, R.E., Greenberg, B.G., Horvitz, D.G. and Abernathy, J.R. (1973). The two alternate questions randomized response model for human surveys. *Journal of the American Statistical Association* **68**, 525–530.

Fox, J.A. and Tracy, P.E. (1984). Measuring associations with randomized response. *Social Science Research* **13**(2), 188–197.

Fox, J.A. and Tracy, P.E. (1986). *Randomized Response: A Method for Sensitive Surveys* (Series: Quantitative Applications in the Social Sciences). SAGE Publications, California.

Franklin, L.A. (1989). Randomized response sampling from dichotomous populations with continuous randomization. *Survey Methodology* **15**, 225–235.

Franklin, L.A. (1998). Randomized response techniques. In *Encyclopedia of Biostatistics* (P. Armitage and T. Colton, eds.), 3696–3703. Wiley, New York.

Gelfand, A.E. and Smith, A.F.M. (1990). Sampling-based approaches to calculating marginal densities. *Journal of the American Statistical Association* **85**, 398–409.

Gelman, A., Carlin, J.B., Stern, H.S. and Rubin, D.B. (1995). *Bayesian Data Analysis*. Chapman & Hall, London.

Geng, Q.X. (2011). Non-randomized response models for sensitive survey questions. Final Year Project for Bachelor of Science in Statistics and Operations Research under the Supervision of Dr. M.L. Tang. Department of Mathematics, Hong Kong Baptist University, Hong Kong.

Gilens, M., Sniderman, P.M. and Kuklinski, J.H. (1998). Affirmative action and the politics of realignment. *British Journal of Political Science* **28**(1), 159–183.

Gjestvang, C.R. and Singh, S. (2006). A new randomized response model. *Journal of the Royal Statistical Society, B* **68**, 523–530.

Goodstadt, M.S. and Gruson, V. (1975). The randomized response technique: A test on drug use. *Journal of the American Statistical Association* **70**, 814–818.

Gould, A.L., Shah, B.V. and Abernathy, J.R. (1969). Unrelated question randomized response techniques with two trials per respondent. In *1969 Social Statistics Section Proceedings of the American Statistical Association*, American Statistical Association, 351–359.

Greenberg, B.G., Abernathy, J.R. and Horvitz, D.G. (1986). Randomized response. In *Encyclopedia of Statistical Sciences* (S. Kotz and N.L. Johnson, eds.), Vol.7, 540–548. Wiley, New York.

Greenberg, B.G., Abul–Ela, A.A., Simmons, W.R. and Horvitz, D.G. (1969). The unrelated question randomized response model: Theoretical framework. *Journal of the American Statistical Association* **64**, 520–539.

Greenberg, B.G., Horvitz, D.G. and Abernathy, J.R. (1974). Comparison of randomized response designs. In *Reliability and Biometry: Statistical Analysis of Life Length* (F. Prochan and R.J. Serfling, eds.), 787–815. SIAM, Philadelphia.

Greenberg, B.G., Kuebler, R.R., Abernathy, J.R. and Horvitz, D.G. (1971). Application of the randomized response technique in obtaining quantitative data. *Journal of the American Statistical Association* **66**, 243–250.

Grewal, I.S., Bansal, M.L. and Singh, S. (2003). Estimation of population mean of a stigmatized quantitative variable using double sampling. *Statistica* **63**(1), 79-88.

Groenitz, H. (2012). A new privacy-protecting survey design for multichotomous sensitive variables. *Metrika*, in press.

Gupta, S., Gupta, B. and Singh, S. (2002). Estimation of sensitivity level of personal interview survey questions. *Journal of Statistical Planning and Inference* **100**, 239–247.

Gupta, S., Shabbir, J. and Sehra, S. (2010). Mean and sensitivity estimation in optional randomized response models. *Journal of Statistical Planning and Inference* **140**, 2870–2874.

Hedayat, A.S. and Sinha, B.K. (1991). *Design and Inference in Finite Population Sampling.* Wiley, New York.

Horvitz, D.G., Greenberg, B.G. and Abernathy, J.R. (1975). Recent developments in randomized designs. In *A Survey of Statistical Design and Linear Models* (J.N. Srivastava, ed.), 271–285. North Holland/American Elsevier Publishing Co., New York.

Horvitz, D.G., Greenberg, B.G. and Abernathy, J.R. (1976). Randomized response: A data-gathering device for sensitive questions. *International Statistical Review* **44**(2), 181–196.

Horvitz, D.G., Shah, B.V. and Simmons, W.R. (1967). The unrelated question randomized response model. In *1967 Social Statistics Section Proceedings of the American Statistical Association*, American Statistical Association, 65–72.

Hu, X. and Batchelder, W.H. (1994). The statistical analysis of general processing tree models with the EM algorithm. *Psychometrika* **59**, 21–47.

Hussain, Z. and Shabbir, J. (2007). Estimation of mean of a sensitive quantitative variable. *Journal of Statistical Research* **41**(2), 83–92.

Imai, K. (2011). Multivariate regression analysis for the item count technique. *Journal of the American Statistical Association* **106**(494), 407–416.

Jann, B. and Brandenberger, L. (2012). An experimental survey measuring plagiarism using the crosswise model: Codebook and documentation. The Institute of Sociology, University of Bern, Bern, Switzerland.

Jann, B., Jerke, J. and Krumpal, I. (2012). Asking sensitive questions using the crosswise model: An experimental survey measuring plagiarism. *Public Opinion Quarterly* **76**(1), 32–49.

Janus, A.L. (2010). The influence of social desirability pressures on expressed immigration attitudes. *Social Science Quarterly* **91**(4), 928–946.

Kim, J.M. and Elam, M.E. (2005). A two-stage stratified Warner's randomized response model using optimal allocation. *Metrika* **61**, 1–7.

Kim, J.M. and Warde, W.D. (2004). A stratified Warner's randomized response model. *Journal of Statistical Planning and Inference* **120**, 155–165.

Kim, J.M. and Warde, W.D. (2005). Some new results on the multinomial randomized response model. *Communications in Statistics—Theory and Methods* **34**, 847–856.

Kim, J.M., Tebbs, J.M. and An, S.W. (2006). Extension of Mangat's randomized-response method. *Journal of Statistical Planning and Inference* **136**, 1554–1567.

Kuk, A.Y.C. (1990). Asking sensitive questions indirectly. *Biometrika* **77**, 436–438.

Kuklinski, J.H., Cobb, M.D. and Gilens, M. (1997). Racial attitudes and the "new south." *Journal of Politics* **59**(2), 323–349.

LaBrie, J.W. and Earleywine, M. (2000). Sexual risk behaviors and alcohol: Higher base rates revealed using the unmatched-count technique. *Journal of Sex Research* **37**(4), 321–326.

Lakshmi, D.V. and Raghavarao, D. (1992). A test for detecting untruthful answering in randomized response procedures. *Journal of Statistical Planning and Inference* **31**(3), 387–390.

Lange, K. (2002). *Mathematical and Statistical Methods for Genetic Analysis* (2nd Ed.). Springer, New York.

Lara, D., Strickler, J., Olavarrieta, C.D. and Ellertson, C. (2004). Measuring induced abortion in Mexico: A comparison of four methodologies. *Sociological Methods & Research* **32**(4), 529–558.

Lavender, J.M. and Anderson, D.A. (2008). A novel assessment of behaviors associated with body dissatisfaction and disordered eating. *Body Image* **5**(4), 399–403.

Lee, R.M. (1993). *Doing Research on Sensitive Topics*. Sage Publications, London.

Lensvelt-Mulders, G.J.L.M., Hox, J.J. and van der Heijden, P.G.M. (2005). How to improve the efficiency of randomized response designs. *Quality & Quantity* **39**, 253–265.

Lensvelt-Mulders, G.J.L.M., Hox, J.J., van der Heijden, P.G.M. and Maas, C.J.M. (2005). Meta-analysis of randomized response research: Thirty-five years of validation. *Sociological Methods & Research* **33**(3), 319–348.

Li, H. and Zhang, K.L. (1998). The progress of social behavior sciences related to HIV and AIDS. *Chinese Journal of Preventive Medicine* **2**, 120–124.

Liu, P.T. and Chow, L.P. (1976). The efficiency of the multiple trial randomized response technique. *Biometrics* **32**, 607–618.

Liu, P.T., Chow, L.P. and Mosley, W.H. (1975). Use of the randomized response technique with a new randomizing device. *Journal of the American Statistical Association* **70**, 329–332.

Liu, Y. and Tian, G.L. (2013a). Multi-category parallel models in the design of surveys with sensitive questions. *Statistics and Its Interface* **6**(1), 137–149.

Liu, Y. and Tian, G.L. (2013b). A variant of the parallel model for sample surveys with sensitive characteristics. *Computational Statistics & Data Analysis* **67**, 115–135.

Mangat, N.S. (1994). An improved randomized response strategy. *Journal of the Royal Statistical Society, B* **56**, 93–95.

Mangat, N.S. and Singh, R. (1990). An alternative randomized response procedure. *Biometrika* **77**(2), 439–442.

Migon, H.S. and Tachibana, V.M. (1997). Bayesian approximations in randomized response model. *Computational Statistics & Data Analysis* **24**, 401–409.

Miller, J.D. (1983). The nominative technique: Method and heroin estimated from the 1982 national survey on drug abuse. Unpublished report of National Institute on Drug Abuse, Rockville, Maryland.

Miller, J.D. (1984). A new survey technique for studying deviant behavior. Unpublished doctoral dissertation, George Washington University.

Miller, J.D. (1985). The nominative technique: A new method of estimating heroin prevalence. In *Self-Report Methods of Estimating Drug Use: Meeting Current Challenges to Validity* (B.A. Rouse, N.J. Kozel and L.G. Richards, eds.), 104–124. National Institute on Drug Abuse Research Monograph 57, Rockville, Maryland.

Monto, M.A. (2001). Prostitution and fellatio. *Journal of Sex Research* **38**(2), 140–145.

Moors, J.J.A. (1971). Optimization of the unrelated question randomized response model. *Journal of the American Statistical Association* **66**, 627–629.

Moshagen, M. (2010). MultiTree: A computer program for the analysis of multinomial processing tree models. *Behavior Research Methods* **42**(1), 42–54.

Moshagen, M., Musch, J. and Erdfelder, E. (2012). A stochastic lie detector. *Behavior Research Methods* **44**, 222–231.

Nayak, T.K. (1994). On randomized response surveys for estimating a proportion. *Communications in Statistics—Theory and Methods* **23**(11), 3303–3321.

Nazuk, A. and Shabbir, J. (2010). A new mixed randomized response model. *International Journal of Business and Social Science* **1**(1), 186–190.

Newcombe, R.G. (1998). Improved confidence intervals for the difference between binomial proportions based on paired data. *Statistics in Medicine* **17**, 2635–2650.

Ng, K.W. (1997). Inversion of Bayes formula: Explicit formulas for uncon- ditional pdf. In *Advances in the Theory and Practice in Statistics—A Volume in Honor of Samuel Kotz* (N.L. Johnson and N. Balakrishnan, eds.), 571–584. Wiley, New York.

Ng, K.W., Tian, G.L. and Tang, M.L. (2011). *Dirichlet and Related Distri- butions: Theory, Methods and Applications*. Wiley, New York.

Ng, K.W., Tang, M.L., Tan, M. and Tian, G.L. (2008). Grouped Dirich- let distribution: A new tool for incomplete categorical data analysis. *Journal of Multivariate Analysis* **99**(3), 490–509.

O'Hagan, A. (1987). Bayes linear estimators for randomized response mod- els. *Journal of the American Statistical Association* **82**, 580–585.

Orwin, R.G. and Boruch, R.F. (1982). RRT meets RDD: Statistical strate- gies for assuring response privacy in telephone surveys. *Public Opinion Quarterly* **46**(4), 560–571.

Ostapczuk, M., Musch, J. and Moshagen, M. (2009). A randomized response investigation of the education effect in attitudes towards foreigners. *European Journal of Social Psychology* **39**, 920–31.

Ostapczuk, M., Musch, J. and Moshagen, M. (2011). Improving self-report measures of medication non-adherence using a cheating detection ex- tension of the randomized-response-technique. *Statistical Methods in Medical Research* **20**, 489–503.

Ostapczuk, M., Moshagen, M., Zhao, Z. and Musch, J. (2009). Assessing sensitive attributes using the randomized response technique: Evi- dence for the importance of response symmetry. *Journal of Educa- tional and Behavioral Statistics* **34**(2), 267–287.

Petróczi, A., Nepusz, T., Cross, P., Taft, H., Shah, S., Deshmukh, N., Schaffer, J., Shane, M., Adesanwo, C., Barker, J., and Naughton, D.P. (2011). New non-randomised model to assess the prevalence of discriminating behaviour: A pilot study on mephedrone. *Substance Abuse Treatment, Prevention, and Policy* **6**, 20.

Pitz, G.F. (1980). Bayesian analysis of random response models. *Psycho- logical Bulletin* **87**, 209–212.

Pollock, K.H. and Bek, Y. (1976). A comparison of three randomized re- sponse models for quantitative data. *Journal of the American Statis- tical Association* **71**, 884–886.

Poole, W.K. (1974). Estimation of the distribution function of a continuous type random variable through randomized response. *Journal of the American Statistical Association* **69**, 1002–1005.

Raghavarao, D. and Federer, W.T. (1979). Block total response as an alternative to the randomized response method in surveys. *Journal of the Royal Statistical Society, B* **41**(1), 40–45.

Rao, C.R. (1973). *Linear Statistical Inference and Its Applications*. Wiley, New York.

Rao, J.N.K. and Sitter, R.R. (1995). Variance estimation under two-phase sampling with application to imputation for missing data. *Biometrika* **82**, 453–460.

Rayburn, N.R., Earleywine, M. and Davison, G.C. (2003). Base rates of hate crime victimization among college students. *Journal of Interpersonal Violence* **18**(10), 1209–1221.

Saha, A. (2007). Optional randomized response in stratified unequal probability sampling—A simulation based numerical study with Kuk's method. *Test* **16**(2), 346–354.

Shimizu, I.M. and Bonham, G.S. (1978). Randomized response technique in a national survey. *Journal of the American Statistical Association* **73**, 35–39.

Singh, S. and Kim, J.M. (2011). A pseudo-empirical log-likelihood estimator using scrambled responses. *Statistics and Probability Letters* **81**, 345–351.

Singh, S., Kim, J.M. and Grewal, I.S. (2008). Imputing and jackknifing scrambled responses. *Metron—International Journal of Statistics* **66**(2), 183–204.

Singh, S., Mangat, N.S. and Singh, R. (1997). Estimation of size and mean of a sensitive quantitative variable for a sub-group of a population. *Communications in Statistics—Theory and Methods* **26**, 1793–1804.

Song, J.J. and Kim, J.M. (2012). Bayesian analysis of randomized response sum score variables. *Communications in Statistics—Theory and Methods* **41**, 1875–1884.

Spurrier, J.D. and Padgett, W.J. (1980). The application of Bayesian techniques in randomized response. *Sociological Methodology* **11**, 533–544.

Strauss, S.M., Rindskopf, D.M. and Falkin, G.P. (2001). Modeling relationships between two categorical variables when data are missing: Examining consequences of the missing data mechanism in an HIV data set. *Multivariate Behavioral Research* **36**(4), 471–500.

Suzuki, T., Takahasi, K. and Sakasegawa, H. (1976). Some notes on randomized response techniques. *Proceedings of the Institute of Statistical Mathematics* **24**(1), 1–13.

Swensson, B. (1974). Combined questions: A new survey technique for eliminating evasive answer bias (I)—Basic theory. Report No. 70 of the Errors in Surveys Research Project. Institute of Statistics, University of Stockholm.

Takahasi, K. and Sakasegawa, H. (1977). A randomized response technique without making use of any randomizing device. *Annals of the Institute of Statistical Mathematics* **29**(1), 1–8.

Tan, M., Tian, G.L. and Ng, K.W. (2003). A non-iterative sampling method for computing posteriors in the structure of EM-type algorithms. *Statistica Sinica* **13**(3), 625–639.

Tan, M., Tian, G.L. and Ng, K.W. (2010). *Bayesian Missing Data Problems: EM, Data Augmentation and Non-iterative Computation.* Chapman & Hall/CRC, Boca Raton.

Tan, M., Tian, G.L. and Tang, M.L. (2009). Sample surveys with sensitive questions: A non-randomized response approach. *The American Statistician* **63**(1), 9–16.

Tang, M.L. and Wu, Q. (2013). Non-randomized response model for sensitive survey with non-compliance. Submitted.

Tang, M.L., Tian, G.L., Tang, N.S. and Liu, Z.Q. (2009). A new non-randomized multi-category response model for surveys with a single sensitive question: Design and analysis. *Journal of the Korean Statistical Society* **38**, 339–349.

Tang, M.L., Wu, Q., Tian, G.L. and Guo, J.H. (2012). Two-sample non-randomized response techniques for sensitive questions. *Communication in Statistics—Theory and Methods*, in press.

Tanner, M.A. (1996). *Tools for Statistical Inference: Methods for the Exploration of Posterior Distributions and Likelihood Functions* (3rd Ed.). Springer, New York.

Tanner, M.A. and Wong, W.H. (1987). The calculation of posterior distributions by data augmentation (with discussions). *Journal of the American Statistical Association* **82**, 528–550.

Tian, G.L. (2012). A new non-randomized response model: The parallel model. *Statistica Neerlandica*, in revision.

Tian, G.L., Ng, K.W. and Geng, Z. (2003). Bayesian computation for contingency tables with incomplete cell-counts. *Statistica Sinica* **13**, 189–206.

Tian, G.L., Tan, M. and Ng, K.W. (2007). An exact non-iterative sampling procedure for discrete missing data problems. *Statistica Neerlandica* **61**, 232–242.

Tian, G.L., Yu, J.W., Tang, M.L. and Geng, Z. (2007). A new non-randomized model for analyzing sensitive questions with binary outcomes. *Statistics in Medicine* **26**(23), 4238–4252.

Tian, G.L., Yuen, K.C., Tang, M.L. and Tan, M. (2009). Bayesian non-randomized response models for surveys with sensitive questions. *Statistics and Its Interface* **2**, 13–25.

Tian, G.L., Tang, M.L., Liu, Z.Q., Tan, M. and Tang, N.S. (2011). Sample size determination for the non-randomized triangular model for sensitive questions in a survey. *Statistical Methods in Medical Research* **20**(3), 159–173.

Tracy, D.S. and Mangat, N.S. (1996). Some developments in randomized response sampling during the last decade—A follow up of review by Chaudhuri and Mukerjee. *Journal of Applied Statistical Science* **4**, 533–544.

Tsuchiya, T. (2005). Domain estimators for the item count technique. *Survey Methodology—A Journal of Published by Statistics Canada* **31**(1), 41–51.

Tsuchiya, T., Hirai, Y. and Ono, S. (2007). A study of the properties of the item count technique. *Public Opinion Quarterly* **71**(2), 253–272.

Unnikrishnan, N.K. and Kunte, S. (1999). Bayesian analysis for randomized responses models. *Sankhyā, B* **61**(3), 422–432.

van den Hout, A. and Klugkist, I. (2009). Accounting for non-compliance in the analysis of randomized response data. *Australian & New Zealand Journal of Statistics* **51**(3), 353–372.

van den Hout, A., Böckenholt, U. and van der Heijden, P.G.M. (2010). Estimating the prevalence of sensitive behavior and cheating with a dual design for direct questioning and randomized response. *Journal of the Royal Statistical Society, C* **59**(4), 723–736.

Warner, S.L. (1965). Randomized response: A survey technique for eliminating evasive answer bias. *Journal of the American Statistical Association* **60**, 63–69.

Warner, S.L. (1971). The linear randomized response model. *Journal of the American Statistical Association* **66**, 884–888.

Warner, S.L. (1986). The omitted digit randomized response model for telephone applications. In *Proceedings of the Social Survey Research Methods*, American Statistical Association, 441–443.

Wiederman, M.W. (1997). The truth must be in here somewhere: Examining the gender discrepancy in self-reported lifetime number of sex partners. *Journal of Sex Research* **34**, 375–386.

Williamson, G.D. and Haber, M. (1994). Models for three-dimensional contingency tables with completely and partially cross-classified data. *Biometrics* **50**, 194–203.

Wimbush, J.C. and Dalton, D.R. (1997). Base rate for employee theft: Convergence of multiple methods. *Journal of Applied Psychology* **82**(5), 756–763.

Winkler, R.L. and Franklin, L.A. (1979). Warner's randomized response model: A Bayesian approach. *Journal of the American Statistical Association* **74**, 207–214.

Wiseman, F., Moriarty, M. and Schafer, M. (1975). Estimating public opinion with the randomized response model. *Public Opinion Quarterly* **39**(4), 507–513.

Wu, J.Q., Zhao, P.F., Lu, H., et al. (1996). *The Progress of Social Sciences in Reproductive Health.* Chinese Population Publishing House, Beijing.

Youn, G. (2001). Perceptions of peer sexual activities in Korean adolescents. *Journal of Sex Research* **38**(4), 352–360.

Yu, J.W., Tian, G.L. and Tang, M.L. (2008). Two new models for survey sampling with sensitive characteristic: Design and analysis. *Metrika* **67**, 251–263.

Yu, J.W., Lu, Y. and Tian, G.L. (2013). A survey design for a sensitive binary variable correlated with another non-sensitive binary variable. *Journal of Probability and Statistics*, Volume 2013, Article ID 827048, 11 pages, http://dx.doi.org/10.1155/2013/827048.

Zdep, S.M. and Rhodes, I.N. (1976). Making a randomized response technique work. *Public Opinion Quarterly* **40**(4), 531–537.

Zou, G.H. (1997). Two-stage randomized response procedures as single stage procedures. *Australian Journal of Statistics* **39**, 235–236.

Author Index

Subject Index

acceptance region, 66
AIDS, 1
algorithm
    DA, 107, 112
    data augmentation, 87, 90, 187
    EM, 35, 51, 56, 83, 95, 147, 179
association, 91, 188, 200

bootstrap approach, 96

Central Limit Theorem, 68, 130, 158
class
    non-sensitive, 80
    sensitive, 121
confidence interval
    asymptotic, 31
    bootstrap, 31, 53, 58, 131, 181
    Clopper–Pearson, 217
    likelihood ratio, 184, 216
    Wald, 30, 52, 58, 130
    Wilson, 30, 52, 58, 130, 183
correlation, 91
critical region, 66

data
    cervical cancer, 250
    HIV, 114
    income and sexual partner, 199
    induced abortion, 36
    latent, 113
    multinomial response, 91
    observed, 58
    sensitive quantitative, 5

sex partner, 88
sexual behavior, 56
DDQ, 24, 121
design
    crosswise, 72, 163
    item count, 10
    Latin square measurement, 3
    non-randomized, 44
    parallel, 163
    survey, 23
    triangular, 46
    unrelated question, 5, 7, 119
    Warner, 3, 24
discriminant, 136, 142
distribution
    asymptotic, 53
    beta, 54, 146
    binomial, 95
    chi-squared, 97
    complete-data posterior, 55
    conditional predictive, 55
    Dirichlet, 86, 106
    exponential, 268
    general finite, 265
    grouped Dirichlet, 245
    hypergeometric, 265
    multinomial, 101
    Poisson, 265
    positively correlated bivariate-beta,
        110
    posterior, 55
    prior, 54